Microstructure and Mechanical Properties of Structural Metals and Alloys

Microstructure and Mechanical Properties of Structural Metals and Alloys

Special Issue Editor

Andrey Belyakov

MDPI • Basel • Beijing • Wuhan • Barcelona • Belgrade

Special Issue Editor
Andrey Belyakov
Belgorod National Research University
Russia

Editorial Office
MDPI
St. Alban-Anlage 66
4052 Basel, Switzerland

This is a reprint of articles from the Special Issue published online in the open access journal *Metals* (ISSN 2075-4701) from 2017 to 2018 (available at: https://www.mdpi.com/journal/metals/special_issues/microstructural_metals_alloys)

For citation purposes, cite each article independently as indicated on the article page online and as indicated below:

LastName, A.A.; LastName, B.B.; LastName, C.C. Article Title. *Journal Name* **Year**, *Article Number, Page Range.*

ISBN 978-3-03897-505-2 (Pbk)
ISBN 978-3-03897-506-9 (PDF)

Cover image courtesy of Andrey Belyakov.

Contents

About the Special Issue Editor

Andrey Belyakov (Ph.D., D.Sc.) has been a Leading Research Associate at the Laboratory of Mechanical Properties of Nanostructured Materials and Superalloys, Belgorod National Research University, Belgorod, Russia, since 2007, after several years of research at Ufa State Aviation Technical University, Institute for Metals Superplasticity Problems (Ufa), University of Electro-Communications (UEC, Tokyo), National Institute for Materials Science (NIMS, Tsukuba), and Tokyo Institute of Technology. His research interests are mainly focused on the mechanisms of microstructural changes in various metallic materials subjected to large-strain plastic deformation, including hot and cold working as well as heat treatments. His scientific expertise comprises deformation behavior and microstructure evolution including static, dynamic, and post-dynamic recrystallization. At the moment, he is the Editorial Board member of both journals *Materials and Metals*. To date, he has published (co-authored) over 200 articles in scientific journals and conference proceedings, with over 2500 citations, h-index: 30.

Editorial

Microstructure and Mechanical Properties of Structural Metals and Alloys

Andrey Belyakov

Laboratory of Mechanical Properties of Nanostructured Materials and Superalloys, Belgorod State University, Belgorod, 308015, Russia; belyakov@bsu.edu.ru; Tel.: +7-4722-585457

Received: 27 August 2018; Accepted: 28 August 2018; Published: 29 August 2018

1. Introduction and Scope

Mechanical properties of polycrystalline structural metals and alloys are significantly affected by their microstructures including phase content, grain/subgrain sizes, grain boundary distribution, dispersed particles, dislocation density, etc. The development of metallic materials with desired structural state results in beneficial combinations of mechanical properties. Specific alloying designs along with a wide variety of thermal, deformation, and many other treatments are used to produce metallic semi-products with favorable microstructures in order to achieve the required properties. Therefore, the studies on structure–property relationships are of a great practical importance. The aim of this special issue is to present the latest achievements in theoretical and experimental investigations of mechanisms of microstructural changes/evolutions in various metallic materials subjected to different processing methods and their effect on mechanical properties.

2. Contributions

The present special issue on the microstructure and mechanical properties of structural metals and alloys collects papers dealing with various aspects of microstructure–property relationships of advanced structural steels and alloys including commercial and novel materials. A total of 22 papers cover a ranger of structural metals and alloys. The major portion of these papers is focused on the mechanisms of microstructure evolution and the mechanical properties of metallic materials subjected to various thermo-mechanical, deformation, or heat treatments [1–12]. Another large portion of the studies is aimed on the elaboration of alloying design of advanced steels and alloys [13–16]. The changes in phase content, transformation, and particle precipitation and their effect on the properties are also broadly presented in this collection [17–21]. In two papers [19,22], particular emphasis is placed on the microstructure/property changes caused by irradiation.

Those readers interested in structural steels may learn much from comprehensive investigations of microstructural changes and their effect on mechanical properties caused by plastic working and heat treatment of diverse steel types [2,7,8,10,12,15,16,19–22]. Two of these papers [7,12] present experimental/simulation results of mechanical behavior of high-Mn TWIP steels, which have recently aroused a great interest among material scientists and engineers because of outstanding strength–ductility combination inherent in such steels. Some crucial features of structure–property relations are detailed for advanced heat resistant [10,15,19,22] and stainless [8,21] steels. Materials scientists working with aluminum alloys may find many interesting results for dispersion strengthening, including processing, structural/precipitation analysis, and mechanical testing [13,17,18]. As a guest editor, I have the pleasure to note that present collection is not limited to such frequently used materials like steel and aluminum alloys. Interested readers will find attractive reports on magnesium [1], nickel [3], titanium [4,5], copper [6,11], and tin [14] alloys. Those who are interested in innovative materials, and their processing and applications, are suggested to take a good look at Ti/TiB metal-matrix composite [5] and high-entropy alloys [9]. The great diversity of materials, which are presented in this special issue, involves various techniques of their production. Worthy of mention are severe plastic deformation [5,9] and welding [8,10] as topics of quickened interest. As a guest editor, I sincerely believe that every reader among materials scientists will find interesting and useful information in the present special issue.

3. Conclusions and Outlook

The papers collected in this special issue clearly reflect the modern research trends in materials science. These fields of specific attention are high-Mn TWIP steels, high-Cr heat resistant steels, aluminum alloys, ultrafine grained materials including those developed by severe plastic deformation, and high-entropy alloys. In spite of great effort in the development of advanced structural metals and alloys, these topics deserve further comprehensive investigations. The engineering and technology progress is closely related with the development of new structural materials with improved mechanical properties. This requires deep knowledge of mechanisms and regularities of microstructural changes during processing and exploitation, as well as clear understanding of microstructure–property relationships. No doubt, structural materials will continuously attract a great interest among materials scientists and engineers.

As a guest editor, I would like to thank all the authors for their valuable contribution to the present special issue; special thanks to the Metals editorial team, in particular to Ms. Hollie Huang, for their assistance and support during the preparation of this special issue.

Conflicts of Interest: The author declares no conflict of interest.

References

1. Che, C.; Cai, Z.; Cheng, L.; Meng, F.; Yang, Z. The Microstructures and Tensile Properties of As-Extruded Mg–4Sm–xZn–0.5Zr (x = 0, 1, 2, 3, 4 wt %) Alloys. *Metals* **2017**, *7*, 281. [CrossRef]
2. Niu, G.; Wu, H.; Zhang, D.; Gong, N.; Tang, D. Study on Microstructure and Properties of Bimodal Structured Ultrafine-Grained Ferrite Steel. *Metals* **2017**, *7*, 316. [CrossRef]
3. Underwood, O.D.; Madison, J.D.; Thompson, G.B. Emergence and Progression of Abnormal Grain Growth in Minimally Strained Nickel-200. *Metals* **2017**, *7*, 334. [CrossRef]
4. Li, K.; Yang, P. The Formation of Strong {100} Texture by Dynamic Strain-Induced Boundary Migration in Hot Compressed Ti-5Al-5Mo-5V-1Cr-1Fe Alloy. *Metals* **2017**, *7*, 412. [CrossRef]
5. Zherebtsov, S.; Ozerov, M.; Stepanov, N.; Klimova, M.; Ivanisenko, Y. Effect of High-Pressure Torsion on Structure and Microhardness of Ti/TiB Metal–Matrix Composite. *Metals* **2017**, *7*, 507. [CrossRef]
6. Wang, H.; Huang, H.; Xie, J. Effects of Strain Rate and Measuring Temperature on the Elastocaloric Cooling in a Columnar-Grained Cu71Al17.5Mn11.5 Shape Memory Alloy. *Metals* **2017**, *7*, 527. [CrossRef]
7. Kalinenko, A.; Kusakin, P.; Belyakov, A.; Kaibyshev, R.; Molodov, D.A. Microstructure and Mechanical Properties of a High-Mn TWIP Steel Subjected to Cold Rolling and Annealing. *Metals* **2017**, *7*, 571. [CrossRef]
8. Nam, T.H.; An, E.; Kim, B.J.; Shin, S.; Ko, W.S.; Park, N.; Kang, N.; Jeon, J.B. Effect of Post Weld Heat Treatment on the Microstructure and Mechanical Properties of a Submerged-Arc-Welded 304 Stainless Steel. *Metals* **2018**, *8*, 26. [CrossRef]
9. Zherebtsov, S.; Stepanov, N.; Ivanisenko, Y.; Shaysultanov, D.; Yurchenko, N.; Klimova, M.; Salishchev, G. Evolution of Microstructure and Mechanical Properties of a CoCrFeMnNi High-Entropy Alloy during High-Pressure Torsion at Room and Cryogenic Temperatures. *Metals* **2018**, *8*, 123. [CrossRef]
10. Liu, X.; Cai, Z.; Yang, S.; Feng, K.; Li, Z. Characterization on the Microstructure Evolution and Toughness of TIG Weld Metal of 25Cr2Ni2MoV Steel after Post Weld Heat Treatment. *Metals* **2018**, *8*, 160. [CrossRef]
11. Chen, X.; Jiang, F.; Jiang, J.; Xu, P.; Tong, M.; Tang, Z. Precipitation, Recrystallization, and Evolution of Annealing Twins in a Cu-Cr-Zr Alloy. *Metals* **2018**, *8*, 227. [CrossRef]
12. Torganchuk, V.; Glezer, A.V.; Belyakov, A.; Kaibyshev, R. Deformation Behavior of High-Mn TWIP Steels Processed by Warm-to-Hot Working. *Metals* **2018**, *8*, 415. [CrossRef]
13. Morozova, A.; Mogucheva, A.; Bukin, D.; Lukianova, O.; Korotkova, N.; Belov, N.; Kaibyshev, R. Effect of Si and Zr on the Microstructure and Properties of Al-Fe-Si-Zr Alloys. *Metals* **2017**, *7*, 495. [CrossRef]

14. Park, Y.; Bang, J.H.; Oh, C.M.; Hong, W.S.; Kang, N. The Effect of Eutectic Structure on the Creep Properties of Sn-3.0Ag-0.5Cu and Sn-8.0Sb-3.0Ag Solders. *Metals* **2017**, *7*, 540. [CrossRef]

15. Fedoseeva, A.; Dudova, N.; Kaibyshev, R.; Belyakov, A. Effect of Tungsten on Creep Behavior of 9%Cr–3%Co Martensitic Steels. *Metals* **2017**, *7*, 573. [CrossRef]

16. Chu, R.; Fan, Y.; Li, Z.; Liu, J.; Yin, N.; Hao, N. Study on the Control of Rare Earth Metals and Their Behaviors in the Industrial Practical Production of Q420q Structural Bridge Steel Plate. *Metals* **2018**, *8*, 240. [CrossRef]

17. Ma, P.; Jia, Y.; Gokuldoss, P.K.; Yu, Z.; Yang, S.; Zhao, J.; Li, C. Effect of Al2O3 Nanoparticles as Reinforcement on the Tensile Behavior of Al-12Si Composites. *Metals* **2017**, *7*, 359. [CrossRef]

18. He, H.; Zhang, L.; Li, S.; Wu, X.; Zhang, H.; Li, K. Precipitation Stages and Reaction Kinetics of AlMgSi Alloys during the Artificial Aging Process Monitored by In-Situ Electrical Resistivity Measurement Method. *Metals* **2018**, *8*, 39. [CrossRef]

19. Yang, Z.; Jin, S.; Song, L.; Zhang, W.; You, L.; Guo, L. Dissolution of M23C6 and New Phase Re-Precipitation in Fe Ion-Irradiated RAFM Steel. *Metals* **2018**, *8*, 349. [CrossRef]

20. Muro, M.; Artola, G.; Gorriño, A.; Angulo, C. Effect of the Martensitic Transformation on the Stamping Force and Cycle Time of Hot Stamping Parts. *Metals* **2018**, *8*, 385. [CrossRef]

21. Paulsen, C.O.; Broks, R.L.; Karlsen, M.; Hjelen, J.; Westermann, I. Microstructure Evolution in Super Duplex Stainless Steels Containing σ-Phase Investigated at Low-Temperature Using In Situ SEM/EBSD Tensile Testing. *Metals* **2018**, *8*, 478. [CrossRef]

22. Li, Q.; Shen, Y.; Zhu, J.; Huang, X.; Shang, Z. Evaluation of Irradiation Hardening of P92 Steel under Ar Ion Irradiation. *Metals* **2018**, *8*, 94. [CrossRef]

Article

The Microstructures and Tensile Properties of As-Extruded Mg–4Sm–xZn–0.5Zr (x = 0, 1, 2, 3, 4 wt %) Alloys

Chaojie Che [1,2], Zhongyi Cai [1,*], Liren Cheng [2,*], Fanxing Meng [2,3] and Zhen Yang [1]

[1]   Roll Forging Research Institute, Jilin University, Changchun 130025, China; chechaojie@163.com (C.C.);
      yangzhen0517@163.com (Z.Y.)
[2]   State Key Laboratory of Rare Earth Resource Utilization, Changchun Institute of Applied Chemistry, CAS,
      Changchun 130022, China; mengfanxing@163.com
[3]   School of Materials Science and Engineering, Jilin University, Changchun 130025, China
*   Correspondence: caizy@jlu.edu.cn (Z.C.); lrcheng@ciac.ac.cn (L.C.);
      Tel.: +86-0431-8509-4340 (Z.C.); +86-0431-8526-2414 (L.C.)

Received: 19 June 2017; Accepted: 19 July 2017; Published: 24 July 2017

Abstract: The microstructures and tensile properties of as-cast and as-extruded Mg–4Sm–xZn–0.5Zr (x = 0, 1, 2, 3, 4 wt %) alloys were systematically investigated by optical microscope, X-ray diffractometer (XRD), scanning electron microscope (SEM) and transmission electron microscope (TEM). Numerous nanoscale dynamic precipitates could be observed in the as-extruded alloys containing high content of Zn, and the nanoscale particles were termed as $(Mg,Zn)_3Sm$ phase. Some basal disc-like precipitates were observed in as-extruded Mg–4Sm–4Zn–0.5Zr alloy, which were proposed to have a hexagonal structure with $a = 0.556$ nm. The dynamic precipitates effectively pinned the motions of DRXed (dynamic recrystallized) grain boundaries leading to an obvious reduction of DRXed grain size, and the tensile yield strength of as-extruded alloy was improved. The as-extruded Mg–4Sm–4Zn–0.5Zr alloy exhibits the best comprehensive mechanical properties at room temperature among all the alloys, and the yield strength, ultimate tensile strength and elongation are about 246 MPa, 273 MPa and 21% respectively.

Keywords: Mg–Sm–Zn–Zr; dynamic precipitation; microstructure; mechanical property

1. Introduction

Magnesium and its alloys have great potential use in the fields of automobile and aerospace due to their low density, good machinability, excellent specific strength and stiffness [1–3]. Rare earth (RE) metals have been proved to play an important role in improving mechanical properties of Mg alloys [4]. There has been a great deal of research on 1Mg–Y [5–8], Mg–Gd [9–13] and Mg–Nd [14] alloys, and some of the alloys exhibit good mechanical properties at elevated temperature. As one of the light rare earth elements, Sm has a maximum solubility of about 5.8 wt %, which even is higher than that of Nd (3.6 wt %) in solid Mg. Moreover, the market price of Sm is much cheaper than that of Nd and Y [15]. It is, therefore, meaningful to produce low-price, heat-resistant Mg–Sm alloys with proper mechanical properties to compete with traditional Mg–RE alloys [16–18].

Recently, some investigations about Mg–Sm–Zn alloys have been carried out. Yuan and Zheng have investigated the microstructures and mechanical properties of Mg–3Sm–0.5Gd–xZn–0.5Zr (x = 0, 0.3, 0.6) alloys [19], and prismatic precipitates (base-centered orthorhombic, $a = 0.64$ nm, $b = 2.223$ nm, $c = 0.521$ nm) and basal precipitate γ' (MgZnRE-containing, plate-shaped, hexagonal, $a = 0.55$ nm, $c = 0.52$ nm) have been observed in peak-aged Mg–3Sm–0.5Gd–0.6Zn–0.5Zr alloy. Xia et al. have investigated the precipitation evolution of Mg–4Sm–xZn–Zr (x = 0, 0.3, 0.6, 1.3) (wt %)

alloys [15,20], and they found that a new precipitate β'_z was observed with the Zn addition increasing, when Zn content was higher than ~1 wt %, the basal γ-series precipitates dominated. However, the reports about Mg–Sm–Zn alloy with high Zn content (>2 wt %) and the wrought Mg–Sm–Zn alloy are hardly found.

As is well known, precipitation strengthening is an important way to strengthen the Mg alloys. In fact, the dynamic precipitation also can occur depending on the alloy composition and solid solution content, especially during hot deformation [21]. E. Dogan et al. [21] found a new dynamic precipitate Φ' in AZ31 alloy during different plastic deformation modes, and the precipitate primarily formed along the grain boundaries of the DRXed grains. Hou et al. [11] found extensive dynamic precipitation in Mg–8Gd–2Y–1Nd–0.3Zn–0.6Zr alloy after hot compression at 350 °C and the strain rate of 0.5 s^{-1}, and they thought the formation of precipitate depended strongly on the stress field. Kabir et al. [22] reported that the dynamic precipitation was mainly stimulated by nucleation of dynamic recrystallized grain, in turn, the dynamic precipitation could suppress dynamic recrystallization and refine the recrystallized grain in Mg–Al–Sn alloys. Due to the high content Zn in the Mg–4Sm–xZn–0.5Zr alloys, it is a reasonable inference that the dynamic precipitation will occur during the hot extrusion process, and that the volume fraction of dynamic precipitates increases with the Zn content increasing.

Therefore, in the present work, we have investigated the deformation behaviors, microstructure and tensile properties of the as-extruded Mg–4Sm–xZn–0.5Zr (x = 0, 1, 2, 3 and 4) (wt %) alloys. Meanwhile, the relationship between dynamic precipitation and DRXed grain size and the effects of Zn addition on microstructure and mechanical properties of Mg–4Sm–xZn–0.5Zr alloy have been discussed.

2. Experimental Procedure

The experimental alloys were prepared from commercial high-purity Mg (>99.9%, wt %) and Zn (>99.9%, wt %), Mg–20%Sm and Mg–30%Zr master alloys by melting in an electrical resistance furnace under the protective gas consisting of SF$_6$ and CO$_2$. The melting alloys were maintained at 780 °C for 30 min and then cast into a steel mold (Φ90 mm × ~500 mm) with a circulatory water cooling system. The chemical compositions of obtained alloy ingots were analyzed by using inductively coupled plasma atomic emission spectrometry (ICP). The actual chemical compositions of as-cast alloys were shown in Table 1.

Table 1. Chemical compositions of the as-cast alloys with different zinc contents.

Nominal Alloys	Composition (wt %)			
Mg–4Sm–xZn–0.5Zr	Mg	Sm	Zn	Zr
Mg–4Sm–0.5Zr	Balance	3.50	-	0.40
Mg–4Sm–1Zn–0.5Zr	Balance	3.74	0.82	0.40
Mg–4Sm–2Zn–0.5Zr	Balance	4.06	1.80	0.51
Mg–4Sm–3Zn–0.5Zr	Balance	4.21	2.79	0.60
Mg–4Sm–4Zn–0.5Zr	Balance	4.10	3.67	0.40

In order to analyze the solidification behavior of the experimental alloys, differential scanning calorimetry (DSC) was carried out using a NETZSCH STA 449F3 system equipped with platinum-rhodium crucibles. Samples weighing approximately 30 mg were heated in a flowing argon atmosphere from 20 °C to 700 °C and held for 5 min before being cooled down to 100 °C. Both the heating and cooling curves were recorded at a controlled rate of 1 °C/min. Before hot extrusion, the as-cast alloys were solution treated at first. Solution treatment was carried out under Ar atmosphere for 10 h at 510 °C according to the DSC curve in Figure 1, and then the ingots were quenched in water of ~70 °C. Before the ingots were extruded, both the alloy ingots and extrusion dies were heated to 360 °C and maintained for 90 min. Then the ingots were hot extruded into rods with the diameter of 15 mm at 360 °C with a ratio of ~30:1.

The as-cast and as-extruded samples were etched with 4% nitric acid ethyl alcohol solution, and then examined using both an Olympus optical microscope and a Hitachi S4800 SEM scanning electron microscope (Hitachi S4800 SEM, Tokyo, Japan) operated at 10 kV. The grain size was measured by the standard linear intercept method using an Olympus stereomicroscope. The phases in the experimental alloys were analyzed by an 18 kW type X-ray diffractometer (Rigaku D/max 2500 PC X-ray Diffractometer, Tokyo, Japan) operated at 40 kV and 40 mA and FEI Tecnai G2 F20 transmission electron microscope (TEM, Hillsboro, OR, USA) at 200 kV. The TEM foils were cut from as-extruded bars perpendicular to extrusion direction. Tensile samples of 20 mm in gauge length, 4 mm in gauge width and 2.5 mm in gauge thickness were machined from as-cast ingots and as-extruded bars for tensile tests. The specimens for tensile tests were cut along the extrusion direction. Tensile tests were carried out with a constant displacement rate of 1.0 mm/min on an electronic universal testing machine (SANS CMT–5105, MN, USA). Mechanical properties were determined from a complete stress-strain curve. The yield strength (YS), ultimate tensile strength (UTS) and fracture elongation were obtained based on the average of three tests.

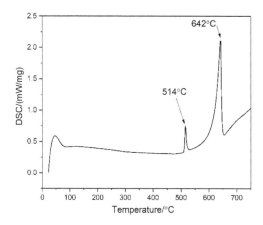

Figure 1. The differential scanning calorimetry (DSC) curve of as-cast Mg–4Sm–4Zn–0.5Zr alloy.

3. Results and Discussion

3.1. Microstructure of As-Cast Alloys

Figure 2 shows the optical micrographs of as-cast Mg–4Sm–xZn–0.5Zr (x = 0, 1, 2, 3, 4 wt %) alloys. It can be observed that all the alloys are composed of α-Mg matrix and network eutectic phase at the grain boundaries, while the grain size varies from one alloy to another. The average grain sizes of five alloys are about 33 ± 3.1 μm, 31 ± 2.0 μm, 40 ± 3.5 μm, 37 ± 2.1 μm and 35 ± 2.3 μm, respectively. The grain sizes of alloys seem to have no specific relationship with the variation of Zn content, which should take relative contents of compounds and actual contents of Zr into consideration comprehensively.

Figure 3 shows the SEM micrographs of as-cast Mg–4Sm–xZn–0.5Zr (x = 0, 1, 2, 3, 4 wt %) alloys. It is obvious that the Mg–4Sm–0.5Zr alloy contains less intermetallic compounds than those Zn-containing alloys. As shown at top-right corner in Figure 3a, a magnifying image of the part surrounded by hollow rectangle shows details of second phase in as-cast Mg–4Sm–0.5Zr alloy. The EDS (Energy Dispersive Spectroscopy) patterns taken from point A in Figure 3a and point D in Figure 3d have been presented in Figure 3f. And the results of EDS are shown in Table 2.

Table 2. The chemical compositions of second phases in Figure 2a,d analyzed by EDS.

Position		Elements			
		Mg K	Sm L	Zn K	Zr L
Point A	Wt %	53.53	43.61	-	2.86
	At %	87.27	11.49	-	1.24
Point D	Wt %	47.78	26.13	23.29	2.79
	At %	77.81	6.88	14.10	1.21

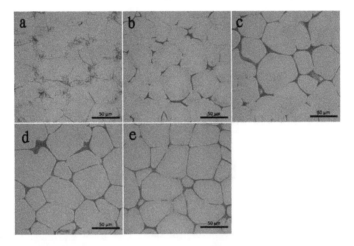

Figure 2. Optical micrographs of as-cast Mg–4Sm–xZn–0.5Zr alloys: (**a**) $x = 0$; (**b**) $x = 1$; (**c**) $x = 2$; (**d**) $x = 3$; (**e**) $x = 4$.

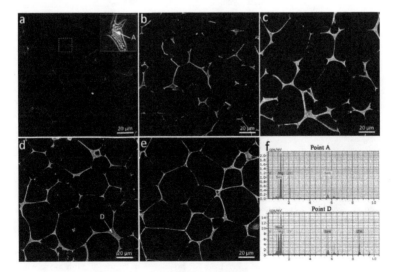

Figure 3. The scanning electron microscope (SEM) micrographs of as-cast Mg–4Sm–xZn–0.5Zr alloys: (**a**) $x = 0$; (**b**) $x = 1$; (**c**) $x = 2$; (**d**) $x = 3$; (**e**) $x = 4$. The hollow rectangle at top-right corner in Figure 2a is the magnifying picture at grain boundary.

Figure 4 shows the XRD patterns of as-cast Mg–4Sm–*x*Zn–0.5Zr (*x* = 0, 1, 2, 3, 4 wt %) alloys. As shown in the nethermost curve in Figure 4, Mg–4Sm–0.5Zr alloy consists of $Mg_{41}Sm_5$ phase and α-Mg phase. While the Zn-containing alloys contain a new compound $(Mg,Zn)_3Sm$ according to the XRD curves. The results of XRD are consistent with the results of EDS in Table 2. The $(Mg,Zn)_3Sm$ phase is similar to $(Mg,Zn)_3RE$ phase reported by Zhang et al. [23] and Yuan et al. [24]. The $(Mg,Zn)_3RE$ phase has a DO_3-type structure with *a* = 0.72 nm. Figure 5a shows a bright field TEM image of $(Mg,Zn)_3Sm$ phase and the sample is taken from the as-cast Mg–4Sm–2Zn–0.5Zr alloy. The SAED (Selected Area Electron Diffraction) pattern in Figure 5b corresponds to the position A in Figure 5a, and the electron beam is parallel to the $[\bar{1}12]$ axis of $(Mg,Zn)_3Sm$ phase. By calculation, the interplanar spacing between $\{220\}_{(Mg,\ Zn)3Sm}$ is about 0.258 nm, $a = 2\sqrt{2} \times 0.258$ nm = 0.729 nm, which agrees well with previous research.

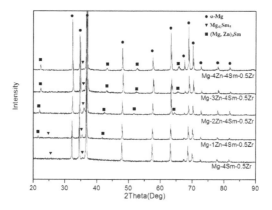

Figure 4. The X-ray diffractometer (XRD) patterns of as-cast Mg–4Sm–*x*Zn–0.5Zr (*x* = 0, 1, 2, 3, 4) (wt %) alloys.

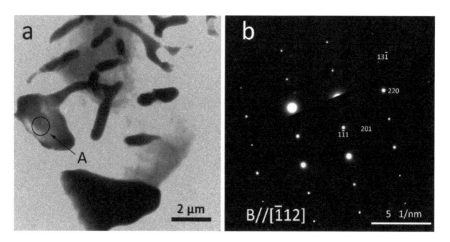

Figure 5. (**a**) A bright field transmission electron microscope (TEM) image of as-cast Mg–4Sm–2Zn–0.5Zr alloy and (**b**) the SAED patterns taken from the position A in Figure 4a, the zone axis is $[\bar{1}12]$.

3.2. Microstructures of Solution-Treated Alloys

Figure 6 shows the optical micrographs of solution-treated alloys at 510 °C for 10 h. Comparing with the as-cast alloys, the average size of grain slightly increased at the T4 state. Although the eutectic compounds at grain boundaries are less than those in as-cast alloys, they do not disappear completely. The solution-treated alloys containing high content of Zn have more eutectic compounds. The XRD pattern of the solution-treated alloys are shown in Figure 7, $Mg_{41}Sm_5$ phases and $(Mg,Zn)_3Sm$ phase still can be detected in solution-treated alloys. The result reveals that the $Mg_{41}Sm_5$ phases and $(Mg,Zn)_3Sm$ phase are thermostable compounds to some degree.

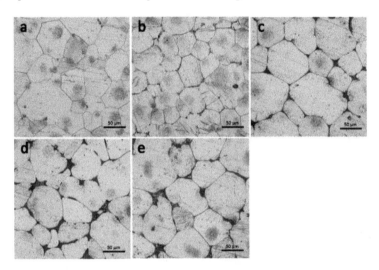

Figure 6. The optical micrographs of solution-treated Mg–4Sm–*x*Zn–0.5Zr alloys: (**a**) *x* = 0; (**b**) *x* = 1; (**c**) *x* = 2; (**d**) *x* = 3; (**e**) *x* = 4.

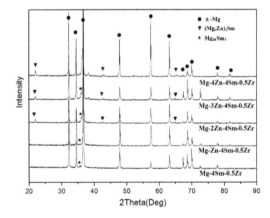

Figure 7. The XRD patterns of solution-treated Mg–4Sm–*x*Zn–0.5Zr (*x* = 0, 1, 2, 3, 4) (wt %) alloys.

3.3. Microstructures of As-Extruded Alloys

Figure 8 shows the SEM micrographs of as-extruded Mg–4Sm–*x*Zn–0.5Zr alloys observed on the longitudinal section along the extrusion direction under different magnification. Here, it is obvious that almost complete dynamic recrystallization has taken place in all the alloys. After hot extrusion, the grains are obviously refined for each alloy. By calculation, the DRXed grain sizes of as-extruded alloys are 4.0 ± 0.3 μm, 5.1 ± 0.2 μm, 6.2 ± 0.3 μm, 4.5 ± 0.2 μm and 2.4 ± 0.3 μm, respectively. Comparing with the other four alloys, the grains of as-extruded Mg–4Sm–4Zn–0.5Zr alloy are obviously finer. As shown in Figure 8a, spherical particles can hardly be observed in as-extruded Mg–4Sm–0.5Zr alloy. With the increasing Zn content, more and more fine and dispersed phases appear in the as-extruded alloys.

As shown in Figure 8, the second phases in as-extruded Zn-containing alloys have various morphologies with different sizes ranging from dozens of nanometers to several micrometers. Some block-shaped particles with diameter of several micrometers can be observed at the grain boundary and their volume fraction is very few. Apart from the big block-shaped particles, a high number density of white particles can be observed in the as-extruded Zn-containing alloys, especially in high Zn-content alloys. The magnifying SEM images show that the fine particles locate both at grain boundaries and at the interior of the grains.

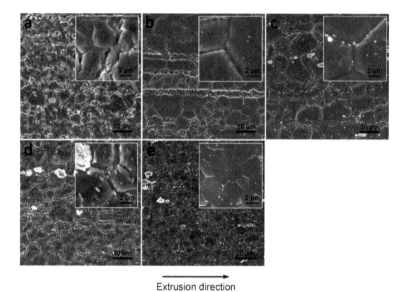

Extrusion direction

Figure 8. The SEM micrographs of as-extruded Mg–4Sm–*x*Zn–0.5Zr alloys observed on the longitudinal section along the extrusion direction: (**a**) *x* = 0; (**b**) *x* = 1; (**c**) *x* = 2; (**d**) *x* = 3; (**e**) *x* = 4.

The XRD patterns in Figure 9 of as-extruded alloys reveal that besides α-Mg, $(Mg,Zn)_3Sm$ and $Mg_{41}Sm_5$ phase can be detected in the as-extruded alloys containing Zn. The results are similar to the XRD patterns of as-cast alloys and solution-treated alloys. Comparing the XRD patterns of as-cast alloys (Figure 4), solution-treated alloys (Figure 7) and as-extruded alloys (Figure 9), an interesting phenomenon can be observed in that the XRD peaks of the $(Mg,Zn)_3Sm$ phase in as-cast alloys and as-extruded alloys shift toward larger angles with increasing Zn content. The phenomenon is not observed in the XRD patterns of solution-treated alloys.

According to Bragg's Law $\lambda = 2d\sin\theta$ (where λ is radiation wavelength, θ is the usual Bragg angle, and d is interplanar crystal spacing), the peaks shift towards larger angle means that the θ has increased with the increasing of Zn content; at the same time, the d should decrease in order to keep the λ constant. Therefore, it can be concluded that the interplanar spacing of $(Mg,Zn)_3Sm$ phase decreases with increasing Zn content in as-cast and as-extruded alloys. Additionally, the interplanar spacing of $(Mg,Zn)_3Sm$ phase remains constant with increasing Zn content in the solution-treated alloys. The decrease of interplanar spacing may be attributed to the internal strain in $(Mg,Zn)_3Sm$ phase caused by redundant Zn in alloys with increasing Zn content in as-cast and as-extruded alloys.

The internal strain is released adequately through solution treatment at 510 °C for 10 h, therefore, the interplanar spacing of $(Mg,Zn)_3Sm$ phase in the solution-treated alloy remains constant.

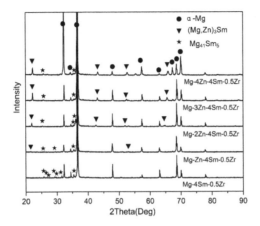

Figure 9. The XRD patterns of solution-treated Mg–4Sm–xZn–0.5Zr (x = 0, 1, 2, 3, 4) (wt %) alloys.

Figure 10 shows a bright field TEM micrograph of ellipsoidal particles with diameter of ~1.5 μm observed in as-extruded Mg–4Sm–2Zn–0.5Zr alloy and corresponding SAED patterns. Figure 10e shows the EDS result taken from the particle in Figure 10a. The EDS result reveals that the atomic ratio of Zn and Sm is ~2.5:1, therefore, the micron-sized particles may be $(Mg,Zn)_3Sm$ phase, which are verified by the SAED patterns. As shown in Figure 10b–d, the SAED patterns are taken from the [001], [$\bar{1}11$] and [$\bar{1}12$] direction, respectively. These particles have relatively big size and irregular shape in general, and should be the fragmented $(Mg,Zn)_3Sm$ phase after hot-extrusion.

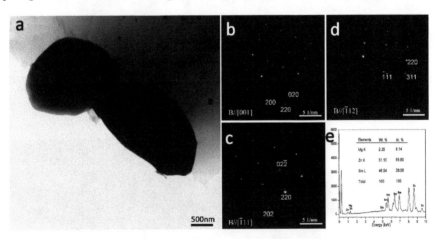

Figure 10. The bright field TEM micrograph of the micron-sized phase observed in as-extruded Mg–4Sm–2Zn–0.5Zr alloy for (**a**) and the corresponding SAED patterns taken along the [001] direction for (**b**), the [$\bar{1}11$] direction for (**c**) and the [$\bar{1}12$] direction for (**d**), respectively. (**e**) The EDS pattern of the micron-sized phase in Figure 10a.

Figure 11 shows a bright field TEM micrograph of a regular polygon particle with a diameter of ~60 nm in as-extruded Mg–4Sm–2Zn–0.5Zr alloy, and the SAED patterns indicate that it is $(Mg,Zn)_3Sm$ phase. The fine regular nano-sized particles should be the dynamic precipitates after hot extrusion.

Besides the block-shaped precipitates, some disc-like precipitates also are observed in the as-extruded Mg–4Sm–4Zn–0.5Zr alloy, as shown in Figure 12a. The disc-like precipitates are about 30 nm wide and 500 nm long, and they all lie along the $(0001)_{\alpha-Mg}$ plane of Mg matrix. Figure 12b shows the HRTEM (High Resolution Transmission Electron Microscopy) image of the basal precipitates taken with electron beam paralleling to the $[2\bar{1}\bar{1}0]_{\alpha-Mg}$ direction of Mg matrix. An examination of these basal precipitate discs using HRTEM reveals that most of them form on several successive $(0001)_{\alpha-Mg}$ planes of the matrix phase. The corresponding FFT (Fast Fourier Transform) pattern is shown at the top-right corner of Figure 12b. The SAED pattern recorded from $[2\bar{4}23]_{\alpha-Mg}$ of Mg matrix regions containing disc-like precipitates is shown in Figure 12c. The ambiguously extra diffraction spots are located at the $1/3(11\bar{2}2)_{\alpha-Mg}$ and $2/3(11\bar{2}2)_{\alpha-Mg}$ positions. This result is similar to the basal precipitates Mg–RE–Zn phase in as-cast Mg–RE–Zn–Zr alloy at T6 state [25]. Therefore, the basal precipitates are proposed to have a hexagonal structure with $a = 0.556$ nm.

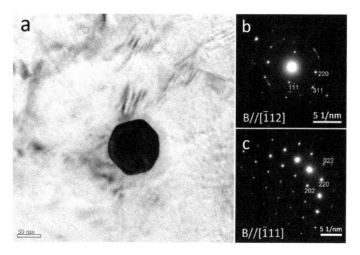

Figure 11. The bright field TEM micrograph of the nanoscale phase observed in as-extruded Mg–4Sm–2Zn–0.5Zr alloy for (**a**), and corresponding SAED patterns taken along the $[\bar{1}12]$ direction for (**b**) and the $[\bar{1}11]$ direction for (**c**).

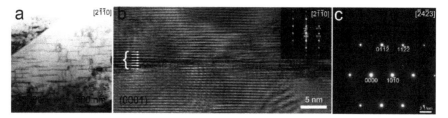

Figure 12. (**a**) The bright field TEM micrograph of disc-like precipitates observed in as-extruded Mg–4Sm–4Zn–0.5Zr alloy, zone axis: $[2\bar{1}\bar{1}0]_{\alpha-Mg}$; (**b**) the HRTEM micrograph of disc-like precipitates observed in as-extruded Mg–4Sm–4Zn–0.5Zr alloy, zone axis: $[2\bar{1}\bar{1}0]_{\alpha-Mg}$; (**c**) the SAED pattern taken from the Mg matrix containing the disc-like precipitates in as-extruded Mg–4Sm–0.5Zr alloy, zone axis: $[2\bar{4}23]_{\alpha-Mg}$.

It is well known that precipitation strengthening is an effective way to strengthen Mg alloy, however, many investigations about precipitation focus on the static precipitation, especially Mg–RE alloys [15,18,26,27]. As mentioned by Kabir [22], the formation of strain-induced precipitates depended on deformation temperature, strain, and strain rate. The dynamic precipitates in Kabir's work mainly distributed at the grain boundaries, while in this paper, the dynamic precipitates can be observed at the grain boundaries and at the interior of the grain, which is consistent with the investigation from Su [28].

The second phase can influence recrystallization, whether fragmented coarse particles or the dynamic precipitates. The effect of second phases on recrystallization depends on their size, spacing and fraction [29], and lies in three aspects: firstly, the stored energy at the positions of the particles increases the driving pressure for recrystallization; secondly, the large particles (≥ 1 µm in diameter) may act as nucleation sites for recrystallization; finally, the closely spaced particles may exert a significant pinning effect on both low and high angle grain boundaries [30]. Some evidence reveals that the particle stimulated nucleation of recrystallization (PSN) may occur during the high temperature deformation [30]. In this work, after hot deformation some static recrystallization has been observed in as-extruded Mg–4Sm–4Zn–0.5Zr alloy, and the nucleation site locate at the large particle, as shown in Figure 13a. In the process of hot deformation massive dispersive particles can effectively pin the migration of boundaries and retard the dynamic recrystallization, which leads to fine DRXed grain in the as-extruded alloy. In Figure 13b the dynamic precipitation can be observed at the grain boundary, as indicated by yellow arrows.

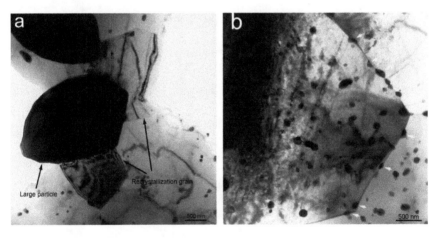

Figure 13. Bright field TEM micrographs taken from as-extruded Mg–4Sm–4Zn–0.5Zr alloy: (**a**) the large particles can promote recrystallization after hot extrusion; (**b**) fine and dispersive particles can pin the grain boundaries.

It is well known that the DRXed grain size of Mg alloy are influenced by several factors, such as deformation temperature, strain rate, deformation degree and the initial grain size [31]. In this work, except the initial grain size, the other factors are almost the same for all the five kinds of alloys. Previous research revealed that the initial grain size was a more sensitivity factor to influence the DRXed grain size in Mg alloy than in other metals [32]. The coarser initial grain always leads to a coarser DRXed grain through hot deformation, and vice versa. On one hand, the finer initial grain provide more grain boundaries to facilitate nucleating of the dynamic recrystallization. On the other hand, with the initial grain decreasing, the strain corresponding to the peak stress also decreases, therefore, the dynamic recrystallization can occur more easily [33]. In the present paper, the variation trend of grain size of as-cast alloys is mainly as same as that of as-extruded alloys with the Zn content increasing. Moreover, the results agree well with the former research. The DRXed grains of as-extruded Mg–4Sm–4Zn–0.5Zr alloy are obviously finer than those of the other alloys, may be attributed to the massive dispersive particles retarding the dynamic recrystallization and pinning the migration of grain boundaries, which agree with the former research from Kabir [22].

3.4. Mechanical Properties

As shown in Figure 14a, when $x \leq 2$, it is obvious that the yield strength and ultimate tensile strength of as-cast alloy increases with increasing Zn content. However, when $2 \leq x \leq 4$, the value of yield strength almost retains a constant of about 135 MPa; at the same time, the elongation has a drop with Zn content increasing. The yield strength σ_y varies with grain size according to the Hall-Petch equation, $\sigma_y = \sigma_0 + k_y d^{-1/2}$, where d is the average grain diameter and σ_0 and k_y are constants for a particular material. The average grain diameters of the as-cast alloys do not change too much, therefore, the grain size reduction is not the main factor to influence strengthening. Solid-solution strengthening plays an important role in strengthening the alloys. The Zn element has a relatively high solid solubility in Mg matrix, hence the yield strength of as-cast alloy increases with Zn content increasing at first, whereas when the Zn content is greater than 2 wt %, the solid solubility of Zn reach extremum at room temperature, which can explain why the value of yield strength almost retains a constant when $x \geq 2$. Moreover, the volume fraction of eutectic compounds increases and the compounds become increasingly coarse, which provides more crack initiations during the tensile test to lead to fracture. Thus, the elongation has an obvious drop with Zn content increasing when $x \geq 2$. The as-cast Mg–4Sm–2Zn–0.5Zr alloy exhibits the best comprehensive mechanical properties at room temperature, and the yield strength, ultimate tensile strength and elongation are 132 MPa, 175 MPa and 8.7%, respectively.

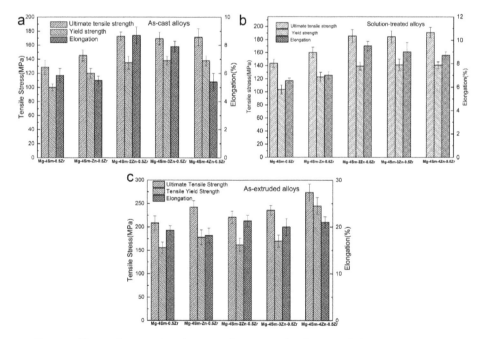

Figure 14. The tensile properties of as-cast, solution-treated and as-extruded Mg–4Sm–xZn–0.5Zr (x = 0, 1, 2, 3, 4 wt %) alloys at room temperature.

Comparing with the tensile properties of as-cast alloys, the ultimate tensile strength and elongation of solution-treated alloys is a bit higher, and the yield strength almost equal to those of as-cast alloys, as shown in Figure 14b. The increasing of elongation and ultimate tensile may be attributed to reduction of the coarse eutectic compounds at grain boundaries. The solution-treated Mg–4Sm–4Zn–0.5Zr alloys exhibits the best comprehensive mechanical properties at room temperature, and the yield strength, ultimate tensile strength and elongation are 191 MPa, 141 MPa and 8.7%.

Figure 14c shows the tensile properties of as-extruded alloys. Comparing with the yield strength and elongation of as-cast alloy, those of as-extruded alloys have clearly improved, which is attributed to the grain refinement after hot extrusion. The fine grains can provide greater total grain boundary to impede dislocation motion. It should be mentioned that grain size reduction improves not only strength, but also the toughness of the alloys. As shown in Figure 7, the volume fraction of second phases increases with Zn content increasing, especially the fine nano-sized phases. It is well known that the operative slip system of Mg is mainly on the basal plane (0001) $\langle 11\bar{2}0 \rangle$ and secondly on vertical face planes $(10\bar{1}0)$ in the direction $\langle 11\bar{2}0 \rangle$ at room temperature. At elevated temperatures, slip also can occur on the $(10\bar{1}1)$ plane in the $\langle 11\bar{2}0 \rangle$ direction [3]. On one hand, the massive fine particles promote recrystallization nucleation decreasing the grain size. On the other hand, the massive fine particles can effectively block slip during tensile test. The as-extruded Mg–4Sm–4Zn–0.5Zr alloy exhibits the best comprehensive mechanical properties at room temperature, and the yield strength, ultimate tensile strength and elongation are about 246 MPa, 273 MPa and 21%, respectively. This result is attributed to the massive dispersed nanoscale particles reducing the DRXed grain size and blocking slip of dislocation effectively in as-extruded Mg–4Sm–4Zn–0.5Zr alloy.

Figure 15 shows the stress-strain curves of as-extruded Mg–4Sm–4Zn–0.5Zr alloy at room temperature and 473 K. It can be seen that the as-extruded Mg–4Sm–4Zn–0.5Zr alloy has a relatively good tensile properties at elevated temperature, which should be attributed to the heat-resistant $(Mg,Zn)_3Sm$ phase with different sizes blocking not only the motion of dislocation but also the slide of the grain boundary.

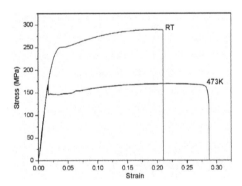

Figure 15. The stress-strain curves of as-extruded Mg–4Sm–4Zn–0.5Zr alloy at room temperature and 473 K.

3.5. Fracture

Figure 16 shows the SEM fractographs of tensile tests of as-cast and as-extruded Mg–4Sm–*x*Zn–0.5Zr alloys at room temperature. It is obvious that the fractographs of as-cast alloys mainly consist of cleavage planes, tear ridges and shallow dimples, revealing the poor ductility of as-cast alloys. The coarse ridges and cleavage planes in Figure 16e correspond to the coarse eutectic compounds in as-cast Mg–4Sm–4Zn–0.5Zr alloy, indicating that excessively coarse eutectic compounds are harmful to the toughness of alloy. The abundant ridges and deep dimples observed in Figure 16f,j reveal the excellent plasticity of as-extruded alloys, which is consistent with the high elongations in Figure 14c.

In short, the fracture mechanism of as-cast alloys is transgranular cleavage fracture, and the ductility of alloys has been improved much by hot extrusion.

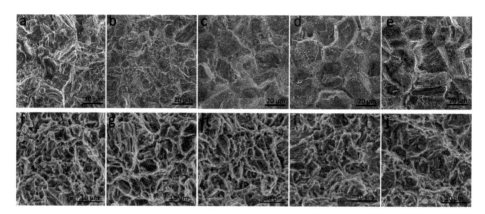

Figure 16. SEM fractographs of tensile tests of as-cast (**a**–**e**) and as-extruded (**f**–**j**) Mg–4Sm–xZn–0.5Zr alloys at room temperature: (**a**,**f**) x = 0; (**b**,**g**) x = 1; (**c**,**h**) x = 2; (**d**,**i**) x = 3; (**e**,**j**) x = 4.

4. Conclusions

(1) The as-cast Mg–4Sm–0.5Zr alloys contains α-Mg matrix and Mg$_{41}$Sm$_5$ phase. The microstructures of as-cast Mg–4Sm–xZn–0.5Zr (x = 1, 2, 3, 4 wt %) alloys mainly consist of α-Mg matrix, Mg$_{41}$Sm$_5$ and (Mg,Zn)$_3$Sm.

(2) The as-cast Mg–4Sm–2Zn–0.5Zr alloy exhibits the best comprehensive tensile properties at room temperature among all the as-cast alloys, and YS, UTS and EL are 132 MPa, 175 MPa and 8.7%, respectively.

(3) f Besides the block-shaped precipitates, some disc-like precipitates also are observed in the as-ex elongation are about 246 MPa, 273 MPa and 21%, respectively, which is attributed to the massive dispersed nanoscale particles effectively reducing the DRXed grain size and blocking slip of dislocation. The dynamic precipitates in as-extruded Mg–4Sm–4Zn–0.5Zr alloy containing basal precipitates having a hexagonal structure with a = 0.556 nm.

Acknowledgments: This project is supported by National Natural Science Foundation of China (Grant No. 51575231).

Author Contributions: Chaojie Che, Fanxing Meng performed the data collection, figures, and data analyses; Chaojie Che wrote the original manuscript; Zhongyi Cai designed the experiments and revised the manuscript; Liren Cheng contributed to data analyses, data interpretation and manuscript revision; Zhen Yang participated in data collection and analyses.

Conflicts of Interest: The author declare no conflict of interest.

References

1. Pan, F.; Yang, M.; Chen, X. A review on casting magnesium alloys: Modification of commercial alloys and development of new alloys. *J. Mater. Sci. Technol.* **2016**, *32*, 1211–1221. [CrossRef]

2. Sarker, D.; Friedman, J.; Chen, D.L. Twin growth and texture evolution in an extruded AM30 magnesium alloy during compression. *J. Mater. Sci. Technol.* **2014**, *30*, 884–887. [CrossRef]

3. Mordike, B.L.; Ebert, T. Magnesium properties-applications-potential. *Mater. Sci. Eng. A* **2001**, *302*, 37–45. [CrossRef]

4. Rokhlin, L.L.; Dobatkina, T.V.; Nikitina, N.I. Constitution and properties of the ternary magnesium alloys containing two rare-earth metals of different subgroups. *Mater. Sci. Forum* **2003**, *419–422*, 291–296. [CrossRef]

5. Nie, J.F.; Muddle, B.C. Precipitation in magnesium alloy WE54 during isothermal ageing at 250 degrees C. *Scr. Mater.* **1999**, *40*, 1089–1094. [CrossRef]
6. Li, D.; Wang, Q.; Ding, W. Characterization of phases in Mg–4Y–4Sm–0.5Zr alloy processed by heat treatment. *Mater. Sci. Eng. A* **2006**, *428*, 295–300. [CrossRef]
7. Antion, C.; Donnadieu, P.; Perrard, F.; Deschamps, A.; Tassin, C.; Pisch, A. Hardening precipitation in a Mg-4Y-3Re alloy. *Acta Mater.* **2003**, *51*, 5335–5348. [CrossRef]
8. Gröbner, J.; Kozlov, A.; Fang, X.Y.; Geng, J.; Nie, J.F.; Schmid-Fetzer, R. Phase equilibria and transformations in ternary Mg-rich Mg–Y–Zn alloys. *Acta Mater.* **2012**, *60*, 5948–5962. [CrossRef]
9. He, S.M.; Zeng, X.Q.; Peng, L.M.; Gao, X.; Nie, J.F.; Ding, W.J. Microstructure and strengthening mechanism of high strength Mg–10Gd–2Y–0.5Zr alloy. *J. Alloys Compd.* **2007**, *427*, 316–323. [CrossRef]
10. Li, D.J.; Zeng, X.Q.; Dong, J.; Zhai, C.Q.; Ding, W.J. Microstructure evolution of Mg–10Gd–3Y–1.2Zn–0.4Zr alloy during heat-treatment at 773k. *J. Alloys Compd.* **2009**, *468*, 164–169. [CrossRef]
11. Hou, X.; Cao, Z.; Zhao, L.; Wang, L.; Wu, Y.; Wang, L. Microstructure, texture and mechanical properties of a hot rolled Mg–6.5Gd–1.3Nd–0.7Y–0.3Zn alloy. *Mater. Des.* **2012**, *34*, 776–781. [CrossRef]
12. Zheng, K.Y.; Dong, J.; Zeng, X.Q.; Ding, W.J. Precipitation and its effect on the mechanical properties of a cast Mg–Gd–Nd–Zr alloy. *Mater. Sci. Eng. A* **2008**, *489*, 44–54. [CrossRef]
13. Nie, J.F.; Gao, X.; Zhu, S.M. Enhanced age hardening response and creep resistance of Mg–Gd alloys containing Zn. *Scr. Mater.* **2005**, *53*, 1049–1053. [CrossRef]
14. Mostafa, A.; Medraj, M. Experimental investigation of the Mg-Nd-Zn isothermal section at 300 degrees C. *Metals* **2015**, *5*, 84–101. [CrossRef]
15. Xia, X.; Sun, W.; Luo, A.A.; Stone, D.S. Precipitation evolution and hardening in Mg-Sm-Zn-Zr alloys. *Acta Mater.* **2016**, *111*, 335–347. [CrossRef]
16. Qi, H.Y.; Huang, G.X.; Bo, H.; Xu, G.L.; Liu, L.B.; Jin, Z.P. Thermodynamic description of the Mg–Nd–Zn ternary system. *J. Alloys Compd.* **2011**, *509*, 3274–3281. [CrossRef]
17. Zheng, X.; Dong, J.; Xiang, Y.; Chang, J.; Wang, F.; Jin, L.; Wang, Y.; Ding, W. Formability, mechanical and corrosive properties of Mg–Nd–Zn–Zr magnesium alloy seamless tubes. *Mater. Des.* **2010**, *31*, 1417–1422. [CrossRef]
18. Saito, K.; Hiraga, K. The structures of precipitates in an Mg-0.5 at %Nd age-hardened alloy studied by HAADF-STEM technique. *Mater. Trans.* **2011**, *52*, 1860–1867. [CrossRef]
19. Yuan, M.; Zheng, Z. Effects of Zn on the microstructures and mechanical properties of Mg–3Sm–0.5Gd–xZn–0.5Zr (x = 0, 0.3 and 0.6) alloy. *J. Alloys Compd.* **2014**, *590*, 355–361. [CrossRef]
20. Xia, X.; Luo, A.A.; Stone, D.S. Precipitation sequence and kinetics in a Mg-4Sm-1Zn-0.4Zr (wt %) alloy. *J. Alloys Compd.* **2015**, *649*, 649–655. [CrossRef]
21. Dogan, E.; Wang, S.; Vaughan, M.W.; Karaman, I. Dynamic precipitation in Mg-3Al-1Zn alloy during different plastic deformation modes. *Acta Mater.* **2016**, *116*, 1–13. [CrossRef]
22. Kabir, A.S.H.; Sanjari, M.; Su, J.; Jung, I.H.; Yue, S. Effect of strain-induced precipitation on dynamic recrystallization in Mg-Al-Sn Alloys. *Mater. Sci. Eng. A Struct.* **2014**, *616*, 252–259. [CrossRef]
23. Zhang, S.; Yuan, G.Y.; Lu, C.; Ding, W.J. The relationship between (mg,zn)3re phase and 14h-lpso phase in Mg-Gd-Y-Zn-Zr alloys solidified at different cooling rates. *J. Alloys Compd.* **2011**, *509*, 3515–3521. [CrossRef]
24. Yuan, M.; Zheng, Z.Q. Effects of heat treatment on microstructure and mechanical properties of Mg-2.6Sm-1.3Gd-0.6Zn-0.5Zr alloy. *Mater. Sci. Technol.* **2014**, *30*, 261–267. [CrossRef]
25. Ping, D.H.; Hono, K.; Nie, J.F. Atom probe characterization of plate-like precipitates in a Mg-Re-Zn-Zr casting alloy. *Scr. Mater.* **2003**, *48*, 1017–1022. [CrossRef]
26. Zheng, J.X.; Zhou, W.M.; Chen, B. Precipitation in Mg-Sm binary alloy during isothermal ageing: Atomic-scale insights from scanning transmission electron microscopy. *Mater. Sci. Eng. A Struct.* **2016**, *669*, 304–311. [CrossRef]
27. Nie, J.F. Precipitation and hardening in magnesium alloys. *Metall. Mater. Trans. A* **2012**, *43*, 3891–3939. [CrossRef]
28. Su, J.; Kaboli, S.; Kabir, A.S.H.; Jung, I.H.; Yue, S. Effect of dynamic precipitation and twinning on dynamic recrystallization of micro-alloyed Mg-Al-Ca alloys. *Mater. Sci. Eng. A* **2013**, *587*, 27–35. [CrossRef]
29. Robson, J.D.; Henry, D.T.; Davis, B. Particle effects on recrystallization in magnesium–manganese alloys: Particle-stimulated nucleation. *Acta Mater.* **2009**, *57*, 2739–2747. [CrossRef]

30. Humphreys, F.J.; Hatherly, M. *Recrystallization and Related Annealing Phenomena*, 2nd ed.; Elsevier: Oxford, UK, 2004.
31. Chen, Z.H.; Xu, F.Y.; Fu, D.F.; Xia, W.J. The dynamic recrystallization of magnesium alloys. *Chem. Ind. Eng. Prog.* **2006**, *25*, 7.
32. Barnett, M.R. Recrystallization during and following hot working of magnesium alloy az31. *Mater. Sci. Forum* **2003**, *419–422*, 503–508. [CrossRef]
33. Barnett, M.R.; Beer, A.G.; Atwell, D.; Oudin, A. Influence of grain size on hot working stresses and microstructures in Mg-3Al-lZn. *Scr. Mater.* **2004**, *51*, 19–24. [CrossRef]

Article

Study on Microstructure and Properties of Bimodal Structured Ultrafine-Grained Ferrite Steel

Gang Niu [1,2], Huibin Wu [1,2,*], Da Zhang [2], Na Gong [2] and Di Tang [1,2]

1 Collaborative Innovation Center of Steel Technology, University of Science and Technology Beijing, Beijing 100083, China; ustbniug@163.com (G.N.); 13120336753@163.com (D.T.)
2 Institute of Engineering Technology, University of Science and Technology Beijing, Beijing 100083, China; claycn@outlook.com (D.Z.); gongnana.cheng@gmail.com (N.G.)
* Correspondence: wuhb@ustb.edu.cn; Tel.: +86-010-6233-2617

Received: 24 June 2017; Accepted: 16 August 2017; Published: 18 August 2017

Abstract: The objective of the study research was to obtain bimodal structured ultrafine-grained ferrite steel with outstanding mechanical properties and excellent corrosion resistance. The bimodal microstructure was fabricated by the cold rolling and annealing process of a dual-phase steel. The influences of the annealing process on microstructure evolution and the mechanical properties of the cold-rolled dual-phase steel were investigated. The effect of bimodal microstructure on corrosion resistance was also studied. The results showed that the bimodal characteristic of ferrite steel was most apparent in cold-rolled samples annealed at 650 °C for 40 min. More importantly, due to the coordinated action of fine-grained strengthening, back-stress strengthening, and precipitation strengthening, the yield strength (517 MPa) of the bimodal microstructure improved significantly, while the total elongation remained at a high level of 26%. The results of corrosion experiments showed that the corrosion resistance of bimodal ferrite steel was better than that of dual-phase steel with the same composition. This was mainly because the Volta potential difference of bimodal ferrite steel was smaller than that of dual-phase steel, which was conducive to forming a protective rust layer.

Keywords: bimodal ferrite steel; ultrafine-grained microstructure; mechanical properties; corrosion resistance

---

1. Introduction

With the increasing demand for lightweight vehicles and reduction in container weight, the steel plates used in vehicles and containers require high strength, which can significantly reduce the thickness and weight of vehicles and containers, and higher plasticity to permit cold forming in the manufacturing process. Good weather resistance is also required to face harsh service environments [1,2]. Dual-phase steel—the main vehicle steel in industry—has the advantages of a high work hardening rate, low yield ratio, and good match of strength and toughness due to the coexistence of martensite and ferrite. However, its corrosion resistance is not ideal because the potential difference between martensite and ferrite is large and abundant defects exist in martensite [3,4]. Single ferrite with few defects possesses remarkable resistance to weathering [5], but its strength is lower than that of dual-phase steel.

Microstructural refinement is one of the most important methods of metallic materials strengthening to yield lightweight components with improved performance [6–8]. In particular, extremely high yield strength was achieved in ultrafine-grained or nanograined steels (UFG/NG), and the yield strength resembled the Hall-Petch relationship [9,10]. However, the ductility of these steels with homogeneous UFG/NG microstructures was considerably lower than those with coarse-grained microstructures. The inferior ductility has severely limited the application of this material. The low ductility was caused by low strain hardening [11–13], which is the result of small grain sizes. In recent years, several research efforts were made to achieve a combination of high strength and reasonable ductility through the creation of heterogeneous microstructures, and it was clearly demonstrated that the bimodal grain size distribution is a simple and effective approach to obtaining heterogeneous microstructures with high strength and sufficient ductility [7,9,14–17]. In addition, the bimodal microstructure is generally composed of a single phase, and there is no obvious potential difference between the two phases [18–20]. In this study, bimodal structured ultrafine-grained ferrite steel is obtained from the dual-phase steel through cold rolling and annealing. It is fabricated by the recrystallization of deformed martensite and deformed ferrite. The main objective of this study is to investigate the microstructure evolution, mechanical properties, and corrosion resistance of bimodal ferrite steel.

2. Materials and Experimental Methods

The chemical composition of low carbon steel was (in wt %): 0.06C, 1.17Si, 1.23Mn, 0.6Cr, 0.24Mo, and 0.025Al, balanced with Fe. Dual-phase steels were prepared by melting in a vacuum induction furnace (Shenyang Jinyan Co., Ltd., Shenyang, China). The cast slabs were reheated at 1200 °C for 4 h and hot-rolled into 6 mm thick plates. Subsequently, the plates were austenitized at 780 °C for 15 min and quenched to room temperature. The quenched microstructure consisted of 74% ferrite and 26% martensite as shown in Figure 1a. Next, the steel sheets were cold rolled with 80% total thickness reduction. The cold-rolled specimen thickness was 1 mm and the deformed microstructure is shown in Figure 1b. Annealing experiments were carried out on these samples (210 mm × 70 mm). The annealing temperature was 650 °C and the holding times were 30 min, 40 min, 50 min, and 70 min.

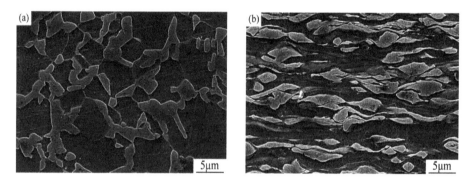

Figure 1. Microstructures of dual-phase steel (**a**) before cold rolling and (**b**) after cold rolling.

The microstructural evolution was analyzed using scanning electron microscopy (ZEISS ULTRA 55, Carl Zeiss, Oberkochen, Germany) and transmission electron microscopy (Tecnai G2 F30 S-TWIN, FEI Company, Hillsboro, OR, USA). The Image Pro-Plus image analysis software (Version 6.0, Media Cybernetics, Inc., Rockville, MD, USA) was applied to determine the grain size of each grain. Tensile tests were carried out at room temperature using a CMT5105 tensile machine (SANS Testing Machine Co., Ltd., Shenzhen, China). Vickers micro-hardness values were measured on an HV-1000 micro-Vickers durometer (300 gf, Shanghai Yanrun instrument factory, Shanghai, China). For SEM examination and Vickers micro-hardness test of the dual-phased microstructure before and after cold rolling (10 mm × 10 mm × 6 mm), and the annealed microstructure (10 mm × 10 mm × 1 mm), specimens were cut from the dual-phase steel's strip before and after cold rolling and the annealed steel's strip, respectively. The specimen surface was ground with emery paper (200-grit to 2000-grit), polished and then etched with 4 vol % natal. Dog-bone-shaped tensile specimens were machined to a gauge length of 25 mm and gauge width of 6 mm. The samples were cut from the dual-phase steel's strip before and after cold rolling and the annealed steel's strip.

The corrosion experiments include neutral salt spray test with 5% NaCl solution (YWX/Q-150 salt fog-box), immersion test, scanning Kelvin probe force microscopy (SKPFM) test, and the electrochemical test. The neutral salt spray was carried out in a salt fog-box at a constant temperature of 50 °C. The immersion test was controlled by a thermostat water bath at 50 °C. The corrosion solutions for the immersion test and the neutral salt spray test were 5% NaCl aqueous solution. The samples for corrosion experiments were cut (50 mm × 15 mm × 1 mm) from the annealed plates (annealed at 650 °C for 40 min) and the dual-phase steel, as mentioned above. All samples were polished to a surface finish of 600-grit with emery paper and ultrasonically cleaned with acetone, dried with clean air and stored in a desiccator before testing. Besides, all experiments were conducted with three groups of contrast samples.

The SKPFM measurements were performed with an atomic force microscope (AFM), MFP-3D infinity (Oxford instruments, Oxford, UK). A Ti/Ir probe was employed for Volta potential measurements. The surfaces of coupons (annealed at 650 °C for 40 min) for the SKPFM test were prepared by grinding to 2000-grit and electropolishing at a voltage of 15 V for 25 sin electrolyte composed of 20 vol % perchloric acid and 80% ethanol. In electrochemical experiments, the exposed surfaces of the specimens for potentiodynamic polarization measurement were 10 mm × 10 mm, with an area of 1 cm^2. All electrochemical test specimens (annealed at 650 °C for 40 min) were enclosed with epoxy resin, leaving a working area of 1 cm^2. Prior to testing, the exposed surface was ground with 200-grit to 2000-grit emery paper. The interfaces between epoxy and sample were sealed with silicone to prevent unwanted crevice corrosion. Potentiodynamic polarization tests were conducted using a CS 310 electrochemical workstation. A three-electrode system was used with the steel specimen as the working electrode, a platinum sheet as the counter electrode, and an Ag/AgCl electrode as the reference electrode. The polarization potential was swept from −0.5 V to 0.8 V vs. the open circuit potential (OCP) at a scan rate of 0.5 mV/s. The frequency range for electrochemical impedance spectroscopy (EIS) was from 10^5 Hz to 10^{-2} Hz. The corrosive electrolyte was 5% NaCl aqueous solution.

3. Results and Discussion

3.1. Effect of Annealing on the Microstructure Evolution of Cold-Rolled Dual-Phase Steel

The cold-rolled microstructure of the dual-phase steel was shown in Figure 1b. It can be seen that the microstructure was fibroid along the rolling direction, and both martensite and ferrite were seriously distorted due to the large deformation. The microstructures of samples annealed at 650 °C for different holding times and the corresponding diagrams of grain size distribution were shown in Figure 2. The grain size measured was that of ferrite. Grains of the coarse grain region and fine grain region were counted separately. The fine grain region was formed by the recrystallization of originally deformed martensite and there were many fine carbides distributed in the fine grain region. However, there was no carbide precipitation in the coarse grain zone. In addition, the micro-hardness of the fine grain region was higher than that of the coarse grain region. Therefore, the annealed microstructures were marked first with a micro-Vickers durometer. To make the statistical results more accurate, three metallographic specimens were prepared for each annealed microstructure and three SEM images were taken from different locations on each metallographic specimen. When the annealing time was 30 min, although preliminary development of the bimodal phenomenon can be seen, the coarse grain region and fine grain region were both fine and the difference between them was not obvious. Furthermore, the grains with diameter ≤200 nm occupied a large proportion of the fine grain region, as shown in Figure 2a,e. When the annealing time was increased to 40 min, the bimodal characteristic was most obvious (Figure 2b). The peak value of the fine grain region was 0.91 μm and the peak value of the coarse grain region was 1.9 μm (Figure 2f). When the annealing time was 50 min, the change of grain size in the coarse grain region was not distinct. However, the grains with diameter ≥1.5 μm increased significantly in the fine grain region (Figure 2g). With the annealing time further increased, the bimodal characteristic gradually weakened (Figure 2d). The difference in grain size between the coarse and fine grain regions gradually reduced and the grain size gradually tended to become uniform, as shown in Figure 2h.

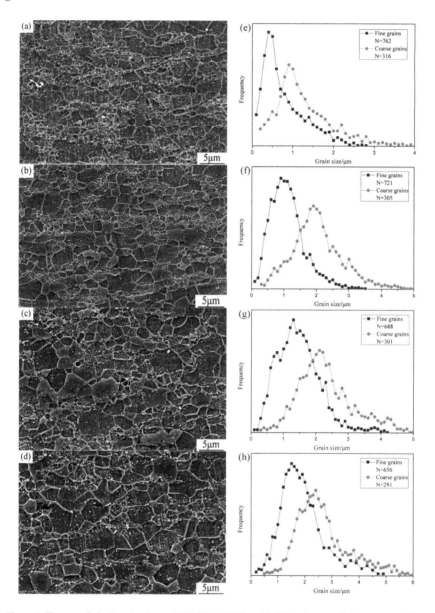

Figure 2. The annealed microstructures at 650 °C and different holding times, and the corresponding diagrams of grain size distribution. Number of the measured grains (N) was given. (**a,e**) 30 min; (**b,f**) 40 min; (**c,g**) 50 min; (**d,h**) 70 min.

The martensite lath structure was preferentially destroyed via the large deformation during the cold rolling process. Hence, the large number of defects and the distortional energy present in the cold-rolled martensite can increase the nucleation rate of recrystallization and effectively refine the annealed microstructure [21,22]. Meanwhile, ferrite carried a much higher plastic deformation as it is softer than martensite. Thus, a large amount of distortional energy was stored in ferrite, which was also beneficial for the recrystallization of the cold-rolled microstructure and favorable for refining the annealed microstructure [7–9]. As shown in Figure 2a, when the annealing time was 30 min, the grains of the original ferrite region almost completed recrystallization. However, there were still incompletely recrystallized grains in the original martensite region. As shown in the white ellipse region of Figure 2a, the original martensite region was still fuzzy and remained slightly distorted. In addition, the Vickers hardness values of microstructures annealed for different times are shown in Figure 4. The microstructure annealed for 30 min retained a higher value of hardness than the subsequent annealing processes. This implied that the recrystallization of grains was not complete when annealed for 30 min. At the beginning of the annealing process, the main driving force (E) for recrystallization comes from the distortional energy [21,23]. Most of the deformation was carried out in the ferrite region due to its soft matrix during the cold rolling process. Hence, the distortional energy of unit volume stored in the martensite region (q_m) was lower than that in the ferrite region (q_f). In addition, there were many precipitates in the martensite region, as shown in Figure 3 (discussed below), which resulted in the restraining force in grain boundary migration. Therefore, the recrystallization of some deformed grains and the growth of some recrystallized grains in favorable positions were both blocked. Thus, the martensite region was gradually transformed to a fine grain region. However, the resistance of grain growth in the primary ferrite region was feeble and the stored distortion energy was higher, so these grains grew quickly and transformed to the coarse grain region.

The microstructure annealed at 650 °C for 40 min is shown in Figure 2b. It can be seen that all grains grew further and recrystallization was completed; grain boundaries in the coarse grains region and fine grains region were both clear. Meanwhile, the regions of fine grains and coarse grains were distributed in strips and formed a lamellar interphase structure, which is consistent with the distribution of cold-rolled martensite and ferrite. Besides, the percentage of fine grain and coarse grain regions were close to those of the original martensite and ferrite regions, respectively. At this condition, the bimodal characteristic of grain distribution was most obvious. When the annealing time was further increased to 50 min and 70 min, the grains of the fine grain region and coarse grain region both grew rapidly due to the more driving force. The Vickers hardness values of these microstructures were low, as shown in Figure 4. Meanwhile, the bimodal characteristics were increasingly obscured as shown in Figure 2c,g and Figure 2d,h. As mentioned earlier, the driving force for recrystallization in the original martensite region (E_m) was lower than that in the original ferrite region (E_f) at the beginning of the annealing process, and the rate of increase R_f was greater than R_m. However, with the increase of annealing time, the deformed grains were gradually replaced with newly undistorted equiaxed grains. Therefore, q_f and q_m decreased gradually and eventually disappeared. When the rate of grain growth in the fine grain region was equal to the rate of grain growth in the coarse grain region, the difference in grain size was the largest and the bimodal characteristic of grain size distribution was most apparent, as shown in Figure 2b,f. The main driving force (E) for grain growth after complete recrystallization is the reduction of grain surface energy. The smaller the grain size is, the higher the surface energy is [24]. The assumption is that all grains are spherical. If R is the grain diameter, let γ be the surface energy per unit surface area of grains. Therefore the driving force for grain growth per unit volume in the martensite region (E_m) and the ferrite region (E_f) are simplified as follows [24].

$$E_m = k\gamma\Delta S/\Delta V = k\gamma dS/dV = 2k\gamma/R_m, \tag{1}$$

$$E_f = 2k\gamma/R_f, \tag{2}$$

where k is the surface energy coefficient. When the annealing time increased further, $E_m > E_f$, the rate of grain growth in the fine grain region exceeded that in the coarse grain region, so the grain sizes of the two regions were gradually consistent.

In addition, the grain growth was also affected by the drag of the second phase particle. The dispersed precipitates in the fine grain region (annealing for 40 min) are shown in Figure 3. A large number of white spots were distributed in the fine-grained region, which presented two kinds of morphology—rectangular and circular—at high magnification. The size of these precipitates was ~50 nm. The analysis results of EDS indicated that these precipitates contained large quantities of elements Cr, Mn, and Mo, as shown in Table 1. During heating of the two-phase region, most of the alloying elements dissolved in austenite due to their high solubilities. And in the subsequent quenching process, it was difficult for most of the alloying elements to diffuse due to the fast cooling rate. After quenching, a large number quantity of alloy elements (Cr and Mo) and carbon were supersaturated in α-Fe. After cold rolling, the martensite lath structure was preferentially destroyed via the large deformation during the cold rolling process and the large number of defects and the distortional energy present in the cold-rolled martensite. Therefore, during the subsequent annealing process, distorted martensite spontaneously recrystallized and the supersaturated carbon and alloy elements precipitated in the form of carbides dispersed in fine grain. Bimodal structure ultrafine-grained ferrite with nano-scale carbides was finally obtained.

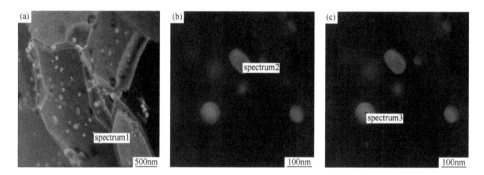

Figure 3. TEM micrographs of precipitates in the fine grain region; (**a**) distribution of precipitates; (**b**) rectangular precipitate; (**c**) circular precipitate.

Table 1. The results of EDS analysis of precipitates and matrix (wt %).

Alloy Element	Spectrum1 (Matrix)	Spectrum2 (Precipitate)	Spectrum3 (Precipitate)
Si	1.5	1.2	0.8
Fe	97.0	81.3	77.7
Mn	1.0	9.0	12.1
Cr	0.5	6.6	7.7
Mo	-	2.0	1.6

3.2. Mechanical Properties of the Annealed Microstructures

The mechanical properties of the cold-rolled samples annealed at 650 °C and different holding times are shown in Figure 4. The yield strength increased first and then decreased slightly with the increase of annealing time, and the elongation increased always. Meanwhile, the yield strength and the elongation of all annealed samples improved relative to the dual-phase steel. Particularly, when the cold-rolled samples were annealed at 650 °C for 40 min, the yield strength (517 MPa) of the bimodal microstructure significantly improved, while the total elongation remained at a high level of 26%, which were attributed to the coordinated action of fine-grained strengthening, back-stress strengthening, and precipitation strengthening [16]. On the one hand, the refinement of bimodal ferrite grains and the increase of dislocation density with tensile strain caused an increase in stress, i.e., fine-grained strengthening, which was caused by the increase in grain boundaries [6,25,26]. On the other hand, the soft lamellae of coarse grains start plastic deformation first during the tensile process. However, they were constrained by the surrounding hard lamellae of the fine grains. Therefore, dislocations in coarse grains were piled up and blocked at lamella interfaces, which were actually grain boundaries. This produced a long-range back stress to increase the difficulty for dislocations to slip in the lamellae of coarse grains until the surrounding lamellae of fine grains started to yield at a larger global strain and to stop the dislocation source from emitting more dislocations [16,27,28]. This means that the soft lamellae constrained by hard lamellae appeared much stronger than when they were not constrained. Besides, there was a large amount of precipitation dispersed in the fine grain region of the bimodal ferrite microstructure, which would induce significant pinning effect on the dislocation motion in tension, i.e., precipitation strengthening [29,30]. In summary, the combined effect of fine-grained strengthening, back-stress strengthening, and precipitation strengthening gave the bimodal ferrite steel reasonable strength, although the microstructure of the bimodal ferrite steel was completely ferrite. Meanwhile, the complete ferrite matrix contributed to prominent elongation.

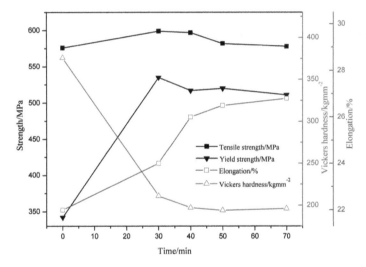

Figure 4. Mechanical properties of cold-rolled specimens annealed at 650 °C and different holding times. The first value of yield strength, tensile strength, and elongation belongs to the dual-phase steel (the Vickers hardness of the dual-phase steel was 194 HV). The first value of Vickers hardness belongs to the dual-phase steel after cold rolling (The values of yield strength, tensile strength, and elongation of the dual-phase steel after cold rolling were 1008 MPa, 1125 MPa, and 4.7%, respectively).

The stress-strain curves of the annealed specimens were also obtained during the tensile process as shown in Figure 5. Yield plateaus appeared in stress-strain curves of the annealed microstructures. Previous studies showed that the occurrence of a yield point needs to satisfy three conditions [31,32]: (i) low initially mobile dislocation density; (ii) rapid increase of dislocation during deformation; and (iii) rate of dislocation slip controlled by the loading stress. The annealed microstructures contained ultrafine-grained ferrite and nanoscale carbides; the recrystallization of ferrite was relatively completed, which caused the great disappearance of matrix dislocation. Because of the pinning effect of the Cottrell atmosphere [33] and precipitates [29,30] on dislocation, the motion of dislocation needed sufficient stress. Once the dislocation was free from pinning, it could move under relatively low stress. Therefore, the continuous appearance of this situation produced a yield platform. Besides, the decrease of grain size was beneficial for producing the yield platform and increasing the extension length of the yield platform [9,34]. Hence, the appearance of the yield plateau in bimodal ferrite steel was mainly due to ultrafine grains of ferrite and numerous nano-scale carbides.

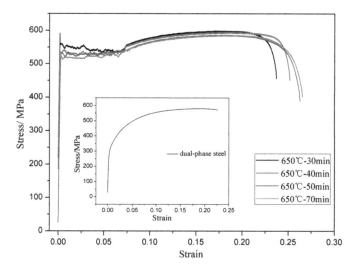

Figure 5. Stress-strain curves of the original dual-phase steel and the specimens annealed at 650 °C for different holding times.

3.3. Corrosion Behavior of Bimodal Ferrite Steel and Dual-Phase Steel

In order to study the corrosion behaviors of bimodal ferrite steel and dual-phase steel, the neutral salt spray, immersion, SKPFM, and electrochemistry tests were used. After neutral salt spray testing for eight days, the rates of weight loss for the two experimental steels are shown in Figure 6. The corrosion rates of both steels increased rapidly at the beginning of corrosion and then remained at a stable level. This is mainly because rust layers were not formed in the early stages and the corrosion resistances of the substrates were poor. In addition, the corrosion rate of bimodal ferrite steel was slightly higher than that of dual-phase steel. It is possible that a micro battery formed between the nanoscale precipitates and substrate, which was likely to promote corrosion. With the increase of corrosion time, a rust layer began to form. The dense and protective rust layer could form easier in the bimodal ferrite steel than in dual-phase steel. This was mostly because a large number of defects existed in martensite, which facilitated the occurrence and development of corrosion [5,6]. Meanwhile, the Figure 7b shows that the Volta potential of martensite was higher than the Volta potential of the surrounding ferrite and that the Volta potential difference between martensite and ferrite was large, which improved the electrochemical activity of the dual-phase microstructure and accelerated the progress of corrosion [19,20]. However, Figure 7a shows that the Volta potential of bimodal ferrite was low and there was no obvious Volta potential difference in bimodal ferrite. It is more difficult to form a dense and protective rust layer on the dual-phase microstructure than the bimodal microstructure. Therefore, when the corrosion time exceeded 72 h, the decrease of corrosion rate in the bimodal ferrite steel was faster due to the massive formation of compacted rust layers that can effectively hinder the permeation of the corrosion medium.

After immersion testing for seven days, the corrosion morphologies of the samples after removing corrosion product are shown in Figure 8. Figure 8a,b show that the corrosion of the two experimental steels were non-uniform, with pitting areas and full corrosion areas. Figure 8c,d show enlargement of the full corrosion area in both steels. Figure 8e,f shows enlargement of the pitting area in both steels. It can be seen that corrosion of dual-phase steel was more serious, and the corrosion depth of dual-phase steel was larger than that of bimodal ferrite steel. As mentioned earlier, there were several defects in the martensite substrate, which accelerated the corrosion of samples, facilitated the extension of corrosion pits in the substrate. And were not conducive to the formation of a compact rust layer. The result of the SKPFM test also showed that the Volta potential difference between martensite and ferrite was large, as shown in Figure 7b. However, Figure 7a implies that the Volta potentials of very few regions were high in bimodal ferrite steel. There was probably a micro battery between the nanoscale precipitates and substrate in the fine grain region.

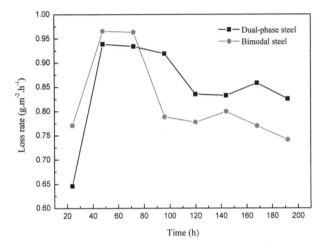

Figure 6. The rate of weight loss for dual-phase steel and bimodal ferrite steel (annealed for 40 min).

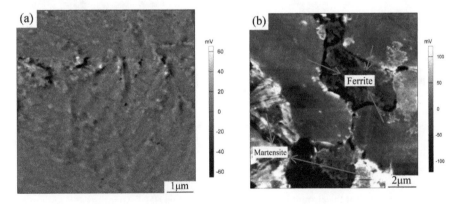

Figure 7. Volta potential images of (**a**) bimodal ferrite steel and (**b**) dual-phase steel.

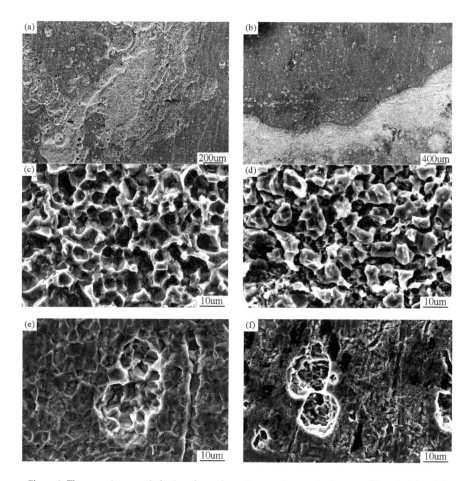

Figure 8. The corrosion morphologies of samples with corrosion product removed (**a,c,e**)—bimodal ferrite steel; (**b,d,f**)—dual-phase steel.

The results of the electrochemistry test are shown in Figure 9. Figure 9a shows that passivation regions were observed in both steels. When the samples were placed in the electrolyte, ion exchange between the sample surface and the solution gradually formed a dynamic balance. When voltage was applied, this dynamic balance was quickly destroyed and the electrode potential increased rapidly, leading to the increase of current density and accelerated the corrosion. However, with the increase in corrosion time, the sample surface was gradually passivated. The electrode potential continued to increase but the current was nearly constant at this stage. This current is known as the passivation current which can be used to characterize the corrosion resistances of metals. The smaller the passivation current is, the better the corrosion resistance is [35]. Hence, the corrosion resistance of bimodal ferrite steel was better than that of dual-phase steel, as shown in Figure 9a. With the further increase of potential, the current increased instantaneously. Here, voltage corresponded to the breakdown voltage of the passivation film, which can be used to characterize the stability of the passivation film. The higher the breakdown potential is, the better the stability of passivation film is [36]. Therefore, the stability of the passivation film in bimodal ferrite steel was better than that of dual-phase steel, as shown in Figure 9a. The electrochemical impedance spectrum (EIS, shown in Figure 9b) results showed that the radius of EIS in the bimodal ferrite steel was larger than that of dual-phase steel. The higher the radius of EIS, the stronger the stability of the passivation film [37–39]. Therefore, the enhancement of corrosion resistance of the bimodal structure ultrafine-grained ferrite steel was mainly achieved by promoting the formation of a dense rust layer and strengthening the stability of the rust layer.

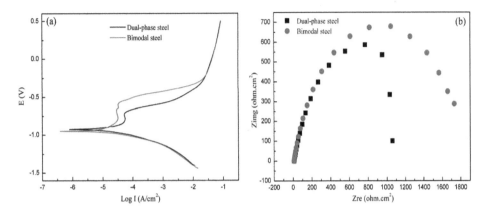

Figure 9. The results of the electrochemistry test: (**a**) polarization curve and (**b**) AC impedance spectroscopy.

4. Conclusions

1. With the annealing time increasing, deformed ferrite and deformed martensite gradually recrystallized. When the samples were annealed at 650 °C for 40 min, a bimodal microstructure was obtained and the fine grain region and coarse grain region were distributed in bands. The peak value of the fine grain region was 0.91 μm, the peak value of coarse grain region was 1.9 μm.

2. Bimodal structured ultrafine-grained ferrite steel had better comprehensive mechanical properties than dual-phase steel. This is mainly because the hard fine grains were embedded in the soft coarse grains to form a lamellar interphase structure. The fine-grained strengthening, back-stress strengthening, and precipitation strengthening produced by the lamellar interphase structure contributed to the excellent match of strength and ductility. Meanwhile, complete ferrite with ultrafine grains was conducive to the appearance of the yield plateau.

3. The neutral salt spray, immersion, SKPFM, and the electrochemistry tests showed that the corrosion resistance of bimodal ferrite steel was better than that of dual-phase steel. The main reason was there were several defects in the martensite substrate and the potential difference between martensite and ferrite was large, which accelerated the corrosion process and was not conducive to the formation of a compact rust layer. On the contrary, the substrate of bimodal ferrite microstructure contained fewer defects and there was no obvious potential difference, which was detrimental to the occurrence and development of corrosion and conducive to forming a protective rust layer. Therefore, the bimodal structured ultrafine-grained ferrite steel possessed excellent corrosion resistance.

Acknowledgments: This research was supported by the National Natural Science Foundation of China (Grant No. 51474031).

Author Contributions: Gang Niu, Huibin Wu and Di Tang conceived and designed the experiments; Gang Niu and Da Zhang performed the experiments; Gang Niu, Huibin Wu, and Na Gong analyzed the data; Da Zhang contributed reagents/materials/analysis tools; Gang Niu wrote the paper; Huibin Wu revised the language in this paper.

Conflicts of Interest: The authors declare no conflict of interest.

References

1. Ghosh, S.; Singh, A.K.; Mula, S.; Chanda, P.; Mahashabde, V.V.; Roy, T.K. Mechanical properties, formability and corrosion resistance of thermomechanically controlled processed Ti-Nb stabilized IF steel. *Mater. Sci. Eng. A* **2017**, *684*, 22–36. [CrossRef]

2. Qu, S.; Pang, X.; Wang, Y.; Gao, K. Corrosion behavior of each phase in low carbon microalloyed ferrite-bainite dual-phase steel: Experiments and modeling. *Corros. Sci.* **2013**, *75*, 67–77. [CrossRef]

3. Wu, H.; Liang, J.; Tang, D.; Liu, X.; Zhang, P.; Yue, Y. Influence of Inclusion on Corrosion Behavior of E36 Grade Low-alloy Steel in Cargo Oil Tank Bottom Plate Environment. *J. Iron Steel Res. Int.* **2014**, *21*, 1016–1021. [CrossRef]

4. Al-Duheisat, S.A.; El-Amoush, A.S. Effect of deformation conditions on the corrosion behavior of the low alloy structural steel girders. *Mater. Des.* **2016**, *89*, 342–347. [CrossRef]

5. Osório, W.R.; Peixoto, L.C.; Garcia, A. Electrochemical corrosion behaviour of a Ti-IF steel and a SAE 1020 steel in a 0.5 M NaCl solution. *Mater. Corros.* **2010**, *61*, 407–411. [CrossRef]

6. Zeng, Y.; Jiang, B.; Li, R.; Yin, H.; Al-Ezzi, S. Grain Refinement Mechanism of the As-Cast and As-Extruded Mg–14Li Alloys with Al or Sn Addition. *Metals* **2017**, *7*, 172. [CrossRef]

7. Zhao, D.; Liu, Y.; Liu, F.; Wen, Y.; Zhang, L.; Dou, Y. ODS ferritic steel engineered with bimodal grain size for high strength and ductility. *Mater. Lett.* **2011**, *65*, 1672–1674.

8. Odnobokova, M.; Belyakov, A.; Kaibyshev, R. Development of nanocrystalline 304L stainless steel by large strain cold working. *Metals* **2015**, *5*, 656–668. [CrossRef]

9. Zhao, M.; Yin, F.; Hanamura, T.; Nagai, K.; Atrens, A. Relationship between yield strength and grain size for a bimodal structural ultrafine-grained ferrite/cementite steel. *Scr. Mater.* **2007**, *57*, 857–860. [CrossRef]

10. Yuan, W.; Panigrahi, S.K.; Su, J.Q.; Mishra, R.S. Influence of grain size and texture on Hall–Petch relationship for a magnesium alloy. *Scr. Mater.* **2011**, *65*, 994–997. [CrossRef]

11. Wei, Y.; Li, Y.; Zhu, L.; Liu, Y.; Lei, X.; Wang, G.; Gao, H. Evading the strength-ductility trade-off dilemma in steel through gradient hierarchical nanotwins. *Nat. Commun.* **2014**, *5*, 3580. [CrossRef] [PubMed]

12. Wu, X.; Yuan, F.; Yang, M.; Jiang, P.; Zhang, C.; Chen, L.; Ma, E. Nanodomained nickel unite nanocrystal strength with coarse-grain ductility. *Sci. Rep.* **2015**, *5*, 11728. [CrossRef] [PubMed]

13. Ueno, H.; Kakihata, K.; Kaneko, Y.; Hashimoto, S.; Vinogradov, A. Nanostructurization assisted by twinning during equal channel angular pressing of metastable 316L stainless steel. *J. Mater. Sci.* **2011**, *46*, 4276–4283. [CrossRef]

14. Wu, H.; Niu, G.; Wu, F.; Tang, D. Reverse-transformation austenite structure control with micro/nanometer size. *Int. J. Miner. Metall. Mater.* **2017**, *24*, 530–537. [CrossRef]

15. Karmakar, A.; Karani, A.; Patra, S.; Chakrabarti, D. Development of bimodal ferrite-grain structures in low-carbon steel using rapid intercritical annealing. *Metall. Mater. Trans. A* **2013**, *44*, 2041–2052. [CrossRef]

16. Wu, X.; Yang, M.; Yuan, F.; Wu, G.; Wei, Y.; Huang, X.; Zhu, Y. Heterogeneous lamella structure unites ultrafine-grain strength with coarse-grain ductility. *Proc. Natl. Acad. Sci. USA* **2015**, *112*, 14501–14505. [CrossRef] [PubMed]

17. Vajpai, S.K.; Ota, M.; Zhang, Z.; Ameyama, K. Three-dimensionally gradient harmonic structure design: An integrated approach for high performance structural materials. *Mater. Res. Lett.* **2016**, *4*, 191–197. [CrossRef]

18. Tan, H.; Jiang, Y.; Deng, B.; Sun, T.; Xu, J.; Li, J. Effect of annealing temperature on the pitting corrosion resistance of super duplex stainless steel UNS S32750. *Mater. Charact.* **2009**, *60*, 1049–1054. [CrossRef]

19. Örnek, C.; Engelberg, D.L. SKPFM measured Volta potential correlated with strain localisation in microstructure to understand corrosion susceptibility of cold-rolled grade 2205 duplex stainless steel. *Corros. Sci.* **2015**, *99*, 164–171. [CrossRef]

20. Nakhaie, D.; Davoodi, A.; Imani, A. The role of constituent phases on corrosion initiation of NiAl bronze in acidic media studied by SEM-EDS, AFM and SKPFM. *Corros. Sci.* **2014**, *80*, 104–110. [CrossRef]

21. Kisko, A.; Hamada, A.S.; Talonen, J.; Porter, D.; Karjalainen, L.P. Effects of reversion and recrystallization on microstructure and mechanical properties of Nb-alloyed low-Ni high-Mn austenitic stainless steels. *Mater. Sci. Eng. A* **2016**, *657*, 359–370. [CrossRef]

22. Wu, H.; Niu, G.; Cao, J.; Yang, M. Annealing of strain-induced martensite to obtain micro/nanometre grains in austenitic stainless. *Mater. Sci. Technol.* **2017**, *33*, 480–486. [CrossRef]

23. Meng, Y.; Sugiyama, S.; Yanagimoto, J. Microstructural evolution during RAP process and deformation behavior of semi-solid SKD61 tool steel. *J. Mater. Process. Technol.* **2012**, *212*, 1731–1741. [CrossRef]

24. Li, W.; Wu, Y.; He, J.; Nieh, T.; Lu, Z. Grain growth and the Hall–Petch relationship in a high-entropy FeCrNiCoMn alloy. *Scr. Mater.* **2013**, *68*, 526–529. [CrossRef]

25. Gutierrez-Urrutia, I.; Zaefferer, S.; Raabe, D. The effect of grain size and grain orientation on deformation twinning in a Fe–22 wt. % Mn–0.6 wt. % C TWIP steel. *Mater. Sci. Eng. A* **2010**, *527*, 3552–3560. [CrossRef]

26. Thijs, L.; Sistiaga, M.L.M.; Wauthle, R.; Xie, Q.; Kruth, J.P.; Van Humbeeck, J. Strong morphological and crystallographic texture and resulting yield strength anisotropy in selective laser melted tantalum. *Acta Mater.* **2013**, *61*, 4657–4668. [CrossRef]

27. Sinclair, C.; Saada, G.; Embury, J. Role of internal stresses in co-deformed two-phase materials. *Philos. Mag.* **2006**, *86*, 4081–4098. [CrossRef]

28. Mughrabi, H. Dislocation wall and cell structures and long-range internal stresses in deformed metal crystals. *Acta Metall.* **1983**, *31*, 1367–1379. [CrossRef]

29. Funakawa, Y.; Shiozaki, T.; Tomita, K.; Yamamoto, T.; Maeda, E. Development of high strength hot-rolled sheet steel consisting of ferrite and nanometer-sized carbides. *ISIJ Int.* **2004**, *44*, 1945–1951. [CrossRef]

30. Kapoor, M.; Isheim, D.; Ghosh, G.; Vaynman, S.; Fine, M.E.; Chung, Y.W. Aging characteristics and mechanical properties of 1600 MPa body-centered cubic Cu and B2-NiAl precipitation-strengthened ferritic steel. *Acta Mater.* **2014**, *73*, 56–74. [CrossRef]

31. Kumar, A.; Singh, S.B.; Ray, K.K. Influence of bainite/martensite-content on the tensile properties of low carbon dual-phase steels. *Mater. Sci. Eng. A* **2008**, *474*, 270–282. [CrossRef]

32. Ucak, A.; Tsopelas, P. Constitutive model for cyclic response of structural steels with yield plateau. *J. Struct. Eng.* **2010**, *137*, 195–206. [CrossRef]

33. Cottrell, A.H.; Bilby, B.A. Dislocation theory of yielding and strain ageing of iron. *Proc. Phys. Soc. Sect. A* **1949**, *62*, 49. [CrossRef]

34. Schulson, E.M.; Weihs, T.P.; Viens, D.V.; Baker, I. The effect of grain size on the yield strength of Ni$_3$Al. *Acta Metall.* **1985**, *33*, 1587–1591. [CrossRef]

35. Balyanov, A.; Kutnyakova, J.; Amirkhanova, N.A.; Stolyarov, V.V.; Valiev, R.Z.; Liao, X.Z.; Zhu, Y.T. Corrosion resistance of ultrafine-grained Ti. *Scr. Mater.* **2004**, *51*, 225–229. [CrossRef]

36. Gurappa, I. Characterization of different materials for corrosion resistance under simulated body fluid conditions. *Mater. Charact.* **2002**, *49*, 73–79. [CrossRef]

37. Liu, C.; Bi, Q.; Leyland, A.; Matthews, A. An electrochemical impedance spectroscopy study of the corrosion behaviour of PVD coated steels in 0.5 N NaCl aqueous solution: Part II.: EIS interpretation of corrosion behavior. *Corros. Sci.* **2003**, *45*, 1257–1273. [CrossRef]

38. Hoseinpoor, M.; Momeni, M.; Moayed, M.H.; Davoodi, A. EIS assessment of critical pitting temperature of 2205 duplex stainless steel in acidified ferric chloride solution. *Corros. Sci.* **2014**, *80*, 197–204. [CrossRef]

39. Ghaffari, M.; Saeb, M.R.; Ramezanzadeh, B.; Taheri, P. Demonstration of epoxy/carbon steel interfacial delamination behavior: Electrochemical impedance and X-ray spectroscopic analyses. *Corros. Sci.* **2016**, *102*, 326–337. [CrossRef]

 *metals*

Article

Emergence and Progression of Abnormal Grain Growth in Minimally Strained Nickel-200

Olivia D. Underwood [1,*], Jonathan D. Madison [2] and Gregory B. Thompson [3]

[1]   Connector & LAC Technology, Sandia National Laboratories, Albuquerque, NM 87185, USA
[2]   Materials Mechanics, Sandia National Laboratories, Albuquerque, NM 87185, USA; jdmadis@sandia.gov
[3]   Metallurgical and Materials Engineering, The University of Alabama, Tuscaloosa, AL 35487, USA; gthompson@eng.ua.edu
*   Correspondence: odunder@sandia.gov; Tel.: +1-505-844-4286

Received: 1 July 2017; Accepted: 23 August 2017; Published: 30 August 2017

Abstract: Grain boundary engineering (GBE) is a thermomechanical processing technique used to control the distribution, arrangement, and identity of grain boundary networks, thereby improving their mechanical properties. In both GBE and non-GBE metals, the phenomena of abnormal grain growth (AGG) and its contributing factors is still a subject of much interest and research. In a previous study, GBE was performed on minimally strained ($\varepsilon < 10\%$), commercially pure Nickel-200 via cyclic annealing, wherein unique onset temperature and induced strain pairings were identified for the emergence of AGG. In this study, crystallographic segmentation of grain orientations from said experiments are leveraged in tandem with image processing to quantify growth rates for abnormal grains within the minimally strained regime. Advances in growth rates are shown to vary directly with initial strain content but inversely with initiating AGG onset temperature. A numeric estimator for advancement rates associated with AGG is also derived and presented.

Keywords: abnormal grain growth; grain boundary engineering; electron backscattered diffraction; growth rate

1. Introduction

Abnormal grain growth (AGG) is a mechanism by which a subset of grains grow at a rate faster than others. AGG is of significance because, like many other grain-scale and sub-grain-scale features such as freckles [1,2], precipitates [2], or dendrite arm spacing [3], AGG has been shown to have a significant effect on properties across many material systems under various loading conditions [4–7]. The exact mechanisms underpinning the occurrence of AGG are still unclear and remain a subject of much interest and research. Though debate on specific mechanisms exists, most authors agree that the formation of AGG is related to low-energy, high-mobility grain boundaries such as $\Sigma 3$ and its variants (e.g., $\Sigma 9$ and $\Sigma 27$) [8–16]. This is of note as Grain Boundary Engineering (GBE) is a specific type of thermomechanical process used to improve material properties by altering their grain boundary network and often resulting in increased frequency of $\Sigma 3$ boundaries [17,18].

In a previous study by the authors, GBE was performed on commercially pure Nickel-200 cold-worked to plastic strains (ε) of 3%, 6%, and 9%, respectively. This pre-strained material was then subjected to cyclic annealing schedules in which the dwell temperature was increased with each cycle. Among these cases, greater initial cold-work resulted in lower AGG onset temperatures. Specifically, for $\varepsilon = 0\%$, 3%, 6%, and 9%, AGG was initially observed at 780 °C, 760 °C, 740 °C, and 720 °C, respectively. Furthermore, in the vicinity of AGG, $\Sigma 3$, $\Sigma 9$, and $\Sigma 27$, boundaries exhibited local maxima either at or in the thermal cycle following the observed onset temperature for AGG [19]. While the kinetics of AGG are hypothesized analytically [8,20–23], it is rather difficult to reliably ascertain abnormal grain growth rates experimentally for a variety of reasons. These challenges include fluctuations in relative size needed to resolve AGG within a given region of interest [24], full discretization of AGG amidst a population of continuously growing grains within a changing global grain size distribution [25], and the practicality of acquiring a reasonable amount of observations within a limited thermal range to support quantification [20].

Additionally, while texture is not a central emphasis in this work, it is noteworthy that other studies have shown that both grain growth and recrystallization kinetics can be additionally influenced by grain orientation and strain path [26–29]. In this work, the effect of varying thermal and mechanical strain routes are examined specifically in relation to grain growth. To that end, one of the aforementioned studies [19] is revisited and careful electron backscatter diffraction (EBSD) experiments, similar to those reported in [30], are performed. In these experiments, however, EBSD is combined with image segmentation based on crystallographic orientation to extract advancement rates associated with initial onset (emergence) and continuing growth (progression) of abnormally large grains in a simple Ni system.

2. Materials and Methods

Commercially pure Ni-200 bars (with dimensions of 152 mm × 6.35 mm × 6.35 mm) were thermomechanically processed using a Stanat Model: TA 215 rolling mill (Stanat Mfg. Co. Inc., Westbury, NY, USA) and a Lindberg/Blue M tube furnace (Lindberg/Mph, Riverside, MI, USA), see Figure 1. Samples were cold rolled using multiple rolling passes to achieve a desired reduction of 0%, 3%, 6%, or 9%, respectively. Once rolled, samples were metallographically polished to a 0.05 μm colloidal silica finish. Focus ion beam (FIB) and micro-hardness indent fiducial markers were then placed on each sample so specific regions of interest (ROIs) could be tracked throughout each stage of the experiment while not impeding the field of view associated with any ROI. At room temperature, EBSD was performed on the rolling direction (RD) plane of the pre-selected ROIs for the 0%, 3%, 6%, and 9% strained cases. Samples were then cyclically and progressively annealed for 30 min at 700 °C, 720 °C, 740 °C, 760 °C, 780 °C, and 800 °C, under a flowing argon-rich or hydrogen-rich atmosphere to prevent oxidation.

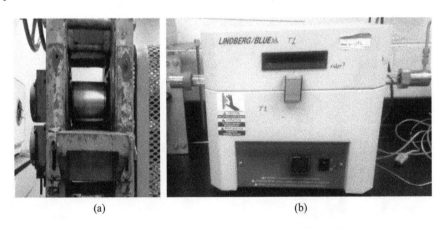

(a)　　　　　　　　　　　　　(b)

Figure 1. Samples were thermomechanically processed using a (a) rolling mill and (b) tube furnace.

After each annealing cycle, see Figure 2, samples were returned to room temperature and EBSD was performed on the same pre-selected ROI to document any changes in the grain population. Operating parameters for the collection of EBSD data utilized the following arrangement: A nominal beam current of 4 nA, a camera binning size of 4 × 4 or 6 × 6, and an indexing step size of 0.5 or 1 μm over multiple cross-sectional areas of 500 × 500 square μm or larger. TexSEM Laboratories (TSL) orientation imaging microscopy analysis by EDAX, Inc. (Mahwah, NJ, USA). was used to analyze

EBSD data and all EBSD maps were filtered using the Neighbor Confidence Index (CI) Correlation followed by Grain CI Standardization cleanup processes, where an average confidence index of >0.1 was used. The reader is directed to reference [19] for additional information regarding the experimental procedure should further detail be desired.

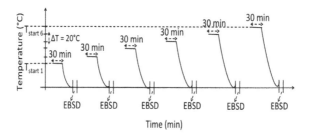

Figure 2. Cyclic annealing schedule with increasing isothermal dwelling anneals used in this study. (Reproduced with permission from [19], © 2016 ASM International and Springer Science + Business Media New York, 2016; EBSD: Electron backscatter diffraction.)

Grain boundary characterization for the unstrained "as-received" Nickel-200 material is shown in Figure 3. The mean grain size diameter, including twin boundary distributions, displayed a near Gaussian distribution with a mean grain size of 7 μm and an area fraction mode of 20% at a grain size of 15 μm, Figure 3a. The polycrystalline material also bore no specific bias in texture or anisotropy in grain morphology but presented a twin-containing equiaxed microstructure, see Figure 3b–e. The orientation maps and pole figure in Figure 3c–e are shown with respect to the RD.

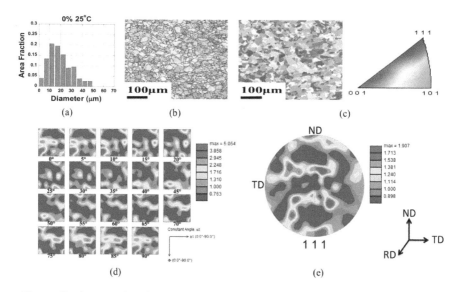

Figure 3. Baseline grain boundary metrics for unstrained, commercially pure Ni-200. (**a**) Grain area fraction distribution; (**b**) electron backscatter diffraction (EBSD) grain contrast map; (**c**) grain orientation map with inverse pole figure (IPF); (**d**) orientation distribution function map and (**e**) pole figure. (Orientations reported with respect to the rolling direction (RD), normal direction (ND) and transverse direction (TD) for (**c**–**e**); Adapted with permission from [19], © 2016 ASM International and Springer Science + Business Media New York, 2016.)

3. Results

For initial induced strains of 0%, 3%, 6%, and 9%, a series of EBSD maps are shown in Figure 4 where a given ROI is maintained along each row. Each cell reveals the EBSD map acquired following a specific cyclic annealing dwell temperature. As can be seen, AGG was observed to initiate at 780 °C for 0% strain; 760 °C for 3% strain; 740 °C for 6% strain; and 720 °C for 9% strain. For the cases in which abnormal grain growth was observed, bimodal distributions for grain size were also seen. Grains occupying the secondary peak of these bimodal distributions, which corresponded to higher mean grain sizes, were considered abnormally large [19]. Quantitatively, these grains corresponded to the upper 30% of their specific grain-size distributions across nearly all cases. So in this way, a consistent numeric threshold was applied for delineation of abnormally large grains. In Figure 4, these grains are outlined in black to assist in clearly identifying their locations and presence.

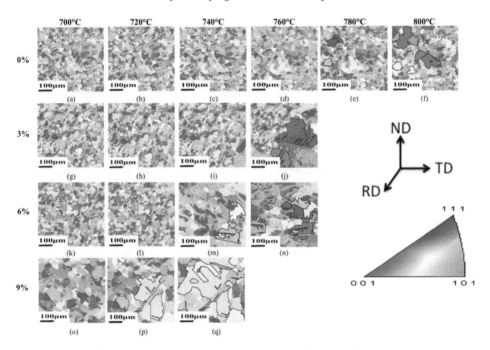

Figure 4. EBSD progression maps for 0% (**a–f**), 3% (**g–j**), 6% (**k–n**), and 9% (**o–q**) induced strain following annealing treatments ranging from 700 °C to 800 °C. Abnormally large grains are outlined in black. (Orientation is shown with respect to the rolling direction (RD); Adapted with permission from [19], © 2016 ASM International and Springer Science + Business Media New York, 2016.)

Inverse pole figure (IPF) maps with respect to the rolling direction (RD) for the ROIs presented in Figure 4 are shown below in Figure 5. The white circles indicate the orientations associated with the abnormally large grains. Minor changes in the texture populations contained within the ROIs are observable at or immediately before the onset of AGG. As might be expected, the population shifts are clearly indicative of the abnormally large grains increasingly occupying significant portions of the ROIs. However, the IPF maps also reveal that among the fields of view investigated, there appears to be no singularly preferred orientation among abnormally large grains in this study.

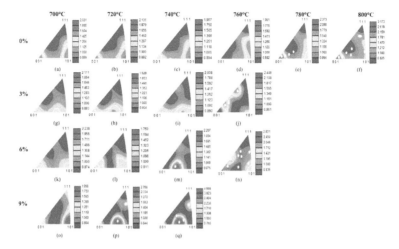

Figure 5. Inverse pole figure maps for 0% (**a–f**), 3% (**g–j**), 6% (**k–n**), and 9% (**o–q**) induced strain following annealing treatments ranging from 700 °C to 800 °C. White circles indicate the location of abnormally large grains. Orientation is shown with respect to the rolling direction (RD).

By utilizing the inherent segmentation available through the differentiation of crystallographic orientations, abnormally large grains were identified, as shown in Figure 4, and then isolated within their fields of view, see Figure 6. These sequestered grains were then used to calculate a local AGG area fraction measurement. By utilizing each successive area fraction, the rates of AGG were determined for both their onset and continued growth behavior. While differences in growth rates for twin and non-twinned grains are not captured here, the experimental quantification of the aggregate development of abnormal grain growth is notable.

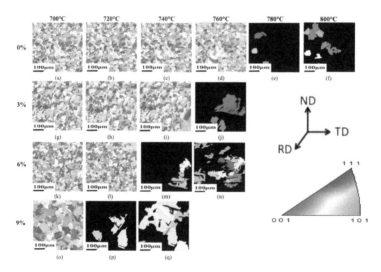

Figure 6. AGG (abnormal grain growth) identification progression maps for 0% (**a–f**), 3% (**g–j**), 6% (**k–n**), and 9% (**o–q**) induced strain following annealing treatments ranging from 700 °C to 800 °C. (Orientation is shown with respect to the rolling direction (RD); Adapted with permission from [19], © 2016 ASM International and Springer Science + Business Media New York, 2016.)

The measured area fractions of abnormally large grains are plotted as a function of annealing temperature associated with their AGG onset, see Figure 7a. These area fractions are also shown as a function of the number of thermal cycles (n) beginning with the cycle prior to AGG, see Figure 7b. The error bars shown depict the area fraction variability associated with a ±10% adjustment to the upper 30% grain size threshold value imposed. As can be observed, the growth rates for all strain levels exhibit a rather consistent two-stage behavior. Where stage 1 appears relatively slow and coincides with the initial observation of AGG, and stage 2 continues rapidly once AGG has clearly presented itself. This more rapid stage is seen to advance at a rate roughly 3 times that of stage 1 for the 0%, 3%, and 9% cases, see Table 1. For convenience, the authors will henceforth refer to stage 1 growth as "emergence" and stage 2 as "progression". Due to challenges associated with reliably identifying the emergence stage for the 3% case, emergence rates are not reported for 3% strain.

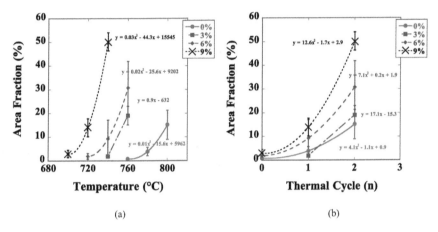

(a)   (b)

Figure 7. Area fraction of abnormally growing grains for 0%, 3%, 6%, and 9% induced strain as a function of (**a**) temperature and (**b**) thermal cycle.

Table 1. Emergence and progression rates for abnormally growing grains.

ε (%)	Emergence	Progression	Progression/Emergence
	Δ (Area Fraction $_{n = 1-0}$)	Δ (Area Fraction $_{n = 2-1}$)	
0	3.03	11.23	3.71
3	-	17.12	-
6	7.34	21.55	2.93
9	10.97	36.25	3.30

As seen in Figure 7, the greater the initial strain content, the higher the rate of abnormal grain growth in both the emergence and progression stages. Alternatively, the higher the AGG onset temperature, the slower the overall AGG advancement. While twinned and non-twinned grains were not differentiated in this analysis, the evolution of grains occupying large regions of cross-sectional area within their physical neighborhood were tracked rather successfully. To provide a more generalized extrapolation of these relative rates, the derivative of the parabolic curve fits in Figure 7b are provided in Figure 8. Here, each data series corresponds to a specific strain content and is denoted by color. The data markers provide the exact values of the calculated differential for up to 10 thermal cycles. Again, no calculated differential values for 3% are included due to the unavailability of a second order polynomial to differentiate.

Figure 8 illustrates a few informative trends. First, AGG in Ni-200 manifests unique growth rates for a given strain irrespective of the number of cyclic annealing exposures it experiences.

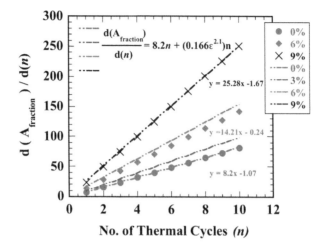

Figure 8. Rate of abnormal grain growth for low strained Ni-200 as a function of the number of thermal annealing cycles.

Stated another way, the initial strain content appears to strongly influence the emergence rate, and this imprint is preserved as abnormal grain growth continues. This estimation can be further simplified and approximated mathematically for all observed strains. This empirical relationship is provided in Equation (1) where:

$$\frac{d\left(A_{fraction}\right)}{d(n)} = 8.2n + \left(\frac{\varepsilon^{2.1}}{6}\right)n. \tag{1}$$

In this instance, n is the quantity of cyclic thermal anneals and ε is the initial plastic strain. These predictive estimates are also included in Figure 8 as dashed lines to assist with their identification in comparison to the calculated differentials. Please note: A numerical prediction provided by Equation (1) is included for an initial 3% strain case by a dashed line, despite the current work's inability to experimentally resolve the emergence rate for 3% strain. However, as can be seen, the predictive agreement is quite reasonable across all strain series shown. Furthermore, Equation (1) also illustrates that higher strain AGG rates can be estimated with a nominal (0%) AGG rate, when further incentivized by the added contribution of the higher strains as provided in the second term of Equation (1). As shown, the numeric approximation must also be accompanied by an appropriate set of prefactors. These values are provided in Equation (1) based on the experiments performed here. This suggests that in the case of consistent AGG advancement from repeated cyclic anneals, at least within the low strain regime, a methodology for the prediction of AGG advancement may be reasonably devised by clearly understanding or merely quantifying the unstrained AGG rates and the relevant scaling factors associated with additional strain content.

4. Discussion

The authors acknowledge that the findings of this work are derived from a limited number of observations within a single and specific lot of a commercially pure Ni-200 material, thermomechanically processed by rolling and exposure to a specific, consistent, and repeated cyclic annealing heat treatment. It should be anticipated that the specific area fraction measurements may be influenced by a number of these factors. Specifically, the most influential factors may be the cross-sectional areas selected for ROIs within each EBSD map, the 20 °C thermal jump between dwells, the 30 min dwell-time interval, and even the utilization of rolling to impart initial plastic strain. While these factors may introduce quantitative variation into specific measures, the authors assert these differences are likely to be insignificant to the fundamental trends observed. It has been shown that differing average grain size and cumulative area fractions can be obtained within the same pre-strained material under cyclic or isothermal annealing by introducing variations in dwell temperature, ramp rate, or hold times. However, once normalized by the local mean grain size, these varying distributions converge to a singular, invariant, cumulative fraction distribution [31]. Likewise, AGG observations could reasonably be expected to follow a similar convergence behavior when multiple observations are normalized to the local mean grain size present. The authors would further assert that the true benefit of this study is held in the observation of self-similar AGG rates being clearly and quantifiably related to initial plastic strain when exposed to equivalent, consistent, and repeated thermal exposures, not the specific area fractions reported in themselves.

Furthermore, the systematic observations across a breadth of strains and temperatures are notable in that AGG is shown to be predictively related to initial strain content. As seen here, initial strain bears a definite and significant influence on not only the emergence rate for AGG, but also, by extension, the continued progression rate. These findings are supported by the work of Decker et al. [32], He et al. [33–35], and Cho et al. [36], who all showed that among instances of self-similar AGG, initial strain content can serve as a significant driving force for grain boundary migration among low-energy, high-mobility grain boundaries [9,36,37]. In this work, this driving force is incorporated into the predictive estimate for AGG advancement as a superposition of a nominal rate (where $\varepsilon = 0$) combined with a weighted contribution from the additional plastic strain present. While good agreement has been achieved here, further experiments, along the lines of those reported herein, could additionally refine these observations and/or develop similar descriptions for other material systems. Such experiments would, however, require identification of self-similar AGG across a collection of strains and temperatures as a pre-requisite.

Two specific studies to further elucidate the AGG advancement rates reported here could include: (1) Capturing regions of interest larger than 500×500 square µm to acquire greater grain populations for measurement, and thereby increase the potential for identification of additional AGG events; or (2) decrease the cyclic annealing thermal or dwell time intervals to values smaller than 20 °C or 30 min, respectively. This would allow the examination of AGG emergence and progression across smaller observation spans. Further studies of this type would either grow the range of instances available for data collection or increase the resolution and granularity over which the AGG phenomena could be observed.

5. Conclusions

(1) Emergence and progression rates for abnormally large grains in minimally strained, commercially pure Ni-200 were observed experimentally by EBSD and quantified using image segmentation based on crystallographic orientation and grain size distributions.

(2) Emergence and progression growth rates were shown to scale directly with increased initial strain content and inversely with onset AGG temperature.

(3) A predictive estimate for the rate of AGG area fraction advancement, as a function of repeated thermal cycles, was determined based on the derivative of the experimental area fraction measures. This estimate has the following form:

$$\frac{d\left(A_{\text{fraction}}\right)}{d(n)} = C_1 n + (C_2 \varepsilon^\alpha) n, \tag{2}$$

where for the experiments performed herein, $C_1 = 8.2$, $C_2 = 1/6$ and $\alpha = 2.1$.

(4)   This numeric estimate indicates two notable implications:

    i.   Area fraction advancement rates for AGG in minimally strained, commercially pure Ni-200 proceed at unique rates for a given initial strain content;

    ii.   This advancement rate can be reasonably approximated across the low strain regime by the superposition of a nominal rate (where $\varepsilon = 0$) combined with the contribution of additional plastic strain (ε), modified by some prefactor value (C_2), and scaled by an appropriate exponential (α).

Acknowledgments: Sandia National Laboratories is a multi-mission laboratory managed and operated by National Technology and Engineering Solutions of Sandia, Limited Liability Company (LLC), a wholly owned subsidiary of Honeywell International, Inc. for the U.S. Department of Energy's National Nuclear Security Administration under contract DE-NA0003525. A subset of experiments for this work were supported by the National Science Foundation under Grant no. DMR-1151109. Compressive rolling was conducted at the Redstone Arsenal in Huntsville, AL, USA and the authors would like to thank Daniel Renner for initial metallographic preparation. Significant electron backscatter diffraction and annealing studies were conducted at the Center for Nanophase Materials Sciences, a U.S. Department of Energy, Office of Science and User Facility in Oak Ridge, TN, USA. The authors benefited greatly from the technical assistance of Donovan Leonard and James Kiggans and, the authors would also like to thank Rodney McCabe from Los Alamos National Laboratory for guidance in EBSD data processing. All other work was performed at Sandia National Laboratories and the authors would like to thank Alice Kilgo for sample preparation, Charles Walker for annealing studies, and Bonnie McKenzie and Joseph Michael for additional EBSD experiments. The late Professor Jeffrey L. Evans is also recognized for his initial motivation of these studies.

Author Contributions: Olivia D. Underwood and Jonathan D. Madison conceived and designed the experiments; Olivia D. Underwood performed the experiments and analyzed the data; Olivia D. Underwood, Jonathan D. Madison, and Gregory B. Thompson wrote the paper.

Conflicts of Interest: The authors declare no conflict of interest.

References

1.   Madison, J.D.; Spowart, J.E.; Rowenhorst, D.J.; Aagesen, L.K.; Thorton, K.; Pollock, T.M. Fluid flow and defect formation in the three-dimensional dendritic structure of nickel-based single crystals. *Metall. Mater. Trans. A* **2012**, *43*, 369–380. [CrossRef]

2.   Pollock, T.M.; Tin, S. Nickel-Based Superalloys for Advanced Turbine Engines: Chemistry, Microstructure, and Properties. *J. Propuls. Power* **2006**, *22*, 361–374. [CrossRef]

3.   Osorio, W.R.; Goulart, P.R.; Santos, G.A.; Neto, C.M.; Garcia, A. Effect of dendritic arm spacing on mechanical properties and corrosion resistance of Al 9 Wt Pct Si and Zn 27 Wt Pct Al alloys. *Metall. Mater. Trans. A* **2006**, *37*, 2525–2538. [CrossRef]

4.   Gabb, T.P.; Kantzos, P.T.; Palsa, B.; Telesman, J.; Gayda, J.; Sudbrack, C.K. *Fatigue Failure Modes of the Grain Size Transition Zone in a Dual Microstructure Disk*; John Wiley & Sons: New York, NY, USA, 2012; pp. 63–72.

5.   Flageolet, B.; Yousfi, O.; Dahan, Y.; Villechaise, P.; Cormier, J. *Characterization of Microstructures Containing Abnormal Grain Growth Zones in Alloy 718*; John Wiley & Sons: New York, NY, USA, 2012; pp. 594–606.

6.   Gabb, T.P.; Kantzos, P.T.; Gayda, J.; Sudbrack, C.K.; Palsa, B. Fatigue resistance of the grain size transition zone in a dual microstructure superalloy disk. *Int. J. Fatigue* **2011**, *33*, 414–426. [CrossRef]

7.   Randle, V.; Coleman, M. Grain growth control in grain boundary engineered microstructures. *Mater. Sci. Forum* **2012**, *715–716*, 103–108. [CrossRef]

8.   Holm, E.A.; Miodownik, M.A.; Rollett, A.D. On abnormal subgrain growth and the origin of recrystallization nuclei. *Acta Mater.* **2003**, *51*, 2701–2716. [CrossRef]

9.   Fang, S.; Dong, Y.P.; Wang, S. The abnormal grain growth of P/M nickel-base superalloy: Strain storage and CSL boundaries. *Adv. Mater. Res.* **2015**, *1064*, 49–54. [CrossRef]

10.   Watson, R.; Preuss, M.; Da Fonseca, J.Q.; Witulski, T.; Terlinde, G.; Buscher, M. Characterization of Abnormal Grain Coarsening in Alloy 718. *MATEC Web Conf.* **2014**, *14*, 1–5. [CrossRef]

11.   Brons, J.G.; Thompson, G.B. A comparison of grain boundary evolution during grain growth in FCC metals. *Acta Mater.* **2013**, *61*, 3936–3944. [CrossRef]

12. Randle, V.; Booth, M.; Owen, G. Time evolution of sigma 3 annealing twins in secondary recrystallized nickel. *J. Microsc.* **2005**, *217*, 162–166.
13. Kazaryan, A.; Wang, Y.; Dregia, S.A.; Patton, B.R. Grain growth in anisotropic systems: Comparison of effects of energy and mobility. *Acta Mater.* **2002**, *50*, 2491–2502. [CrossRef]
14. Upmanyu, M.; Hassold, G.N.; Kazaryan, A.; Holm, E.A.; Wang, Y.; Patton, B.; Srolovitz, D.J. Boundary mobility and energy anisotropy effects on microstructural evolution during grain growth. *Int. Sci.* **2002**, *10*, 201–216.
15. Homma, H.; Hutchinson, B. Orientation dependence of secondary recrystallization in silicon-iron. *Acta Mater.* **2003**, *51*, 3795–3805. [CrossRef]
16. Lin, P.; Palumbo, G.; Harase, J.; Aust, K.T. Coincident site lattice (CSL) grain boundaries and goss texture development in Fe-3% Si Alloy. *Acta Mater.* **1996**, *44*, 4677–4683. [CrossRef]
17. Watanabe, T. An approach to grain boundary design for strong and ductile polycrystals. *J. Glob.* **1984**, *11*, 47–84.
18. Randle, V. Twinning-related grain boundary engineering. *Acta Mater.* **2004**, *52*, 4067–4081. [CrossRef]
19. Underwood, O.D.; Madison, J.D.; Martens, R.L.; Thompson, G.B.; Welsh, S.L.; Evans, J.L. An examination of abnormal grain growth in low strain nickel-200. *Metallogr. Microstruct. Anal.* **2016**, *5*, 302–312. [CrossRef]
20. Randle, V.; Horton, D. Grain growth phenomena in nickel. *Scr. Metall. Mater.* **1994**, *31*, 891–895. [CrossRef]
21. Hillert, M. On the theory of normal and abnormal grain growth. *Acta Metall.* **1965**, *13*, 227–238. [CrossRef]
22. Holm, E.A.; Hassold, G.N.; Miodownik, M.A. On misorientation distribution evolution during anisotropic grain growth. *Acta Mater.* **2001**, *49*, 2981–2991. [CrossRef]
23. Rollett, A.D.; Srolovitz, D.J.; Anderson, M.P. Simulation and theory of abnormal grain growth-anisotropic grain boundary energies and mobilities. *Acta Metall.* **1989**, *37*, 1227–1240. [CrossRef]
24. Rollett, A.D.; Mulins, W.W. On the growth of abnormal grains. *Scr. Mater.* **1997**, *36*, 975–980. [CrossRef]
25. Thompson, C.V.; Frost, H.J.; Spaepen, F. The relative rates of secondary and normal grain growth. *Acta Metall.* **1987**, *35*, 887–890. [CrossRef]
26. Sztwiertnia, K. Recrystallization textures and the concept of oriented growth revisited. *Mater. Lett.* **2014**, *123*, 41–43. [CrossRef]
27. Jensen, D.J. Growth rates and misorientation relationships between growing nuclei/grains and the surrounding deformed matrix during recrstalization. *Acta Metall. Mater.* **1995**, *43*, 4117–4129. [CrossRef]
28. Akhiani, H.; Nezakat, M.; Sonboli, A.; Szpunar, J. The origin of annealing texture in a cold-rolled incoloy 800H/Ht after different strain paths. *Mater. Sci. Eng.* **2014**, *619*, 334–344. [CrossRef]
29. Akhiani, H.; Nezakat, M.; Szpunar, J.A. Evolution of deformation and annealing textures in incoloy 800H/HT via different rolling paths and strains. *Mater. Sci. Eng.* **2014**, *614*, 250–263. [CrossRef]
30. Wilson, A.W.; Madison, J.D.; Spanos, G. Determining phase volume fraction in steels by electron backscattered diffraction. *Sci. Mater.* **2001**, *45*, 1335–1340. [CrossRef]
31. Sahay, S.S.; Malhotra, C.P.; Kolkhede, A.M. Accelerated grain growth behavior during cyclic annealing. *Acta Mater.* **2003**, *51*, 339–346. [CrossRef]
32. Decker, R.F.; Rush, A.I.; Dano, A.G.; Freeman, J.W. *Abnormal Grain Growth in Nickel-Base Heat-Resistant Alloys*; University of North Texas: Denton, TX, USA, 1957.
33. He, G.; Tan, L.; Liu, F.; Huang, L.; Huang, Z.; Jiang, L. Unraveling the formation mechanism of abnormally large grains in an advanced polycrystalline nickel base superalloy. *J. Alloys Compd.* **2017**, *718*, 405–413. [CrossRef]
34. He, G.; Liu, F.; Huang, L.; Huang, Z.; Jiang, L. Controlling grain size via dynamic recrystallization in an advanced polycrystalline nickel base superalloy. *J. Alloys Compd.* **2017**, *701*, 909–919. [CrossRef]
35. He, G.; Tan, L.; Liu, F.; Huang, L.; Huang, Z.; Jiang, L. Revealing the role of strain rate during multi-pass compression in an advanced polycrystalline nickel base superalloy. *Mater. Charact.* **2017**, *128*, 123–133. [CrossRef]
36. Cho, Y.K.; Yoon, D.Y.; Henry, M.F. The Effects of deformation and pre-heat treatment on abnormal grain growth in RENÉ 88 superalloy. *Metall. Mater. Trans. A* **2001**, *32*, 3077–3090. [CrossRef]
37. Bozzolo, N.; Agnoli, A.; Souai, N.; Bernacki, M.; Loge, R.E. Strain induced abnormal grain growth in nickel base superalloys. *Mater. Sci. Forum* **2013**, *753*, 321–324. [CrossRef]

Article

The Formation of Strong {100} Texture by Dynamic Strain-Induced Boundary Migration in Hot Compressed Ti-5Al-5Mo-5V-1Cr-1Fe Alloy

Kai Li and Ping Yang *

School of Materials Science and Engineering, University of Science and Technology Beijing, Beijing 100083, China; likai09260926@126.com
* Correspondence: yangp@mater.ustb.edu.cn; Tel.: +86-010-8237-6968

Received: 6 September 2017; Accepted: 26 September 2017; Published: 3 October 2017

Abstract: The microstructure and texture evolution of Ti-5Al-5Mo-5V-1Cr-Fe alloy during hot compression were investigated by the electron backscatter diffraction technique. The results reveal that two main texture components containing <100> and <111> fiber textures form after the hot compression. The fraction of each component is mainly controlled by deformation and strain rate. Dynamic strain-induced boundary migration (D-SIBM) is proved to be the reason that <100>-oriented grains grow towards <111>-oriented grains. The <100>-oriented grains coarsen with the increasing <100> texture intensity. Dynamic recrystallization (DRX) occurs under a low strain rate and large deformation. The DRX grains were detected by the method of grain orientation spread. The DRX grains reserve a <100> fiber texture similar to the deformation texture; however, DRX is not the main reason causing the formation of a strong <100> texture, due to its low volume fraction.

Keywords: hot compression; dynamic recovery; dynamic recrystallization; texture

1. Introduction

Titanium and titanium alloys are widely used for aviation, aerospace, marine, and other special applications owing to their low density, high strength, and good fatigue and corrosion performance. Thermomechanical processing is usually used for titanium alloys to improve their mechanical properties by changing their microstructure. The microstructure and mechanical properties of titanium alloys are sensitive to processing parameters such as deformation temperature and strain rate [1,2]. The hard processing characterization of titanium alloy is also a problem during industrial manufacture [3]. Many titanium alloys are hot processed in the β single region to obtain a homogeneous microstructure.

Textures of the β phase in many titanium alloys have been investigated by many researchers [4–6]. Different texture components occur under different processing parameters, and the evolution of texture would determine the mechanical properties of titanium alloys [7]. Kou et al. [8] revealed the texture revolution of Ti-15Mo-3Al-2.7Nb-0.2Si alloy during hot rolling. With increasing rolling reduction, the texture of the sample changed evidently and the final texture of rolling was a weak Goss texture in the β phase. Kim et al. [9] obtained a well-developed {001} <110> texture in Ti-22Nb-6Ta alloy under cold rolling and heat treatment at 873 K, and a {112} <110> recrystallization texture was developed after the heat treatment at 1173 K. Some specific textures may cause heterogeneity or poor fatigue performance; as a result, revealing the regularity of texture evolution for controlling its production is apparently important. Previously, we studied a strong <100> texture in Ti-5Al-5Mo-5V-1Cr-1Fe forged bar, illustrating that repeated recovery caused by the formation of large <100>-oriented grains [10].

Dynamic recovery of the β phase is quite common during hot deformation, and recrystallization occurs under some special deformation processes [11,12]. Dynamic recrystallization of a hot-rolled titanium and its effects were studied by Chen et al. [13]. They concluded that the weakening of the texture was associated with the rotation of the dynamic recrystallization grains towards the preferred slip systems, leading to large misorientations between them. The texture of dynamically recovered β phase differs greatly from the dynamic recrystallization β phase. Li et al. [14] studied the effect of dynamic restoration on texture evolution. A strong <001> fiber texture develops where only dynamic recovery (DRC) occurs, and the deformation texture is weakened to a large extent after recrystallization. The texture of compressed molybdenum always has a similar characteristic with β titanium, because of the same body-centered cubic structure. Sophie et al. [15] observed strong <100> fiber texture and <111> fiber texture in compressed samples, and the volume fraction of <100> texture increased with increasing true strain and deformation temperature. Although the dynamic recovery and dynamic recrystallization during hot deformation have been investigated by various techniques, the effect of each mechanism on texture evolution remains unclear. Moreover, the formation of strong <100> fiber texture under a low strain rate and high deformation temperature needs to be revealed from its microstructure evolution.

This study focused on investigating the growth mechanism of <100>-oriented grains from low strain to high strain. The orientation characterizations of Ti-5Al-5Mo-5V-1Cr-1Fe alloy under different strain rates and compression temperatures were studied by the electron backscatter diffraction (EBSD) technique. The dynamic recrystallization grains were selected by the grain orientation spread (GOS) method to reveal the effect of dynamic recovery and recrystallization on texture evolution.

2. Materials and Experimental Procedures

The Ti-5Al-5Mo-5V-1Cr-1Fe alloy used in this study was obtained from a hot forged bar with a diameter of 350 mm. The composition of the alloy was 5.23 Al, 4.85 Mo, 4.93 V, 0.92 Cr, 1.17 Fe, and balance Ti. The β → α + β transus temperature of Ti-5Al-5Mo-5V-1Cr-1Fe was approximately 1143 K.

Cylindrical samples 6 mm in diameter and 12 mm in height were manufactured from the as-received Ti-5Al-5Mo-5V-1Cr-1Fe titanium bar. Before compression, each sample was heated to 1223 K and held for 5 min to obtain homogeneous β phase grains. The compression was accomplished using a Gleeble-1500 machine (DSI, St. Paul, MN, USA). The heating method was resistance heating with thermocouple to measure the temperature. And there was weak temperature fluctuation during the compression. The strain rates were 0.01 s^{-1}, 0.1 s^{-1}, 1 s^{-1} and the height reductions were 20%, 40%, 60%, and 80%. All the samples were quenched by water to retain the deformed microstructure, except those which were slowly cooled to achieve recrystallization. Two other samples under 60% compression with 1 s^{-1} at 1153 K were slowly cooled by 1 K/s to obtain static recrystallization.

The section along and perpendicular to the compressing direction (CD) of the compressed samples was prepared for microstructure and texture determination. For EBSD examination (Oxford Instruments, London, UK), the samples were electropolished with a polishing solution of 5% perchloric acid and 95% ethyl alcohol using at 30 V for 30 s at room temperature. An EBSD system (Channel 5), mounted on an Ultra55 scanning electron microscope (ZEISS, Oberkochen, Germany), was applied to reveal the orientation feature and texture evolution under compression. The work distance for the EBSD test was 16 mm. The tested area was about 2 mm × 1.5 mm and the number of pixels was about 120,000 for each samples. The recrystallization grains were determined through the GOS method, which shows the orientation spread of all the grains. They were considered recrystallization grains once the GOS reached below 2°.

3. Results and Discussions

3.1. Initial Microstructure

There is no strong texture component in the as-received forged Ti-5Al-5Mo-5V-1Cr-1Fe alloy bar (forged 10 times around the transus point, with α + β two phases), shown in the {100} pole figure in

Figure 1a. The microstructure of Ti-5Al-5Mo-5V-1Cr-1Fe alloy after being held at 1223 K for 5 min is shown in Figure 1b. The initial grains of the alloy are equiaxed with a size in the range of 80–100 μm.

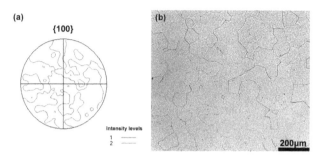

Figure 1. {100} Pole figure (**a**) of the as-received Ti-5Al-5Mo-5V-1Cr-1Fe alloy bar and original microstructure; (**b**) after solution treatment.

3.2. Texture Evolution during Hot Compression

Deformation parameters including the strain and strain rate exert an important effect on the texture components during hot compression around β transus temperature. Figure 2 shows the EBSD test results of the samples compressed at different strain rates of 1 s^{-1}, 0.1 s^{-1}, and 0.01 s^{-1} by reductions of 20%, 40%, 60%, and 80% at 1153 K. Figure 3 shows the fractions of two main fiber textures including <100> and <111> fiber textures. The two main texture components of the sample with 1 s^{-1} strain rate nearly retained the same volume fraction after 40% compression. The intensity of the <100> fiber texture increased with decreasing strain rate. In particular, the sample compressed with 80% reduction under 0.01 s^{-1} strain rate only showed <100> texture. Thus, it can be deduced that at the initial stage of compression, the strain rate has a weak effect on the texture evolution. With increasing deformation, low strain rate may be beneficial for the development of <100> texture, and <111> texture gradually weakened or even vanished.

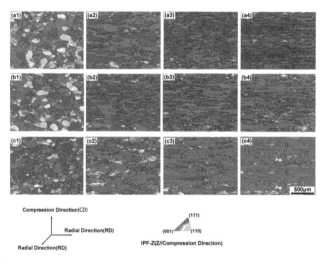

Figure 2. Orientation maps with IPF (inverse pole figure) colors by EBSD (electron backscatter diffraction) under different parameters: (**a1**)–(**a4**) are under 1 s^{-1} with 20%, 40%, 60%, and 80% reduction; (**b1**)–(**b4**) are under 0.1 s^{-1} with 20%, 40%, 60%, and 80% reduction; (**c1**)–(**c4**) are under 0.01 s^{-1} with 20%, 40%, 60%, and 80% reduction; the compressing temperature is 1153 K.

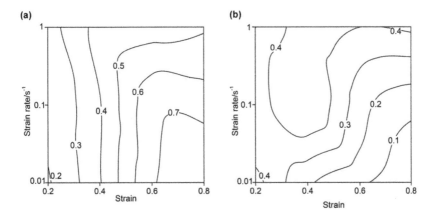

Figure 3. Fractions of <100> and <111> fiber textures under different strains and strain rates corresponding to Figure 2, the compressing temperature is 1153 K: (**a**) <100> fraction distribution; (**b**) <111> fraction distribution.

3.3. Dynamic Recrystallization during Hot Compression

Dynamic recrystallization has an important effect on the microstructure and texture evolution during hot compression. Li et al. [14] considered that DRX (dynamic recrystallization) enhanced with increasing strain rate as the dislocations accumulate rapidly, providing sufficient driving force for the transformation of the low angle boundaries into the high angle boundaries. Dynamic recrystallization grains under different compressing parameters were selected through the GOS value [16,17]. It is identified that the GOS of the recrystallized grains are below 2°. The EBSD mappings ranking the GOS of the samples corresponding to Figure 2 are shown in Figure 4. At higher strain rate, dynamic recrystallization could hardly happen and the microstructure of the samples is in the completely deformed state. The volume fraction of dynamic recrystallization distinctly rises with increasing deformation and decreasing strain rate. As recrystallization is a time-consuming process, the time is not enough for dynamic recrystallization under the condition of high strain rate. The size of the recrystallized grain is larger in lower strain rate compressed samples than in higher strain rate compressed samples. Figure 5 shows the orientation characterization of the recrystallization grains. The nucleation of dynamic recrystallization is almost located at <100>-oriented grain boundary. The misorientation inside the <100>-oriented grains increases with the rotation of each grain during compression. After full recovery, the dislocations in the <100>-oriented grains tangle together and sub-grains form inside the grains with the accumulation of dislocation clusters. Sub-grains at the grain boundary begin to grow, and their deviation from the original grain increases further and some nucleation of dynamic recrystallization then forms at the grain boundary. The recrystallized grains generate from the <100>-oriented grains and, as a result, the new dynamic recrystallized grains present a <100> recrystallization texture inherited from the initial texture. However, the <100> texture intensity is lower than the compressed grains which have not recrystallized. The dynamic recrystallization proportion in sample c4 has an obvious increase compared to other samples, and the dynamic recrystallization grain fraction is only 15.5%. As a result, dynamic recrystallization is beneficial for the formation of <100> fiber texture to a certain extent; however, it is not the main reason for this formation because of its low volume fraction.

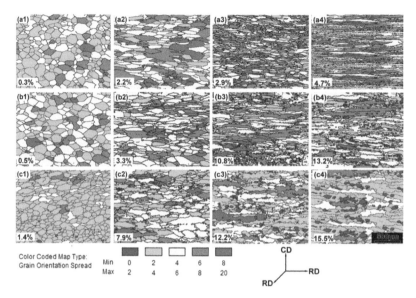

Figure 4. Grain orientation spread of the samples with different parameters: (**a1**)–(**a4**) are under 1 s^{-1} with 20%, 40%, 60%, and 80% reduction; (**b1**)–(**b4**) are under 0.1 s^{-1} with 20%, 40%, 60%, and 80% reduction; (**c1**)–(**c4**) are under 0.01 s^{-1} with 20%, 40%, 60%, and 80% reduction; the compressing temperature is 1153 K; the blue grains are selected as the recrystallization grains.

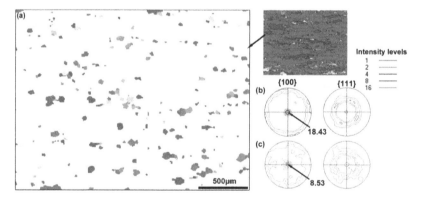

Figure 5. Orientation characterization of recrystallization grain of sample c3: (**a**) orientation distribution map of recrystallized grains with IPF colors; (**b**) {100} and {111} pole figures of the deformed grains; (**c**) {100} and {111} pole figures of the recrystallized grains.

3.4. D-SIBM (Dynamic Strain-Induced Boundary Migration) about <100>-Oriented Grains

Although dynamic recrystallization would occur during hot compression, deformed grains with a highly dynamic recovered state often dominate the evolution of the compressed microstructure. Warchomicka et al. [18] revealed that the main dynamic restoration mechanism is dynamic recovery for the hot compressed Ti55531 alloy at all the deformation parameters. Dynamic recovery could not solely change the deformation texture, and dynamic recrystallization is not enough to change the deformation texture because of its low volume fraction, as discussed in Section 3.2. Strain-induced boundary migration (SIBM) is an important mechanism during the recrystallization of deformed metals [19]. Nucleation would be initiated at the grain boundaries with the SIBM due to different stored energies and dislocation densities between <100>-oriented grains and <111>-oriented grains. The orientation of new grain may be different than either of the initial grains or be similar to one of the neighboring grains. The stored energy is partially released by dynamic recovery, but is not enough for the nucleation of dynamic recrystallization. The storage energy and dislocation density still remain different between <100>-oriented grains and <111>-oriented grains during hot compression. Primig et al. [15] considered that the <111>-oriented grains have a larger Taylor factor than <100> grains in molybdenum alloy when {101} and {112} slip systems are taken into account for the calculation in bcc (body-centered cubic) metals. As a result, <111>-oriented grains have a higher stored energy during compression. The bcc titanium alloy has the same crystal structure as the molybdenum alloy. Therefore, it provided a driving force for the grain boundary migration from <100>-oriented grains to <111>-oriented grains rather than nucleation in the grain boundary. As a result, the <100>-oriented grains grow and the <111>-oriented grains are gradually merged in a grain growth process during hot compression. Therefore, dynamic strain-induced boundary migration (D-SIBM) is proposed to illustrate the microstructure evolution. D-SIBM is the main mechanism along with deformation and dynamic recovery of the microstructure, which is different than SIBM during static recrystallization. In the case of low strain rate, dislocations are eliminated because of its rapid recovery rate. Orientation mappings have been tested by the EBSD method to observe the dynamic process of the migration from the <100>-oriented grains to <111>-oriented grains during compression as shown in Figure 6. The grains are at an initially deformed state towards <100> and <111> orientations with the reduction of 40%. With increasing deformation, it can be concluded that the orientations of the deformed grains rotate to <111> or <100> orientations from their initial orientations. As a result, orientations of the deformed grains become stable around <100> and <111>. Meanwhile, the grain boundary migrates more markedly from <100>-oriented grains to <111>-oriented grains, as shown with black arrows in Figure 6. A bowed grain boundary is found in the samples with 40% reduction, and D-SIBM mechanism is proven to be the main path for the growth of <100> grains and the decrease in <111> grains. Eventually, the <100>-oriented fiber texture occupies over 80%, and the <111> fiber texture nearly vanishes when the compressing reduction reaches 80%, as shown in Figure 3b.

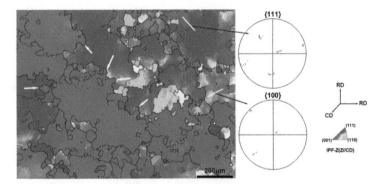

Figure 6. EBSD map with IPF colors of a sample at a strain rate of 0.01 s^{-1} and 40% reductions at 1153 K perpendicular to CD (compressing direction); bowed grain boundaries during migration are shown with white arrows; variation in orientation of deformed grains are shown with {111} and {100} pole figures.

3.5. Recrystallization after Hot Compression

When the compressed sample at a strain rate of 1 s^{-1} is held at the compressing temperature or treated at a low cooling rate, the sample would recrystallize. To study the process of recrystallization, a partially recrystallized microstructure was obtained at a 1 K/s cooling rate after compression, shown in Figure 7a,b. A weak <100> texture was retained for the partial recrystallized grains. The location of the new grains is almost at the tip of the deformed grains, and can provide more stored energy for recrystallization. The microstructure and texture of the fully recrystallized sample after compression are shown in Figure 7c,d. It can be clearly indicating that the texture of recrystallization is nearly in a random orientation, which is quite different from the deformed texture. Only an extremely weak <100> texture is preserved, because the <100>-oriented grains may be restored to a certain degree during compression. As a result, there is hardly texture inheritance during recrystallization.

The <100> texture is common in a compressed β-titanium alloy, and it is harmful to the mechanical properties because of its large <100>-oriented grains and the low elasticity modulus from the <100> direction. As the mechanism of static recrystallization is different from dynamic recrystallization, texture is also different under the two conditions. Deformation with a relatively high strain rate and static recrystallization during the heat treatment would weaken the deformation texture and refine the grain size.

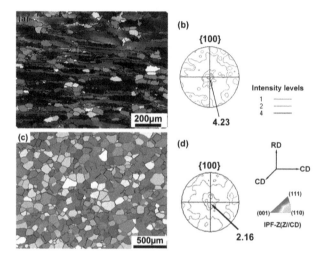

Figure 7. Orientation maps with IPF colors and {100} pole figure of partial recrystallization and complete recrystallization of the sample cooled by 1 K/s after compression with a reduction of 60% at 1153 K and 1 s^{-1} strain rate; (**a,b**) cooled to 1123 K and then water quenched, partial recrystallization; (**c,d**) cooled to 1073 K and then water quenched; complete recrystallization; the colored grains are recrystallized grains and the grey grains are deformed grains.

4. Conclusions

(1) Dynamic recrystallization under different parameters during hot deformation was investigated. Low strain rate and large deformation could induce the formation of dynamic recrystallization. Time is a critical factor causing the formation of this type of dynamic recrystallization. The texture after dynamic recrystallization is the <100> fiber texture, similar to the deformed texture.

(2) Dynamic strain-induced boundary migration was discovered during the low strain rate compression and is considered as the main mechanism causing the formation of a strong <100> texture during compression at a high temperature and low strain rate. The increasing temperature and strain rate of hot compression could promote the migration of <100>-oriented grains towards <111>-oriented grains. As a result, the <100> texture would be strengthened and the <111> texture would be weakened.

(3) To control the texture of the BCC titanium alloys, high strain rate and recrystallization after compression should be applied during hot deformation, which would be beneficial to eliminate the strong <100> texture. Nucleation of static recrystallization in the hot-compressed samples

forms at the tip of the deformed <111>-oriented grains in the samples with $1\ s^{-1}$ strain rate because of the inhomogeneous strain inside these grains. The texture after complete recrystallization is a weak <100> texture, close to the random orientation distribution.

Acknowledgments: This work was supported by National Natural Science Foundation of China (No. 51771024).

Author Contributions: Kai Li and Ping Yang conceived and designed the experiments; Kai Li and Ping Yang performed the experiments and analyzed the data; Kai Li wrote the paper.

Conflicts of Interest: The authors declare no conflict of interest.

References

1. Markovsky, P.E.; Matviychuk, Y.V.; Bondarchuk, V.I. Influence of grain size and crystallographic texture on mechanical behavior of TIMETAL-LCB in metastable β-condition. *Mater. Sci. Eng. A* **2013**, *559*, 782–789.
2. Fan, J.K.; Kou, H.C.; Lai, M.J.; Tang, B.; Chang, H.; Li, J.S. Hot deformation mechanism and microstructure evolution of a new near β titanium alloy. *Mater. Sci. Eng. A* **2013**, *584*, 121–132.
3. Razavi, S.M.J.; Ferro, P.; Berto, F. Fatigue assessment of Ti-6Al-4V circular notched specimens produced by selective laser melting. *Metals* **2017**, *7*, 291. [CrossRef]
4. Raghunathan, S.L.; Dashwood, R.J.; Jackson, M.; Vogel, S.C.; Dye, D. The evolution of microtexture and macrotexture during subtransus forging of Ti-10V-2Fe-3Al. *Mater. Sci. Eng. A* **2008**, *488*, 8–15.
5. Bai, X.F.; Zhao, Y.Q.; Zhang, Y.S.; Zeng, W.D.; Yu, S.; Wang, G. Texture evolution in TLM titanium alloy during uniaxial compression. *Mater. Sci. Eng. A* **2013**, *588*, 29–33.
6. Obasi, G.C.; Fonseca, D.; Quinta, G.; Rugg, D.; Preuss, M. The effect of beta grain coarsening on variant selection and texture evolution in a near-beta Ti alloy. *Mater. Sci. Eng. A* **2013**, *576*, 272–279.
7. Banumathy, S.; Mandal, R.K.; Singh, A.K. Texture and anisotropy of a hot rolled Ti-16Nb alloy. *J. Alloys Compd.* **2010**, *500*, L26–L30.
8. Kou, H.C.; Chen, Y.; Tang, B.; Cui, Y.W.; Sun, F.; Li, J.S.; Xue, X.Y. An experimental study on the mechanism of texture evolution during hot-rolling process in a β titanium alloy. *J. Alloys Compd.* **2014**, *603*, 23–27.
9. Kim, H.Y.; Sasaki, T.; Okutsu, K.; Kim, J.I.; Inamura, T.; Hosoda, H.; Miyazaki, S. Texture and shape memory behavior of Ti-22Nb-6Ta alloy. *Acta Mater.* **2006**, *54*, 423–433.
10. Li, K.; Yang, P. Investigation of microstructure and texture of beta phase in a forged TC18 titanium alloy bar. *Acta Metall. Sin.* **2014**, *50*, 707–714.
11. Wang, K.X.; Zeng, W.D.; Zhao, Y.Q.; Lai, Y.J.; Zhou, Y.G. Dynamic globularization kinetics during hot working of Ti-17 alloy with initial lamellar microstructure. *Mater. Sci. Eng. A* **2010**, *527*, 2559–2566.
12. Hua, K.; Xue, X.Y.; Kou, H.C.; Fan, J.K.; Tang, B.; Li, J.S. Characterization of hot deformation microstructure of a near beta titanium alloy Ti-5553. *J. Alloys Compd.* **2014**, *615*, 531–537.
13. Chen, Y.; Li, J.S.; Tang, B.; Kou, H.C.; Xue, X.Y.; Cui, Y.W. Texture evolution and dynamic recrystallization in a beta titanium alloy during hot-rolling process. *J. Alloys Compd.* **2015**, *618*, 146–152.
14. Li, L.; Luo, J.; Yan, J.J.; Li, M.Q. Dynamic globularization and restoration mechanism of Ti-5Al-2Sn-2Zr-4Mo-4Cr alloy during isothermal compression. *J. Alloys Compd.* **2015**, *622*, 174–183.
15. Primig, S.; Leitner, H.; Knabl, W.; Lorich, A.; Clemens, H.; Stickler, R. Textural evolution during dynamic recovery and static recrystallization of molybdenum. *Metall. Mater. Trans. A* **2012**, *43*, 4794–4805.
16. Wright, S.I.; Nowell, M.M.; Field, D.P. A review of strain analysis using electron backscatter diffraction. *Microsc. Microanal.* **2011**, *17*, 316–329.
17. Biswas, S.; Kim, D.I.; Suwas, S. Asymmetric and symmetric rolling of magnesium: Evolution of microstructure, texture and mechanical properties. *Mater. Sci. Eng. A* **2012**, *550*, 19–30.
18. Warchomicka, F.; Poletti, C.; Stockinger, M. Study of the hot deformation behaviour in Ti-5Al-5Mo-5V-3Cr-1Zr. *Mater. Sci. Eng. A* **2011**, *528*, 8277–8285.
19. Humphreys, F.J.; Hatherly, M. *Recrystallization and Related Annealing Phenomena*, 2nd ed.; Elsevier: Oxford, UK, 2004; pp. 251–257.

Article

Effect of High-Pressure Torsion on Structure and Microhardness of Ti/TiB Metal–Matrix Composite

Sergey Zherebtsov [1], Maxim Ozerov [1,*], Nikita Stepanov [1], Margarita Klimova [1] and Yulia Ivanisenko [2]

[1] Laboratory of Bulk Nanostructured Materials, Belgorod State University, Belgorod 308015, Russia; Zherebtsov@bsu.edu.ru (S.Z.); stepanov@bsu.edu.ru (N.S.); klimova_mv@bsu.edu.ru (M.K.)

[2] Karlsruhe Institute of Technology, Institute of Nanotechnology, 76021 Karlsruhe, Germany; julia.ivanisenko@kit.edu

* Correspondence: ozerov@bsu.edu.ru; Tel.: +7-919-223-8528

Received: 19 October 2017; Accepted: 13 November 2017; Published: 16 November 2017

Abstract: Effect of high-pressure torsion (HPT) at 400 °C on microstructure and microhardness of a Ti/TiB metal–matrix composite was studied. The starting material was produced by spark plasma sintering of a mixture of a pure Ti and TiB_2 (10 wt %) powders at 1000 °C. The microstructure evolution during HPT was associated with an increase in dislocation density and substructure development that resulted in a gradual microstructure refinement of the Ti matrix and shortening/redistribution of TiB whiskers. After five revolutions, a nanostructure with (sub) grain size of ~30 nm was produced in Ti matrix. The microhardness increased with strain attaining the value ~520 HV after five revolutions. The contribution of different hardening mechanisms into the hardness of the Ti/TiB metal–matrix composite was quantitatively analyzed.

Keywords: metal–matrix composite; high-pressure torsion; microstructure evolution; microhardness

1. Introduction

Due to high strength-to-density ratio, excellent corrosion resistance and good biocompatibility titanium and titanium alloys are attractive for various applications, including the aerospace, automotive, chemical and biomedical industries [1]. However, relatively low strength, hardness and wear resistance limit the application of titanium and low-alloyed titanium alloys. One of the effective methods to improve their strength-related mechanical properties is creating a metal–matrix composite (MMC) by inserting ceramic fibers or particles into the Ti matrix [2–7]. In this case, high strength and stiffness of ceramic reinforcements combine with good toughness provided by a metal matrix. Some improvements of wear resistance and high temperature properties of MMCs (metal–matrix composites) can also be expected [6,8–10]. Titanium alloys can be reinforced by TiB_2, TiN, B_4C, ZrC, TiB, TiC, and Al_2O_3 [6]. Among various reinforcements, TiB seems to be the most attractive option because it has a close to titanium density, high Young's modulus and reasonable stability at the processing temperatures [11,12]. In addition, TiB creates minimal residual stresses due to similarity of thermal expansion coefficient and good crystallographic interfaces with the titanium matrix [11–13]. The effect of TiB on the strength properties of Ti-based matrix was either as good as, if not better than, most widely used reinforcements such as TiC or Al_2O_3 [7,14].

Fully dense bulk Ti/TiB specimens can be obtained in situ during the spark plasma sintering (SPS) process through the $TiB_2 + Ti \rightarrow Ti + 2TiB$ reaction [13]. The TiB crystals have a whisker-like shape with very small (down to nano-range) diameter [13]. Due to high heating rate and high pressures, consolidation during the SPS occurs at relatively low temperatures and within short time intervals. Therefore, in contrast to many other sintering treatments, noticeable microstructure coarsening can be avoided during the SPS. However, high strength of Ti/TiB MMCs is usually accompanied by low ductility at room temperature [11,15].

Thermomechanical treatment can considerably improve ductility and strength and decrease ductile-to-brittle transition temperature of MMCs [15–18]. However, the microstructure evolution of MMCs during deformation under different conditions has not been studied comprehensively so far [19]. One of the promising ways to modify the microstructure of metallic materials considerably is severe plastic deformation (SPD) [20], which usually resulted in microstructure refinement to the nanoscale region. Although SPD of various materials by different methods has been a subject of intensive investigations during few last decades [20–22], SPD of MMCs was studied insufficiently. Recent studies on the influence of high pressure torsion (HPT) on microstructure and mechanical properties of various MMCs composites, including Ti/Al$_2$O$_3$ and Ti/TiO$_2$ [23,24], have shown that SPD basically resulted in grain refinement of the matrix material, homogeneous distribution of the second phase particles and diminished particle size [25], thereby enhancing both strength and ductility of the MMCs [26]. It should be noted that the majority of known investigations of microstructure evolution during SPD were conducted using MMCs reinforced with carbon-based or oxide particles of nearly equiaxed shape [22–26]. Much less attention was paid to SPD of MMCs reinforced by elongated particles or fibers, such as Ti/TiB. The processing should obviously break the TiB whiskers with high aspect ratio and reduce the grain size of the Ti matrix. However, no information on this topic was found in the literature.

In this work, a Ti/TiB MMC was produced by the spark plasma sintering using a Ti-10 wt % TiB$_2$ powder mixture at a temperature of 1000 °C. Then the sintered composite was subjected to HPT at 400 °C with a number of revolutions in the interval 1–5. Microstructure evolution of the composite was comprehensively (i.e., both Ti matrix and features of TiB whiskers were investigated) studied using SEM (Scanning Electron Microscope) and TEM (Transmission Electron Microscope) and microhardness measurements were used to estimate the effect of HPT on mechanical properties.

2. Materials and Procedure

Commercial Ti powders (wt % of impurities: 0.07 N, 0.05 C, 0.34 H, 0.34 (Fe + Ni) and 0.1 Si; and TiB$_2$ (wt % of impurities: 0.04 O, 0.04 C, 0.02 Fe) were used as the raw materials. In both cases, the particles had an irregular shape; the average particles sizes of the Ti and TiB$_2$ powders were ~25 and ~4 μm, respectively. A mixture of Ti powder with 10 wt % of TiB$_2$ was prepared using a Retsch RS200 vibrating cup mill (RETSCH, Haan, Germany) for 1 h in ethanol at the milling rotation speed of 700 rpm. Specimens of Ti/TiB metal–matrix composite measured 19 mm diameter and 20 mm height were produced through the SPS process under vacuum at 1000 °C and 40 MPa for 5 min using a Thermal Technology SPS10-3 machine (Thermal Technology, LLC, Santa Rosa, CA, USA).

Disks measured 0.7 mm thickness and 10 mm diameter were cut from the Ti/TiB specimens using the electric-discharge machine (Sodick Inc., Schaumburg, IL, USA) and then deformed by HPT in a Bridgman anvil type unit using a custom-built computer-controlled device (Klement GmbH, Lang, Austria) at 400 °C and 6 GPa with a speed of 1 rpm. The number of revolutions was N = 1, 2 or 5. The corresponding shear strain level γ can be calculated as [22]:

$$\gamma = \frac{2\pi N r}{h} \tag{1}$$

where N is the number of revolutions, r is the radius and h is the thickness of the specimen. The temperature of deformation was chosen based on preliminary experiments; at lower temperatures the MMC did not possess enough ductility for the SPD treatment.

X-ray diffraction (XRD) analysis was done for the shear plane of specimens using an ARL-Xtra diffractometer (Thermo Fisher Scientific, Portland, OR, USA) with CuK$_\alpha$ radiation. The dislocation density was estimated as broadening of XRD peaks using Williamson-Hall method [27]. A JEOL JEM-2100 transmission electron microscope (TEM; JEOL, Tokyo, Japan) and a Quanta 200 scanning electron microscope (SEM; Thermo Fisher Scientific, Portland, OR, USA) were used for microstructure examination. Structure and mechanical properties were studied in the mid-thickness of the specimens in the axial cross-section (using SEM and microhardness) or in the shear plane (using TEM). SEM analysis and microhardness measurements were done at the distance of 0, or ~5 mm from the center; the observation point for TEM was spaced ~1.5 mm from the edge of the disc.

Specimens for SEM analysis were mechanically polished in water with different SiC papers and colloidal silica suspension; the final size of Al_2O_3 abrasive was 0.05 μm. Etching was carried out with Kroll's reagent (95% H_2O, 3% HNO_3, 2% HF). Thin-foil specimens for TEM characterization were prepared by mechanical thinning followed by electropolishing on a twin-jet TENUPOL-5 at 29 V and at −35 °C using an electrolyte containing 60 mL perchloric acid, 600 mL methanol and 360 mL butanol.

Vickers microhardness was determined under a load of 1 kg for 10 s. The reported hardness values were the average of at least 10 measurements.

3. Results

3.1. Initial Microstructure

According to XRD data (Figure 1), the Ti/TiB MMC consisted of hcp (hexagonal close-packed) α-Ti, TiB_2 with a hexagonal lattice, and TiB with an orthorhombic lattice. The volume fractions of the phases in the initial condition were 78.6% of Ti, 19% of TiB and 2.4% of TiB_2.

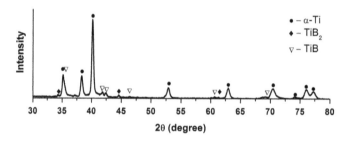

Figure 1. XRD (X-ray diffraction) patterns of the Ti/TiB MMC (metal–matrix composite) sintered at 1000 °C.

The microstructure of Ti/TiB MMC was rather complicated to analyze since there were no obvious links between the images obtained by either optical or scanning electron microscopy in an unetched state using a BSE backscattered electrons; Thermo Fisher Scientific, Portland, OR, USA) detector (Figure 2a) or in an etched state (Figure 2b–e). On the unetched surface, relatively large light particles of different shape and size (5–10 μm) can be seen (Figure 2a). However, in the etched condition, the microstructure of the Ti/TiB MMC consisted of TiB whiskers heterogeneously distributed within the Ti matrix; some residual TiB_2 particles were also observed (shown by arrows in Figure 2b). The variation in density of the TiB whiskers and the presence of the remnants of TiB_2 particles were seen as dark-gray or light-gray areas on the unetched surface (areas #1–3 in Figure 2a vs. Figure 2b and corresponding enlarged images in Figure 2c–e). The microhardness for areas #1–3 in Figure 2 was found to be 420 ± 15 HV, 580 ± 25 HV and 940 ± 175 HV, respectively; i.e., an increase in density of TiB whiskers resulted in an increase in hardness.

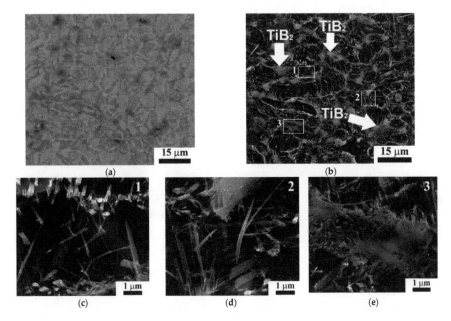

Figure 2. Microstructure of the Ti/TiB composite sintered at 1000 °C: (**a**) unetched; or (**b–e**) etched surface. Images (**c–e**) correspond to areas #1–3, respectively, in (**a,b**).

TEM examination also revealed the TiB whiskers heterogeneously distributed in the Ti matrix (Figure 3a). In the majority of the microstructures, a very high dislocation density was observed, probably due to a large number of the TiB particles. Individual grains cannot be distinguished in the microstructure, however the size of areas with relatively low dislocation density (i.e., the space between areas with a large number of the TiB particles and high dislocation density) was ~1–1.5 μm. The TiB whiskers had an irregular hexagonal shape (Figure 3b) with sides parallel to the (100), (101) and $(10\bar{1})$ planes [28]. Many stacking faults were observed in the (100) plane of the TiB whisker. Due to the presence of the orientation relationship (OR) between the TiB particles and the Ti matrix (which is usually described as $(10\bar{1}0)_\alpha//(100)_{TiB}$ and $[01\bar{1}0]_\alpha//[0\bar{1}1]_{TiB}$ [29]), the interphase Ti/TiB boundaries are very clear without noticeable internal stresses. The transversal size of the TiB whiskers varied in a wide interval from tens to few hundred nanometers with the average value of 63 ± 35 nm.

Figure 3. TEM (Transmission Electron Microscope) images of: (**a**) the initial microstructure of the Ti/TiB composite; and (**b**) a cross-section of a TiB whisker.

3.2. Microstructure Evolution during High-Pressure Torsion

The influence of HPT on the microstructure depended essentially on the distance from the specimen center (Figure 4). In the central part, the microstructure after one revolution of HPT did not change noticeably. At the edge of specimens ($\gamma \approx 45$), a microstructure consisting of dark wavy bands (elongated regions with a different density of the TiB whiskers) and black spots (remnants of TiB_2 surrounded by a "brushes" of TiB whiskers, Figure 2b) was revealed (Figure 4a). Quite similar microstructure was observed in the center of specimens after five revolutions (Figure 4b). Severe deformation ($\gamma \approx 224$) achieved at the edge of specimens after five revolutions (Figure 4c,d) formed quite homogeneous microstructure with numerous inclusions of TiB debris of different size.

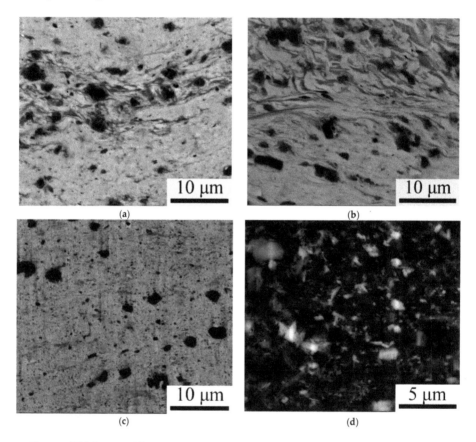

Figure 4. SEM (Scanning Electron Microscope) images ((**a–c**) unetched; and (**d**) etched) of the Ti/TiB MMC microstructure after HPT (high-pressure torsion): (**a**) one revolution at the edge of the specimen; (**b**) five revolutions in the center of the specim; (**c,d**) five revolutions at the edge of the specimen.

The apparent length of the whiskers dropped by a factor of ~5.5 during the first revolution of HPT and then decreased slower (by ~55%) while the number of rotation increased from 1 to 5 (Figure 5). The length-to-diameter aspect ratio of the TiB whiskers approached values of ~5 at the final stages HPT. For comparison, in the initial condition the aspect ratio was ~47. It should be noted that XRD analysis (not shown) did not reveal any noticeable changes in the fraction of the constitutive phases.

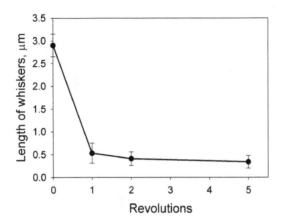

Figure 5. Apparent length of TiB whiskers in the Ti/TiB MMC after HPT.

Meanwhile, TEM analysis showed considerable refinement of the titanium matrix of the composite as a result of HPT (Figures 6a and 7a). Deformation to $\gamma \approx 31$ (1 revolution) resulted in the formation of a cellular microstructure with a very high dislocation density (Figure 7b). Dislocation density measured by XRD increased from 1.2×10^{15} m^{-2} in the initial condition to 1.7×10^{15} m^{-2} after one revolution. The boundaries of cells were rather wide and diffuse; however, some (sub)grains with clean thin boundaries and reduced dislocation density can also be seen in the microstructure. The average size of cells and (sub)grains was found to be ~90 nm after one revolution of HPT (Figure 7a). The interphase Ti/TiB boundaries were blurred due to high internal stresses caused by the high dislocation density in the vicinity of the interfaces, however no sign of cracks or voids formation along the interfaces was observed.

During further deformation, the microstructure almost did not change qualitatively. However, increase in strain to $\gamma = 62$ (two revolutions) and to $\gamma = 157$ (five revolutions) decreased the size of cells and (sub)grain to 55 and 34 nm, respectively, and increased dislocation density to 4.0×10^{15} m^{-2} and 8.2×10^{15} m^{-2}, respectively (Figures 6b,c and 7).

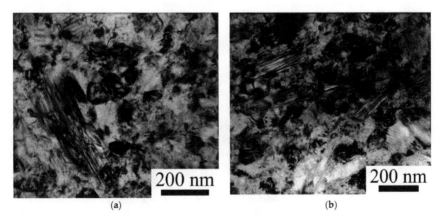

(a)   (b)

Figure 6. *Cont.*

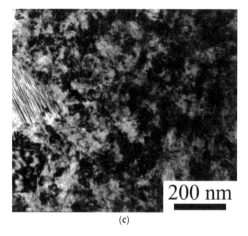

(c)

Figure 6. TEM images of microstructure of the Ti/TiB MMC after HPT: (**a**) one revolution; (**b**) two revolutions; and (**c**) five revolutions.

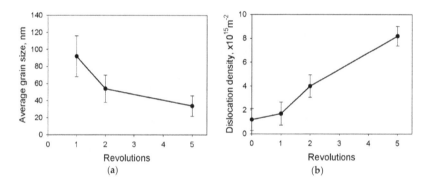

Figure 7. Microstructure parameters of the Ti/TiB MMC after HPT depending on the number of revolutions: (**a**) average (sub)grain size; (**b**) dislocation density.

3.3. Microhardness

Figure 8a shows the microhardness evolution at the center and the edge of the disks during HPT. The microhardness at the edge of the specimen increased with strain more intensively than that in the center. The maximum microhardness attained after five revolutions ($\gamma \approx 224$) was ~510 HV and 485 HV at the edge and in the center, respectively. Both values were higher than those in the initial condition; however, the increment did not exceed 10%. In comparison with the microhardness of commercially pure titanium (Ti of 99.4% purity) subjected to HTP at room temperature (~300 HV [30]), the reinforcement with TiB particles and further particles refinement due to HPT yield ~80% hardening. It should be noted that the difference in hardness between the center and edge of the specimen increased during HPT. Usually the hardness of the center of specimens "caught up" the hardness of the edge after a certain strain [22]. However, in our case, five revolutions did not result in hardness saturation at the edge of the specimen yet; that is why hardening in the center occurred slower.

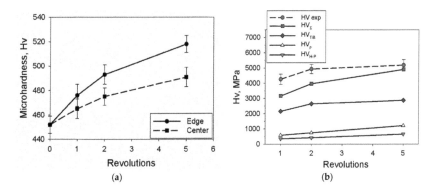

Figure 8. Microhardness evolution of the Ti/TiB MMC during HPT: (**a**) Experimental data; and (**b**) contributions of various mechanism into hardness. Experimental data for the edge of the specimen was used in (**b**) as a reference.

4. Discussion

The results of the present work show considerable changes in microstructure as a result of one to five revolutions during HPT at 400 °C. The main changes were associated with the formation of a nanostructure in the titanium matrix and refinement/redistribution of the TiB particles. However, it should be noted that the microstructure was highly heterogeneous through the specimen; in the center of the discs, the microstructure remained moderately deformed even after five revolutions (Figure 3b). This is an obvious result of a strong gradient of the imposed strain along the radius of the discs (see Equation (1)) and is typical of the HPT process [22].

Formation of a nanocrystalline microstructure after HPT is well documented for various metallic materials [22] including MMCs with different matrixes and reinforcements [22]. The microstructure evolution during HPT at room temperature is usually associated with the formation of low-angle grain boundaries at low strains and transformation of some of these low-angle subgrain boundaries into high-angle grain boundaries at higher strains giving rise to a considerable microstructure refinement. Hughes and Hansen [31] ascribed this process to gradual transformation of geometrically necessary boundaries, which separate microvolumes with different combinations of slip systems, from low-angle subboundaries into high-angle grain boundaries. At elevated temperatures due to the development of thermoactivated processes such a fragmented structure can transform into grain one through continuous dynamic recrystallization (cDRX). In the investigated composite, cDRX was found to operate during uniaxial compression at 700 °C [19]. Therefore, the formation of new very small (34–90 nm, Figure 7a) grains with relatively low dislocation density and well-defined boundaries (Figure 6) during HPT at 400 °C can also be attributed to the occurrence of cDRX. It worth noting that the grain size attained in the commercial pure titanium (Grade 4) during HPT at the same temperature was found to be much larger (approximately 250 nm) [29]. The formation of very small grains in the Ti/TiB MMCs during HPT can be associated with a higly constrained deformation due to the present of a large number of TiB whiskers and very high dislocation density in the initial condition. All these factors resulted in higher flow stress of the composite in comparison with that of commercial pure titanium. Since the size of recrystallized grains (D) depends on the flow stress (σ_s) as $\sigma_s = KD^{-N}$, where K and N are constants [32], an increase in flow stress during deformation should lead to the formation of much smaller recrystallized grains, which is in agreement with the obtained result.

Another effect of severe plastic deformation on the microstructure of Ti/TiB MMC was a considerable (by ~6 times) decrease in the length of the TiB whiskers. This change occurred in the very beginning of the HPT (Figure 4); further straining resulted in more homogeneous redistribution of the TiB particles in the Ti matrix. The latter effect was mentioned by Bachmaier and Pippan [25] with respect to various MMCs.

More interesting is to consider the effect of the observed features on mechanical properties. The microstructural changes occurring during HPT were accompanied by a noticeable rise in hardness (Figure 8a). The contributions of the most relevant hardening mechanisms in hardness of the composite can be expressed as:

$$HV_\Sigma = HV_0 + HV_\rho + HV_{H-P} + HV_{TiB} \tag{2}$$

where HV_0 denotes the friction stress, HV_ρ is the substructure hardening, HV_{H-P} is the Hall–Petch hardening and HV_{TiB} is the precepetation hardening by debris of TiB.

The substructure hardening HV_ρ can be expressed as:

$$HV_\rho = M\alpha Gb\sqrt{\rho} \tag{3}$$

where M is the average Taylor factor, α is a constant, G is the shear modulus, b is the Burgers vector and ρ is the dislocation density. Typical values of M and α accepted for approximate calculations are 3 and 0.5, respectively. The Hall–Petch contribution to the strength is typically of the form:

$$HV_{H-P} = K_y\, d^{-1/2} \tag{4}$$

where K_y is the Hall–Petch coefficient and d is the grain size. The precipitation hardening HV_{TiB} (Orowan strengthening) can be calculated using the formula [33]:

$$HV_{TiB} = \frac{MGb}{2.36\pi} \ln(\frac{0.57DS^{\frac{1}{3}}}{b}) \frac{1}{(0.92.V^{-\frac{1}{3}} - 1.14)DS^{\frac{1}{3}}} \tag{5}$$

where D is the diameter of TiB whiskers, V is the volume fraction of TiB, and S is aspect ratio of TiB whiskers.

Some input parameters for the titanium matrix were taken from [34]: K_y = 0.3 MPa·m$^{1/2}$, HV_0 = 496 MPa and b = 2.9 × 10^{-10} m. The value of the shear stress G = 130 GPa for the Ti/TiB MMC was taken from [11]. The grain size d, dislocation density ρ, and particles diameter X were taken from Figures 4 and 6a,b, respectively. The volume fraction of TiB determined using X-ray analysis was 19%.

In the case of composite materials, some contribution to the total strength can also be expected from [5]: (i) solid solution strengthening by precipitation of interstitial carbon, oxygen and nitrogen atoms in the Ti lattice; (ii) prismatic punching of dislocations at the Ti/TiB interface due to thermal mismatch between the Ti and TiB; and (iii) load transfer from the Ti matrix to TiB by an interfacial shear stress. The total strengthening effect by interstitial C, O and N atoms in titanium, according to Ref. [35], was found to be 160 MPa (this value was included into HV_0). The solubility of boron in titanium is very low (<0.001 at %) [6] therefore the effect of boron atoms on strengthening was neglected. Due to very small difference in coefficient of thermal expansions between Ti and TiB the effect of prismatic punching of dislocations at Ti/TiB interface on strength was too small to mention. According to Refs. [33,36], the effect of load transfer is much lower than that of Orowan strengthening for the aspect ratio smaller than 10 (in our case ~5) so that this factor was also not taken into account.

The calculated contributions of different hardening mechanisms into the total hardness are shown in Figure 8b. It is seen that the Hall–Petch hardening has the lowest effect on strength. Substructure hardening gives a somewhat higher contribution. The contribution of these two hardening mechanisms expectably increased with strain due to the microstructure refinement and increase in dislocation density (Figure 7). Meanwhile, hardening due to the TiB precipitates gives a much more pronounced effect which is more than twice of that for the combined effect of Hall–Petch and substructure hardenings. Precipitation hardening also increased during deformation due to shortening of the reinforcements (Figure 5). More homogeneous distribution of the TiB particles can also give some effect in the mechanical properties, however more likely it can result in an increase in ductility [16,17,22] rather than in a pronounced strengthening.

The overall effect of various strengthening mechanisms fits well with the experimental result, suggesting that the precipitation hardening contribution is much more important than the substructure and Hall–Petch hardenings. Therefore, the properties of MMCs are most probably associated with the morphology and distribution of reinforcements rather than with the properties of the matrix. This statement is in agreement with a number of investigations on the contribution of various strengthening mechanisms after SPD. In various precipitation-hardened alloys (Al-Mg-Sc [37], Cu-Cr-Zr [38] or Mg-Y-Nd-Zr [39]), the effect of Orowan strengthening was equal to or higher than the total contribution of grain-size and substructure strengthenings.

An attractive application of Ti/TiB MMCs with high hardness can be associated with production of medical instruments, in particular cutting tools, because this material has some undeniable advantages over the "medical" steel: possibility of application in a magnetic field, low specific gravity and high biocompatibility.

5. Conclusions

Microstructure evolution and microhardness of Ti/TiB metal–matrix composite were studied during high-pressure torsion (HPT) at 400 °C. The following conclusions were made:

(1) Processing by HPT produced a microstructure with (sub)grains of ~34 nm after five revolutions ($\gamma \approx 157$). The microstructure evolution was associated with an intensive increase in dislocation density and substructure development, resulting in a gradual microstructure refinement of the Ti matrix and shortening/redistribution of TiB whiskers.

(2) The microhardness increased with strain attaining the maximum value (~520 HV) at the edge of the disk after five revolutions. Analysis of contributions of different hardening mechanisms into the hardness of the Ti/TiB metal–matrix composite shows that an increase in hardness can mostly be ascribed to a contribution of precipitation hardening. The combined effect of substructure and Hall–Petch hardening give approximately two times lower contribution.

Acknowledgments: The authors gratefully acknowledge the financial support from the Russian Science Foundation (Grant Number 15-19-00165). The authors are grateful to the personnel of the Joint Research Centre, Belgorod State University for their assistance with the instrumental analysis.

Author Contributions: Sergey Zherebtsov, Yuliya Ivanisenko and Nikita Stepanov conceived and designed the experiments. Maxim Ozerov, Margarita Klimova and Yuliya Ivanisenko performed the experiments. Maxim Ozerov, Nikita Stepanov and Sergey Zherebtsov analyzed the data and wrote the paper.

Conflicts of Interest: The authors declare no conflict of interest.

References

1. Leyens, C.; Peters, M. *Titanium and Titanium Alloys. Fundamentals and Applications*; Wiley-VCH: Weinheim, Germany, 2003; pp. 1–499.
2. Saito, T.; Furuta, T.; Yamaguchi, T. Development of low cost titanium matrix composite. In *Advances in Titanium Metal Matrix Composites, the Minerals, Metals and Materials Society*; Froes, F.H., Storer, J., Eds.; TMS: Warrendale, PA, USA, 1995; pp. 33–44.
3. Radhakrishna Bhat, B.V.; Subramanyam, J.; Bhanu Prasad, V.V. Preparation of Ti-TiB-TiC & Ti-TiB composites by in-situ reaction hot pressing. *Mater. Sci. Eng. A* **2002**, *325*, 126–130.
4. Li, S.; Sun, B.; Imaia, H.; Kondoh, K. Powder metallurgy Ti-TiC metal matrix composites prepared by in situ reactive processing of Ti-VGCFs system. *Carbon* **2013**, *61*, 216–228. [CrossRef]
5. Munir, K.S.; Zheng, Y.; Zhang, D.; Lin, J.; Li, Y.; Wen, C. Improving the strengthening efficiency of carbon nanotubes in titanium metal matrix composites. *Mater. Sci. Eng. A* **2017**, *696*, 10–25. [CrossRef]
6. Godfrey, T.M.T.; Goodwin, P.S.; Ward-Close, C.M. Titanium particulate metal matrix composites—Reinforcement, production methods, and mechanical properties. *Adv. Eng. Mater.* **2000**, *2*, 85–91. [CrossRef]

7. Lindroos, V.K.; Talvitie, M.J. Recent advances in metal matrix composites. *J. Mater. Process. Technol.* **1995**, *53*, 273–284. [CrossRef]

8. Yamamoto, T.; Otsuki, A.; Ishihara, K.; Shingu, P.H. Synthesis of near net shape high density TiB/Ti composite. *Mater. Sci. Eng. A* **1997**, *239–240*, 647–651. [CrossRef]

9. Kumari, S.; Prasad, N.E.; Chandran, K.S.R.; Malakondaiah, G. High-temperature deformation behavior of Ti-TiB$_w$ in-situ metal-matrix composites. *JOM* **2004**, *5*, 51–55. [CrossRef]

10. Tsang, H.T.; Chao, C.G.; Ma, C.Y. Effects of volume fraction of reinforcement on tensile and creep properties of in-situ TiBTi MMC. *Scr. Mater.* **1997**, *37*, 1359–1365. [CrossRef]

11. Morsi, K.; Patel, V.V. Processing and properties of titanium–titanium boride (TiB$_w$) matrix composites—A review. *J. Mater. Sci.* **2007**, *42*, 2037–2047. [CrossRef]

12. Ravi Chandran, K.S.; Panda, K.B.; Sahay, S.S. TiB$_w$-reinforced Ti composites: Processing, properties, application, prospects, and research needs. *JOM* **2004**, *56*, 42–48. [CrossRef]

13. Feng, H.; Zhou, Y.; Jia, D.; Meng, Q.; Rao, J. Growth mechanism of in situ TiB whiskers in spark plasma sintered TiB/Ti metal matrix composites. *Cryst. Growth Des.* **2006**, *6*, 1626–1630. [CrossRef]

14. Da Silva, A.A.M.; Dos Santos, J.F.; Strohaecker, T.R. Microstructural and mechanical characterisation of a Ti6Al4V/TiC/10p composite processed by the BE-CHIP method. *Comp. Sci. Technol.* **2005**, *65*, 1749–1755. [CrossRef]

15. Ozerov, M.; Stepanov, N.; Kolesnikov, A.; Sokolovsky, V.; Zherebtsov, S. Brittle-to-ductile transition in a Ti-TiB metal-matrix composite. *Mater. Lett.* **2017**, *187*, 28–31. [CrossRef]

16. Gaisin, R.A.; Imayev, V.M.; Imayev, R.M. Effect of hot forging on microstructure and mechanical properties of near α titanium alloy/TiB composites produced by casting. *J. Alloys Compd.* **2017**, *723*, 385–394. [CrossRef]

17. Imayev, V.; Gaisin, R.; Gaisina, E.; Imayev, R.; Fecht, H.-J.; Pyczak, F. Effect of hot forging on microstructure and tensile properties of Ti-TiB. *Mater. Sci. Eng. A* **2014**, *609*, 34–41. [CrossRef]

18. Huang, L.; Cui, X.; Geng, L.; Fu, Y. Effects of rolling deformation on microstructure and mechanical properties of network structured TiB$_w$/Ti composites. *Trans. Nonferr. Met. Soc. China* **2012**, *22*, 79–83. [CrossRef]

19. Ozerov, M.; Klimova, M.; Kolesnikov, A.; Stepanov, N.; Zherebtsov, S. Deformation behavior and microstructure evolution of a Ti/TiB metal-matrix composite during high-temperature compression tests. *Mater. Des.* **2016**, *112*, 17–26. [CrossRef]

20. Valiev, R.Z.; Islamgaliev, R.K.; Alexandrov, I.V. Bulk nanostructured materials from severe plastic deformation. *Prog. Mater. Sci.* **2000**, *45*, 103–189. [CrossRef]

21. Valiev, R.Z.; Estrin, Y.; Horita, Z.; Langdon, T.G.; Zechetbauer, M.J.; Zhu, Y.T. Producing bulk ultrafine-grained materials by severe plastic deformation. *JOM* **2006**, *58*, 33–39. [CrossRef]

22. Zhilyaev, A.P.; Langdon, T.G. Using high-pressure torsion for metal processing: Fundamentals and applications. *Prog. Mater. Sci.* **2008**, *53*, 893–979. [CrossRef]

23. Stolyarov, V.V.; Zhu, Y.T.; Lowe, T.C.; Islamgaliev, R.K.; Valiev, R.Z. Processing nanocrystalline Ti and its nanocomposites from micrometer-sized Ti powder using high pressure torsion. *Mater. Sci. Eng. A* **2000**, *282*, 78–85. [CrossRef]

24. Edalati, K.; Iwaoka, H.; Horita, Z.; Konno, M.; Sato, T. Unusual hardening in Ti/Al$_2$O$_3$ nanocomposites produced by high-pressure torsion followed by annealing. *Mater. Sci. Eng. A* **2011**, *529*, 435–441. [CrossRef]

25. Bachmaier, A.; Pippan, R. Generation of metallic nanocomposites by severe plastic deformation. *Int. Mater. Rev.* **2013**, *58*, 41–62. [CrossRef]

26. Islamgaliev, R.K.; Buchgraber, W.; Kolobov, Y.R.; Amirkhanov, N.M.; Sergueeva, A.V.; Ivanov, K.V.; Grabovetskaya, G.P. Deformation behavior of Cu-based nanocomposite processed by severe plastic deformation. *Mater. Sci. Eng. A* **2001**, *319–321*, 872–876. [CrossRef]

27. Williamson, G.K.; Hall, W.H. X-ray line broadening from filed aluminium and wolfram. *Acta Metall.* **1953**, *1*, 22–31. [CrossRef]

28. Ozerov, M.; Klimova, M.; Vyazmin, A.; Stepanov, N.; Zherebtsov, S. Orientation relationship in a Ti/TiB metal-matrix composite. *Mater. Lett.* **2017**, *186*, 168–170. [CrossRef]

29. Islamgaliev, R.K.; Kazyhanov, V.U.; Shestakova, L.O.; Sharafutdinov, A.V.; Valiev, R.Z. Microstructure and mechanical properties of titanium (Grade 4) processed by high-pressure torsion. *Mater. Sci. Eng. A* **2008**, *493*, 190–194. [CrossRef]

30. Edalati, K.; Matsubara, E.; Horita, Z. Processing pure Ti by high-pressure torsion in wide ranges of pressures and strain. *Metall. Mater. Trans. A* **2009**, *40*, 2079–2086. [CrossRef]

31. Hughes, D.; Hansen, N. Microstructure and strength of nickel at large strains. *Acta Mater.* **2000**, *48*, 2985–3004. [CrossRef]
32. Humphreys, F.; Hatherly, M. *Recrystallization and Related Annealing Phenomena*, 2nd ed.; Elsevier: Oxford, UK, 2004; pp. 1–605.
33. Chen, B.; Shen, J.; Ye, X.; Jia, L.; Li, S.; Umeda, J.; Takahashi, M.; Kondoh, K. Length effect of carbon nanotubes on the strengthening mechanisms in metal matrix composites. *Acta Mater.* **2017**, *140*, 317–325. [CrossRef]
34. Frost, H.J.; Ashby, M.F. *Deformation-Mechanism Maps*; Pergamon Press: Oxford, UK, 1982; pp. 1–166.
35. Conrad, H. Effect of interstitial solutes on the strength and ductility of titanium. *Prog. Mater. Sci.* **1981**, *26*, 123–403. [CrossRef]
36. Kelly, A.; Tyson, W.R. Tensile properties of fibre-reinforced metals: Copper/Tungsten and copper/molybdenum. *J. Mech. Phys. Solids* **1965**, *13*, 329–350. [CrossRef]
37. Harrell, T.J.; Topping, T.D.; Wen, H.; Hu, T.; Schoenung, J.M.; Lavernia, E.J. Microstructure and strengthening mechanisms in an ultrafine grained Al-Mg-Sc alloy produced by powder metallurgy. *Metall. Mater. Trans. A* **2014**, *45*, 6329–6343. [CrossRef]
38. Mishnev, R.; Shakhova, I.; Belyakov, A.; Kaibyshev, R. Deformation microstructures, strengthening mechanisms, and electrical conductivity in a Cu-Cr-Zr alloy. *Mater. Sci. Eng. A* **2015**, *629*, 29–40. [CrossRef]
39. Lukyanova, E.A.; Martynenko, N.S.; Shakhova, I.; Belyakov, A.N.; Rokhlin, L.L.; Dobatkin, S.V.; Estrin, Y.Z. Strengthening of age-hardenable WE43 magnesium alloy processed by high pressure torsion. *Mater. Lett.* **2016**, *170*, 5–9. [CrossRef]

Article

Effects of Strain Rate and Measuring Temperature on the Elastocaloric Cooling in a Columnar-Grained Cu$_{71}$Al$_{17.5}$Mn$_{11.5}$ Shape Memory Alloy

Hui Wang, Haiyou Huang * and Jianxin Xie

Key Laboratory for Advanced Materials Processing of the Ministry of Education, Institute for Advanced Materials and Technology, University of Science and Technology Beijing, Beijing 100083, China; wangh9130@163.com (H.W.); jxxie@ustb.edu.cn (J.X.)
* Correspondence: huanghy@mater.ustb.edu.cn; Tel.: +86-010-6233-2253

Received: 22 October 2017; Accepted: 18 November 2017; Published: 27 November 2017

Abstract: Solid-state refrigeration technology based on elastocaloric effects (eCEs) is attracting more and more attention from scientists and engineers. The response speed of the elastocaloric materials, which relates to the sensitivity to the strain rate and measuring temperature, is a significant parameter to evaluate the development of the elastocaloric material in device applications. Because the Cu-Al-Mn shape memory alloy (SMA) possesses a good eCE and a wide temperature window, it has been reported to be the most promising elastocaloric cooling material. In the present paper, the temperature changes (ΔT) induced by reversible martensitic transformation in a columnar-grained Cu$_{71}$Al$_{17.5}$Mn$_{11.5}$ SMA fabricated by directional solidification were directly measured over the strain rate range of 0.005–0.19 s^{-1} and the measuring temperature range of 291–420 K. The maximum adiabatic ΔT of 16.5 K and a lower strain-rate sensitivity compared to TiNi-based SMAs were observed. With increasing strain rate, the ΔT value and the corresponding coefficient of performance (COP) of the alloy first increased, then achieved saturation when the strain rate reached 0.05 s^{-1}. When the measuring temperature rose, the ΔT value increased linearly while the COP decreased linearly. The results of our work provide theoretical reference for the design of elastocaloric cooling devices made of this alloy.

Keywords: shape memory alloy; columnar grain; Cu-Al-Mn; elastocaloric effect; strain rate; measuring temperature

---

1. Introduction

The elastocaloric effect (eCE) refers to the thermal response of a given material to external uniaxial stress, which is commonly quantified by the isothermal entropy change (ΔS) and adiabatic temperature change (ΔT). Elastocaloric refrigeration based on eCEs, which is a new type of solid-state refrigeration, has drawn significant attention in recent years owing to its higher coefficient of performance (COP), its lower device costs, and its eco-friendliness as a promising alternative to conventional vapor compression, as well as to its downscaling ability for microcooling [1–5]; it is regarded as the most applicable solid-state refrigeration technology [6]. Shape memory alloys (SMAs) can undergo a reversible martensitic transformation (MT) by applying external stress, which is accompanied by a large latent heat absorption and release. These features make them important elastocaloric materials and thus draw momentous attention.

The refrigeration capability (RC) of a given elastocaloric material can be evaluated by directly measuring the adiabatic temperature change of the phase transformation or deformation. Some related research indicated that a high ΔT value of more than 10 K has been directly measured in Ti-Ni- [7,8], Ni-Fe- [9] and Cu-based SMAs [10], which has laid a good material foundation for the development and application of elastocaloric refrigeration technology. At present, an increasing number of promising elastocaloric refrigeration devices are being developed [11]. In addition to evaluating the RC, the responding speed of SMAs is also a significant factor that influences the system cooling efficiency of the solid refrigeration system. Some researchers have carried out a series of research on the influence of strain rates on the eCE in Ti-Ni-based SMAs, such as $Ti_{49.1}Ni_{50.5}Fe_{0.4}$ foils and $Ti_{50.4}Ni_{49.6}$ films by Ossmer et al. [12,13], $Ti_{51.1}Ni_{48.9}$ wires by Tušek et al. [8], $Ti_{55.4}Ni_{44.6}$ wires by Tobush et al. [14], a $Ti_{47.25}Ni_{45}Cu_5V_{2.75}$ block by Schmidt et al. [15], and so on. Compared with the Ti-Ni-based SMAs, Cu-based SMAs also have a large ΔT value, which has been proved by Xu et al. They found a ΔT value of 12–13 K in columnar-grained Cu-Al-Mn SMAs [10], covering a wide temperature range of more than 100 K, which, combined with a low applying stress [10,16] and low material costs, make the columnar-grained Cu-Al-Mn SMAs a promising material for solid-state refrigeration. In this paper, over the strain rate from 0.005 to 0.19 s^{-1} and the test temperature range from 291 to 393 K, the temperature changes induced by reversible MT of columnar-grained $Cu_{71}Al_{17.5}Mn_{11.5}$ SMAs have been systematically measured, the effects of the strain rate ($\dot{\varepsilon}$) and measuring temperature (T_A) on the eCEs have been obtained, and the influence mechanism is discussed. The results of this study can provide a theoretical reference for the design and application of elastocaloric cooling devices that are made of Cu-based SMAs.

2. Materials and Methods

A $Cu_{71}Al_{17.5}Mn_{11.5}$ ingot with a columnar-grained microstructure was prepared by directional solidification [17]. At first, the ingot was annealed at 1073 K for 5 min followed by quenching into ice water to obtain a single β_1 phase. Then, the ingot was aged at 473 K for 15 min to stabilize the MT temperatures. Dog bone-shaped tensile samples with a gauge size of 25 mm × 6 mm × 2 mm were cut out from the ingot, with their longitudinal direction along the solidification direction (SD). The surface of the tensile specimen was polished with 500–3000 # sandpaper and then subjected to mechanical polishing and electrolytic polishing. The texture characterization was conducted by electron back-scattered diffraction (EBSD) with a minimum misorientation resolution of 2°. The electrolytic polishing solution was as follows: 250 mL of H_3PO_4, 250 mL of alcohol, 50 mL of glycerol, 5 g of urea, and 500 mL of H_2O, with a voltage of 10 V and duration from 80 to 120 s. The transformation temperatures and latent heat were determined by NETZSCH 404F3 using differential scanning calorimetry (DSC, Mettler-Toledo, Zurich, Switzerland) under a nitrogen inert gas flow of 10 mL·s^{-1} with a heating/cooling rate of 10 K/min. Tensile tests were conducted on a Mechanical Testing System (MTS) testing machine (Wister Industrial Equipment cooperation Limited, Shenzhen, China) equipped with a thermostatic chamber. During the tensile test, all samples were loaded at a constant strain rate of 0.05 s^{-1} to the maximum strain of 10% and were then unloaded at different strain rates to zero stress. In this work, the measuring temperature range was from 291 to 420 K, and the unloading strain rate range was from 0.005 to 0.19 s^{-1}. The temperature of the sample in the tensile cycle was monitored with a K-type thermocouple welded on the center of the sample surface. The authors used a set of self-built equipment for data acquisition and a MATLAB program (v7.0, MathWorks, Natick, MA, USA, 2012) to display the measured temperature.

3. Results and Discussion

The $Cu_{71}Al_{17.5}Mn_{11.5}$ SMA sample was composed of a single austenite phase β_1 with L2$_1$ crystallographic structure at room temperature. When a stress applied in the sample was higher than the critical stress of MT, a stress-induced MT of $\beta_1 \rightarrow \beta_1'$ occurred and the β_1' martensite had an 18R ordered structure. The phase structure and transformation behavior have been determined by X-ray diffraction (XRD, SmartLab, Rigaku Corporation, Matsubara-cho, Akishima-shi, Tokyo, Japan) and transmission electron microscopy (TEM, F20, FEI, Hillsboro, OR, USA) operated at 200 kV at room temperature in previous work by authors and other researchers [18–22].

The DSC curve during the heating/cooling process of the as-quenched sample is shown in Figure 1. An exothermic peak corresponding to the MT during cooling and an endothermic peak corresponding to the reverse austenitic transformation during heating can be clearly observed. The MT and reverse transformation temperatures can be obtained from the curve: the MT starting temperature was M_s = 247 K, the MT finishing temperature was M_f = 235 K, the reverse transformation starting temperature was A_S = 253 K and the reverse transformation finishing temperature was A_f = 265 K; the thermal hysteresis was 18 K.

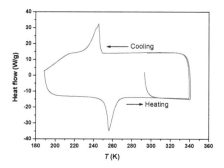

Figure 1. Differential scanning calorimetry (DSC) curve of the columnar-grained $Cu_{71}Al_{17.5}Mn_{11.5}$ shape memory alloy (SMA; heating/cooling rate of 10 K/min).

The EBSD orientation map and the inverse pole figures (Figure 2) illustrate that the columnar-grained Cu-Al-Mn sample has a strong <001> oriented texture along the SD (solidification direction).

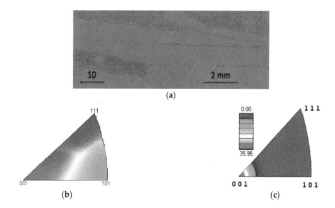

Figure 2. (a) Electron back-scattered diffraction (EBSD) quasi-colored orientation map along solidification direction (SD); (b) the reference stereographic triangle; (c) inverse pole figure along the SD, which illustrate that the columnar-grained $Cu_{71}Al_{17.5}Mn_{11.5}$ SMA sample has a strong <001>-oriented texture along the SD.

The tensile stress-strain curves for the columnar-grained $Cu_{71}Al_{17.5}Mn_{11.5}$ SMA are shown in Figure 3; these were tested at the same loading strain rate $\dot{\varepsilon}_l$ of 0.05 s^{-1} while gradually increasing the unloading strain rates $\dot{\varepsilon}_u$ from 0.005 to 0.19 s^{-1}. Specifically, nine strain rates of 0.005, 0.01, 0.05, 0.07, 0.096, 0.1, 0.13, 0.15, and 0.19 s^{-1} were chosen. It should be noted that in order to avoid the experimental deviation from different samples, the test was carried out using the same sample at each measuring temperature. For all the test conditions in Figure 3, when the measuring temperature was ≤393 K, an almost 100% strain recovery could be observed after the samples underwent a loading strain of 10%, indicating an excellent superelasticity of the columnar-grained $Cu_{71}Al_{17.5}Mn_{11.5}$ SMA. With increasing tensile cycles, both the critical stresses of MT and reverse transformation were decreased gradually. Because the loading strain rate is constant, the stress of MT decreasing in loading processes can be attributed to the influence of cycle numbers, which may be related to the fatigue effect. However, the stress of reverse transformation in unloading processes may be affected by loading processes, strain rates or cycle numbers.

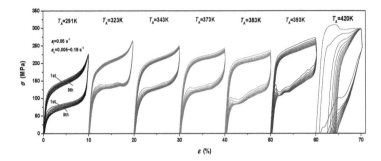

Figure 3. Stress-strain curves for the columnar-grained $Cu_{71}Al_{17.5}Mn_{11.5}$ shape memory alloys (SMAs) with different unloading strain rates ($\dot{\varepsilon}_u$ = 0.005–0.19 s^{-1}) at different measuring temperatures.

In this work, the influence of loading processes can be excluded as a result of the same loading strain and loading strain rate. In order to clarify the influence of cycle numbers or strain rate on the unloading curves, a comparative tensile test with 10 cycles was carried on at a constant unloading rate of 0.05 s^{-1}, and the stress-strain curves are shown in Figure 4a. Comparing the stress-strain curve tested at 291 K in Figure 3 and the curve in Figure 4a, the reduction of the transformation stress has closed relations with the increase in cycle numbers. In other words, the strain rate has little effect on the transformation stress of columnar-grained $Cu_{71}Al_{17.5}Mn_{11.5}$ SMA, which is different from the phenomenon that was observed in Ti-Ni alloys [14,15]. For Ti-Ni alloys, with increasing strain rates, the stress-strain loop enlarges clearly; that is, the loading stress increases and unloading stress decreases. The strain rate sensitivity of the transformation stress in SMAs is related to the capability of stress relaxation, which is caused by martensite nucleation and growth during the MT process. Under constant strain conditions, the MT of Ti-Ni SMAs requires 3 min to induce a stress relaxation of 50 MPa [15]. In other words, when the strain rate exceeds 0.02 s^{-1} of Ti-Ni SMAs, the stress will clearly increase as a result of stress relaxation hysteresis. Figures 3 and 4a indicate that the phenomenon of transformation stress increase is not observed in columnar-grained $Cu_{71}Al_{17.5}Mn_{11.5}$ SMAs until the strain rate reaches 0.19 s^{-1}, which means that the stress relaxation capacity of columnar-grained $Cu_{71}Al_{17.5}Mn_{11.5}$ SMAs is more than 10 times that of Ti-Ni SMAs. Therefore, the strain rate sensitivity of columnar-grained $Cu_{71}Al_{17.5}Mn_{11.5}$ SMAs is dramatically lower than that of Ti-Ni SMAs.

In order to study the effect of measuring temperature on tensile stress-strain curves of the columnar-grained $Cu_{71}Al_{17.5}Mn_{11.5}$ SMA, a series of tensile cycle tests were carried out between the temperature range from 291 to 420 K, and the results are subsequently drawn in Figure 3. A comparison of these stress-strain curves indicates that as the measuring temperature rises, the MT stress increases gradually. When the measuring temperature reaches 420 K, the recoverable strain of the sample decreases rapidly with the increase in the number of tensile cycles. The microstructure observation found that a large number of dislocations were observed in the sample, indicating that the tensile stress was high enough to start the dislocation slip. This phenomenon implies the deformation mechanism begins to change from superelastic deformation induced by phase transformation to permanent plastic deformation caused by a dislocation slip when the sample deforms at 420 K, which can be defined as the critical temperature of stress-induced MT in columnar-grained $Cu_{71}Al_{17.5}Mn_{11.5}$ alloys. Therefore, the upper limit of the temperature window of the eCE of columnar-grained $Cu_{71}Al_{17.5}Mn_{11.5}$ alloys is less than 420 K. In other words, the temperature window of the columnar-grained $Cu_{71}Al_{17.5}Mn_{11.5}$ alloy is 265–393 K. The width of the temperature window is 128 K.

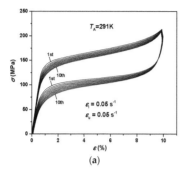 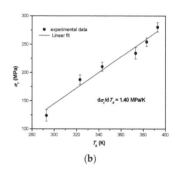

(a)   (b)

Figure 4. (a) Stress-strain curves with constant loading rate ($\dot{\varepsilon}_1 = 0.05\ s^{-1}$) at 291 K; (b) T_A dependence of σ_c for the columnar-grained $Cu_{71}Al_{17.5}Mn_{11.5}$ shape memory alloy (SMA).

On the basis of the first cycle of the measured stress–strain curve in the temperature range of 291–393 K, the MT critical stress (σ_c) as a function of T_A is plotted in Figure 4b. Figure 4b indicates that σ_c linearly increases with the increase of measuring temperature in the range of 291–393 K, and $d\sigma_c/dT_A = 1.40\ MPa/K$ for the columnar-grained $Cu_{71}Al_{17.5}Mn_{11.5}$ alloy, which can be determined by linear fitting.

Figure 5 indicates the temperature change of the columnar-grained $Cu_{71}Al_{17.5}Mn_{11.5}$ SMA samples in loading-unloading cycles at different measuring temperatures. It can be seen from Figure 5a–d that at a constant strain rate ($0.05\ s^{-1}$), the ΔT values measured from both loading and unloading processes for different cycles are almost unchanged in the temperature range of 291–393 K, which indicates that no evident influence of cycle numbers on ΔT was observed after 10 tensile cycles at a constant strain rate; this implies a good stability of the eCE in the columnar-grained $Cu_{71}Al_{17.5}Mn_{11.5}$ SMA. When the strain rate is less than $0.05\ s^{-1}$, $|\Delta T|$ increases with increasing $\dot{\varepsilon}_u$. When the strain rate reaches $0.05\ s^{-1}$ or above, $|\Delta T|$ achieves saturation. For instance, at $T_A = 291\ K$, when $\dot{\varepsilon}_u$ increases from 0.005 to $0.05\ s^{-1}$, ΔT changes from 7.53 to 11.21 K. After $\dot{\varepsilon}_u$ reaches $0.05\ s^{-1}$, the ΔT values remain constant within the range of 11.21–11.51 K. For different T_A values, the variation trends of the T vs. t curves from the samples with increasing $\dot{\varepsilon}_u$ are similar.

In addition, the phenomenon of temperature irreversibility can also be found from Figure 5; that is, the absolute values of the temperature change between loading and unloading are not equal. For example, when the strain rate is $0.05\ s^{-1}$ at 291 K, the loading temperature rises are 14.8–15.3 K (Figure 5a) and 13.7–13.9 K (Figure 5b), while the unloading temperature drops are 13.2–13.5 K (Figure 3a) and 11.2 K (Figure 5b). The absolute value of the unloading temperature drop is 1.6–2.7 K, which is lower than the loading temperature rise. Irreversibility is caused by the existence of the hysteresis area during the loading-unloading process. The formation of the hysteresis area is directly related to the frictional origin during the transformation. These frictions include the interfacial friction between martensite and austenite, the interaction between the phase interface and grain boundary, or other defects [12,14]. When the measuring temperature reached 420 K, as a result of the occurrence of irreversible deformation caused by a dislocation slip, the return strain gradually reduced with the increasing stretching cycles, resulting in a gradually reduced temperature change, as shown in Figure 5e.

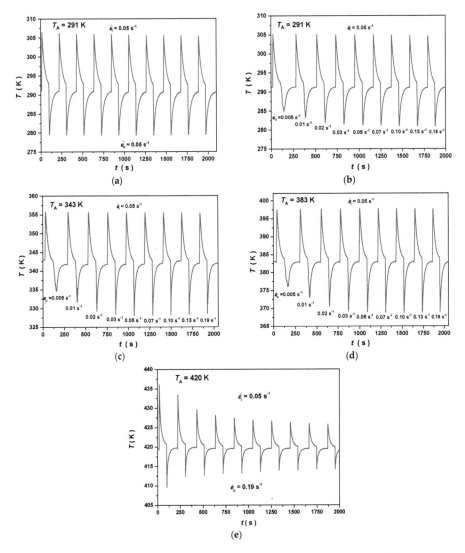

Figure 5. Temperature-time profiles with a specific unloading strain rate: (a) $\dot{\varepsilon}_u = 0.05\ \text{s}^{-1}$, $T_A = 291$ K; (e) $\dot{\varepsilon}_u = 0.19\ \text{s}^{-1}$, $T_A = 420$ K and with various unloading strain rates ($\dot{\varepsilon}_u = 0.005$–$0.19\ \text{s}^{-1}$) at $T_A = 291$ K (b); $T_A = 343$ K (c); and $T_A = 383$ K (d) for the columnar-grained $Cu_{71}Al_{17.5}Mn_{11.5}$ shape memory alloy (SMA) samples ($\dot{\varepsilon}_l = 0.05\ \text{s}^{-1}$ for all samples).

On the basis of the *T-t* profiles (four typical profiles are shown in Figure 5), a high $|\Delta T|$ value of 11.3–16.5 K can be obtained in the columnar-grained $Cu_{71}Al_{17.5}Mn_{11.5}$ SMA during unloading processes, covering a wide temperature range of more than 100 K. The entropy change of the phase transformation associated with the elastocaloric cooling can be estimated as

$$\Delta S \approx -\frac{\Delta T}{T_A} C_p \tag{1}$$

where C_p is the heat capacity, measured to be 455 J/kg·K for the $Cu_{71}Al_{17.5}Mn_{11}$ SMA [10]. Strictly speaking, when calculating the adiabatic temperature change or entropy change, because of the coexistence of two phases during transformation, we should consider that $C_p = xC_p^A + (1 - x)C_p^M$, where x is the fraction of the authentic phase, and C_p^A and C_p^M are the heat capacity of the authentic phase and martensitic phase, respectively. However, C_p^A and C_p^M are approximately equal for SMAs [23]. Therefore, in the actual measurement and calculation, we supposed $C_p \approx C_p^A \approx C_p^M$. In this paper, we experimentally measured the C_p^A value and used it in the calculation. According to the ΔT values measured above, the maximum ΔS can be estimated to be 19.5 J/(kg·K). Additionally, ΔS can also be calculated from the Clausius-Clapeyron equation:

$$\Delta S = -\frac{1}{\rho}\frac{d\sigma_c}{dT_A}\varepsilon \qquad (2)$$

where ρ is the density and ε is the transformation strain. For the columnar-grained $Cu_{71}Al_{17.5}Mn_{11.5}$ SMA, $d\sigma_c/dT_A$ and ε can be determined to be 1.40 MPa/K from Figure 4b and 8.3% from Figure 3, and $\rho = 7.40 \times 10^3$ kg/m^3 [24]. The calculated ΔS value from the Clausius-Clapeyron equation is about 15.7 J/(kg·K), which is smaller than the calculation results based on ΔT. In addition, the theoretical maximum of the isothermal entropy change also can be calculated to be 25.0 J/kg·K by the latent heat of phase transformation determined by DSC measurement [10]. Therefore, the entropy change estimated on the basis of the experimental data in this paper approached ~78% of its theoretical value. It is generally believed that the latent heat of phase transformation determined by the DSC method corresponds to a completed entropy change from 100% phase transformation without any loss, thus often called the theoretical entropy change. The entropy change value estimated on the basis of the experimentally measured ΔT value is less than the theoretical entropy change, because of the existence of internal frictions induced by the phase interface frictions and the interactions between phase migration and defects. In addition, in the actual tensile deformation processes, incomplete MT may be another probable cause of the smaller entropy change value estimated by stress-strain curves. According to the above discussion, the energy loss induced by internal frictions is about 5.5 J/(kg·K) for the columnar-grained $Cu_{71}Al_{17.5}Mn_{11.5}$ SMA.

The $|\Delta T|$ value and the COP for unloading processes in the columnar-grained $Cu_{71}Al_{17.5}Mn_{11.5}$ SMA as a function of $\dot{\varepsilon}$ and T_A are summarized in Figure 6. The COP of the material, which describes the cooling efficiency, is defined by the ratio of cooling power (ΔQ) to input work (ΔW) [25]:

$$COP = \Delta Q/\Delta W \qquad (3)$$

where ΔQ can be estimated from the latent heat, which is $\Delta T_{ad} \times C_p$ (ΔT_{ad} is the adiabatic temperature change), and ΔW can be obtained by integrating the area enclosed by the stress hysteresis loop (Figure 3). Figure 6a indicates that all $|\Delta T|$ vs. $\dot{\varepsilon}$ curves show a similar variation trend. At first, the $|\Delta T|$ value increases with increasing $\dot{\varepsilon}$. When the strain rate reaches 0.05 s^{-1}, the $|\Delta T|$ value remains almost unchanged, implying it achieves saturation. The critical strain rate $\dot{\varepsilon}_c$ corresponding to the onset of $|\Delta T|$ saturation is less than that of Ti-Ni alloy, which was reported as 0.2 s^{-1} [13]. In other words, it is easier to reach near-adiabatic conditions using the columnar-grained $Cu_{71}Al_{17.5}Mn_{11.5}$ SMA. In the process of MT (reverse MT), the homogeneity of martensite (austenite) nucleation and extension is a noteworthy factor to influence $\dot{\varepsilon}_c$. The more homogeneously the MT occurs in the sample, the lower the $\dot{\varepsilon}_c$ value [12]. The columnar-grained Cu-Al-Mn SMA has a homogeneous contribution of stress/strain in the whole sample as a result of a high deformation and transformation compatibility among grains [26]. Therefore, the martensite (austenite) can nucleate and grow homogeneously in the columnar-grained samples. Furthermore, the Cu-based SMAs have a higher thermal conductivity compared to the Ti-Ni alloy, which also helps to obtain a uniform temperature distribution within a very short period of time. The above two reasons show that the columnar-grained Cu$_{71}$Al$_{17.5}$Mn$_{11.5}$ SMA has a low $\dot{\varepsilon}_c$, which can reduce the design difficulty of refrigeration devices.

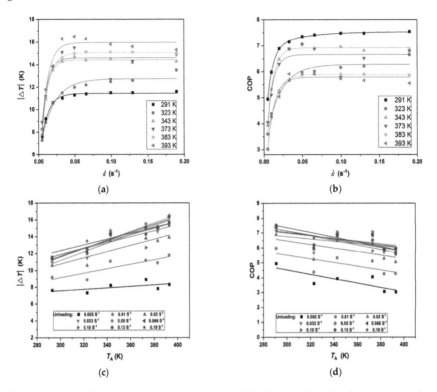

Figure 6. The $|\Delta T|$ and coefficient of performance (COP) values for unloading process in the columnar-grained Cu$_{71}$Al$_{17.5}$Mn$_{11.5}$ shape memory alloy (SMA) as a function of $\dot{\varepsilon}$ and T_A: (**a**) $|\Delta T|$ vs. $\dot{\varepsilon}$; (**b**) COP vs. $\dot{\varepsilon}$; (**c**) $|\Delta T|$ vs. T_A; (**d**) COP vs. T_A.

Figure 6b indicates that the variation trend of COP with increasing strain rate is consistent with $|\Delta T|$, while the variations of $|\Delta T|$ and COP with T_A are the opposite, as shown in Figure 6c,d. For refrigeration device or system design, a higher $|\Delta T|$ and COP are expected. For the columnar-grained Cu$_{71}$Al$_{17.5}$Mn$_{11.5}$ SMA, in order to achieve a stable and high refrigeration capability and COP, the applied strain rates should be more than 0.05 s^{-1}.

4. Conclusions

In summary, the effects of strain rates and the measuring temperature on the elastocaloric cooling in a columnar-grained Cu$_{71}$Al$_{17.5}$Mn$_{11.5}$ SMA were experimentally investigated over the strain rate range of 0.005–0.19 s^{-1} and the measuring temperature range of 291–393 K. With the increasing stain rate, the ΔT and COP values of the alloy increase firstly and then achieve saturation when the strain rate reaches 0.05 s^{-1}, which indicates a lower strain rate sensitivity of refrigeration capability compared to Ti-Ni-based SMAs (about 0.2 s^{-1}). The relatively low strain rate sensitivity of the columnar-grained Cu$_{71}$Al$_{17.5}$Mn$_{11.5}$ SMA is attributed to two reasons. The first is a high deformation and phase transformation compatibility among grains in the columnar-grained microstructure, which causes a homogenous stress/strain distribution in the entire samples. The other is a high thermal conductivity of Cu-based SMAs, which also helps to obtain a uniform temperature distribution in a very short period of time. In addition, a maximum adiabatic ΔT value of 16.5 K with corresponding

ΔS of 19.5 J/(kg·K) and a wide operational temperature window ($\omega_T > 100$ K) were directly measured in the experiment. The above results demonstrate that the columnar-grained $Cu_{71}Al_{17.5}Mn_{11.5}$ SMA is a promising candidate of elastocaloric materials with the advantages of a high refrigeration capability in a wide strain-rate range and operational temperature window, which are beneficial for the design and application to refrigeration devices.

Acknowledgments: This work was supported by the National Key Research and Development Program of China (Grant No. 2016YFB0700500) and the National Natural Science Foundation of China (Grant No. 51574027). Hui Wang thanks the State Key Laboratory of Technologies in Space Cryogenics Propellants, Technical Institute of Physics and Chemistry, Chinese Academy of Sciences for providing experimental equipment.

Author Contributions: Hui Wang and Haiyou Huang conceived and designed the experiments; Hui Wang performed the experiments; Hui Wang and Haiyou Huang analyzed the data and wrote the paper; Haiyou Huang and Jianxin Xie supported the writing of the paper. All authors participated in the discussions of the results.

Conflicts of Interest: The authors declare no conflict of interest.

References

1. Moya, X.; Kar-Narayan, S.; Mathur, N.D. Caloric materials near ferroic phase transitions. *Nat. Mater.* **2014**, *13*, 439–450. [CrossRef] [PubMed]
2. Gschneidnerjr, K.A.; Pecharsky, V.K.; Tsokol, A.O. Recent developments in magnetocaloric materials. *Rep. Prog. Phys.* **2005**, *68*, 1479–1539. [CrossRef]
3. Mischenko, A.S.; Zhang, Q.; Scott, J.F.; Whatmore, R.W.; Mathur, N.D. Giant electrocaloric effect in thin-film $PbZr_{0.95}Ti_{0.05}O_3$. *Science* **2006**, *311*, 1270–1271. [CrossRef] [PubMed]
4. Mañosa, L.; González-Alonso, D.; Planes, A.; Bonnot, E.; Barrio, M.; Tamarit, J.L.; Aksoy, S.; Acet, M. Giant solid-state barocaloric effect in the Ni-Mn-In magnetic shape-memory alloy. *Nat. Mater.* **2010**, *9*, 478–481. [CrossRef] [PubMed]
5. Bonnot, E.; Romero, R.; Mañosa, L.; Vives, E.; Planes, A. Elastocaloric effect associated with the martensitic transition in shape-memory alloys. *Phys. Rev. Lett.* **2008**, *100*, 125901. [CrossRef] [PubMed]
6. Goetzler, W.; Zogg, R.; Young, J.; Johnson, C. *Energy Savings Potential and RD&D Opportunities for Non-Vapor-Compression HVAC Technologies*; U.S. Department of Energy (Navigant Consulting Inc.): Washington, DC, USA, 2014.
7. Cui, J.; Wu, Y.; Muehlbauer, J.; Hwang, Y.; Radermacher, R.; Fackler, S.; Wuttig, M.; Takeuchi, I. Demonstration of high efficiency elastocaloric cooling with large ΔT using NiTi wires. *Appl. Phys. Lett.* **2012**, *101*, 1175–1178. [CrossRef]
8. Tušek, J.; Engelbrecht, K.; Mikkelsen, L.; Pryds, N. Elastocaloric effect of Ni-Ti wire for application in a cooling device. *J. Alloys Compd.* **2015**, *117*, 10–19. [CrossRef]
9. Pataky, G.J.; Ertekin, E.; Sehitoglu, H. Elastocaloric cooling potential of NiTi, Ni_2FeGa, and CoNiAl. *Acta Mater.* **2015**, *96*, 420–427. [CrossRef]
10. Xu, S.; Huang, H.Y.; Xie, J.; Takekawa, S.; Xu, X.; Omori, T.; Kainuma, R. Giant elastocaloric effect covering wide temperature range in columnar-grained $Cu_{71.5}Al_{17.5}Mn_{11}$ shape memory alloy. *APL Mater.* **2016**, *4*, 106106. [CrossRef]
11. Qian, S.; Geng, Y.; Wang, Y.; Radermacher, R. A review of elastocaloric cooling: Materials, cycles and system integrations. *Int. J. Refrig.* **2016**, *64*, 1–19. [CrossRef]
12. Ossmer, H.; Miyazaki, S.; Kohl, M. The elastocaloric effect in TiNi-based foils. *Mater. Today Proc.* **2015**, *2*, S971–S974. [CrossRef]
13. Ossmer, H.; Lambrecht, F.; Gültig, M.; Chluba, C.; Quandt, E.; Kohl, M. Evolution of temperature profiles in Ti-Ni films for elastocaloric cooling. *Acta Mater.* **2014**, *81*, 9–20. [CrossRef]
14. Tobushi, H.; Shimeno, Y.; Hachisuka, T.; Tanaka, K. Influence of strain rate on superelastic properties of TiNi shape memory alloy. *Mech. Mater.* **1998**, *30*, 141–150. [CrossRef]
15. Schmidt, M.; Schütze, A.; Seelecke, S. Elastocaloric cooling processes: The influence of material strain and strain rate on efficiency and temperature span. *Appl. Phys. Lett.* **2016**, *4*, 10–19. [CrossRef]
16. Manosa, L.; Jarque-Farnos, S.; Vives, E.; Planes, A. Large temperature span and giant refrigerant capacity in elastocaloric Cu-Zn-Al shape memory alloys. *Appl. Phys. Lett.* **2013**, *103*, 211904. [CrossRef]

17. Liu, J.L.; Huang, H.Y.; Xie, J.X. Superelastic anisotropy characteristics of columnar-grained Cu-Al-Mn shape memory alloys and its potential applications. *Mater. Des.* **2015**, *85*, 211–220. [CrossRef]
18. Xu, S.; Huang, H.; Xie, J.; Kimura, Y.; Xu, X.; Omori, T.; Kainuma, R. Dynamic recovery and superelasticity of columnar-grained Cu-Al-Mn shape memory alloy. *Metals* **2017**, *7*, 141. [CrossRef]
19. Liu, J.L.; Chen, Z.H.; Huang, H.Y.; Xie, J.X. Microstructure and superelasticity control by rolling and heat treatment in columnar-grained Cu-Al-Mn shape memory alloy. *Mater. Sci. Eng. A* **2017**, *696*, 315–322. [CrossRef]
20. Dutkiewicz, J.; Kato, H.; Miura, S.; Messerschmidtet, U.; Bartsch, M. Structure changes during pseudoelastic deformation of CuAlMn single crystals. *Acta Mater.* **1996**, *44*, 4597–4609. [CrossRef]
21. Kato, H.; Dutkiewicz, J.; Miura, S. Superelasticity and shape memory effects in Cu-23 at. % Al-7 at. % Mn alloy single crystals. *Acta Metall. Mater.* **1994**, *42*, 1359–1365. [CrossRef]
22. Ahlers, M. Martensite and equilibrium phases in Cu-Zn and Cu-Zn-Al alloys. *Prog. Mater. Sci.* **1986**, *30*, 135–186. [CrossRef]
23. Mañosa, L.; Planes, A. Materials with giant mechanocaloric effects: Cooling by strength. *Adv. Mater.* **2017**, *29*, 1603607. [CrossRef] [PubMed]
24. Sutou, Y.; Koeda, N.; Omori, T.; Kainuma, R.; Ishida, K. Effect of aging on bainitic and thermally induced martensitic transformations in ductile Cu-Al-Mn-based shape memory alloys. *Acta Mater.* **2009**, *57*, 5748–5758. [CrossRef]
25. Ossmer, H.; Chluba, C.; Krevet, B.; Quandt, E.; Rohde, M.; Kohl, M. Elastocaloric cooling using shape memory alloy films. *J. Phys. Conf. Ser.* **2013**, *476*, 012138. [CrossRef]
26. Liu, J.L.; Huang, H.Y.; Xie, J.X. The roles of grain orientation and grain boundary characteristics in the enhanced superelasticity of $Cu_{71.8}Al_{17.8}Mn_{10.4}$ shape memory alloys. *Mater. Des.* **2014**, *64*, 427–433. [CrossRef]

Article

Microstructure and Mechanical Properties of a High-Mn TWIP Steel Subjected to Cold Rolling and Annealing

Alexander Kalinenko [1], Pavel Kusakin [1], Andrey Belyakov [1], Rustam Kaibyshev [1] and Dmitri A. Molodov [2,*]

[1]   Laboratory of Mechanical Properties of Nanostructured Materials and Superalloys,
      Belgorod State University, Pobeda 85, Belgorod 308015, Russia; lexs4@mail.ru (A.K.);
      kusakin@bsu.edu.ru (P.K.); belyakov@bsu.edu.ru (A.B.); rustam_kaibyshev@bsu.edu.ru (R.K.)
[2]   Institute of Physical Metallurgy and Metal Physics, RWTH Aachen University, Kopernikusstraße 14,
      Aachen 52056, Germany
*   Correspondence: molodov@imm.rwth-aachen.de; Tel.: +49-241-8026873

Received: 20 November 2017; Accepted: 14 December 2017; Published: 18 December 2017

Abstract: The structure–property relationship was studied in an Fe-18Mn-0.6C-1.5Al steel subjected to cold rolling to various total reductions from 20% to 80% and subsequent annealing for 30 min at temperatures of 673 to 973 K. The cold rolling resulted in significant strengthening of the steel. The hardness increased from 1900 to almost 6000 MPa after rolling reduction of 80%. Recovery of cold worked microstructure developed during annealing at temperatures of 673 and 773 K, resulting in slight softening, which did not exceed 0.2. On the other hand, static recrystallization readily developed in the cold rolled samples with total reductions above 20% during annealing at 873 and 973 K, leading to fractional softening of about 0.8. The recrystallized grain size depended on annealing temperature and rolling reduction; namely, it decreased with a decrease in the temperature and an increase in the rolling reduction. The mean recrystallized grain size from approximately 1 to 8 μm could be developed depending on the rolling/annealing conditions. The recovered and fine grained recrystallized steel samples were characterized by improved strength properties. The yield strength of the recovered, recrystallized, and partially recrystallized steel samples could be expressed by a unique relationship taking into account the fractional contributions from dislocation and grain size strengthening into overall strength.

Keywords: high-Mn TWIP steel; cold rolling; annealing; recovery; recrystallization; strengthening

---

1. Introduction

High-Mn austenitic TWIP/TRIP steels have aroused a great interest among materials scientists and engineers because of their outstanding strength-ductility combinations [1–4]. These steels are considered as the most promising materials for various structural applications in automobile and building industries [5,6]. In general, the mechanical properties of steel semi-products depend on their microstructure, including dislocation substructures, which can be substantially varied by using appropriate regimes/conditions for the applied thermo-mechanical treatment. Therefore, studies on the microstructure/substructure evolution during thermo-mechanical treatment are of great practical importance.

Presently, the most efficient processing methods for production of large-scale steel products involve plate or caliber rolling combined with some heat treatment. The deformation microstructures that develop in steels subjected to rolling depend sensitively on the processing conditions [7–9]. Cold rolling is accompanied by strain hardening, the rate of which gradually decreases with increasing the total rolling strain [10]. The well-developed subgrains can be obtained by warm rolling that is accompanied by dynamic recovery. Dynamic and/or post-dynamic recrystallization during hot working may lead to the uniform microstructure with the mean grain size depending on deformation conditions [9]. Regarding high-Mn austenitic steels, such steels are commonly characterized by a low stacking fault energy (SFE), which controls the development of deformation twinning and/or martensitic transformation leading to twinning- and/or transformation-induced plasticity [11,12]. On the other hand, low SFE hampers any dislocation rearrangements and, thus, slows down the dislocation recovery processes. Therefore, high-Mn austenitic steels are hardly susceptible to dynamic recovery under warm rolling conditions, and the work hardened microstructures remain almost unchanged up to high temperatures sufficient for recrystallization development [13,14].

High-Mn austenitic steels exhibit significant strain hardening owing to increasing the dislocation density during cold-to-warm working [13–15]. The dislocation strengthening leads to remarkable increase in both the yield strength and the ultimate tensile strength [15]. On the other hand, strengthening by cold-to-warm working is generally accompanied by a degradation of plasticity. The total elongation may drop to a few percent after large strain cold working [16,17]. The mechanical properties of work hardened high-Mn steels can be improved by an appropriate heat treatment. Beneficial combination of the strength and ductility can be obtained by large strain cold rolling followed by a recrystallization annealing. The combination of large strain deformation with subsequent annealing may result in the development of special microstructures. Those are partially recrystallized microstructures, where recrystallized grains are surrounded by work hardened portions, and/or ultrafine grained microstructures. Both partially recrystallized and completely recrystallized ultrafine grained microstructures may provide useful strength-ductility combinations.

The aim of the present paper is to study the effect of the rolling strain and subsequent annealing at elevated temperatures on the development of recovery and recrystallization in an advanced 18% Mn austenitic steel. Contributions of the dislocation strengthening and the grain size strengthening into the yield strength of the steel with various microstructures including recovered, partially recrystallized, and ultrafine grained ones are particularly addressed.

2. Materials and Methods

An Fe-18%Mn-0.6%C-1.5%Al steel was hot rolled to a total reduction of 80% and annealed at a temperature of 1423 K for 1 h followed by air cooling. Then, the steel samples were cold rolled to total rolling reductions of 20%, 40%, 60%, and 80% at room temperature. These cold rolled steel samples were annealed at various temperatures in the range of 673 to 973 K for 30 min.

The structural investigations were carried out on the sample sections parallel to the normal direction (ND), using a Nova Nanosem 450 scanning electron microscope (SEM, FEI, Hillsboro, OR, USA) equipped with electron back-scatter diffraction (EBSD) analyzer incorporating orientation imaging microscopy (OIM, EDAX Inc., Mahwah, NJ, USA). The SEM specimens were prepared by electro-polishing using a solution of 90% acetic acid and 10% of perchloric acid at a voltage of 20 V. The OIM maps of $100 \times 100\ \mu m^2$ and $50 \times 50\ \mu m^2$ for the cold rolled and annealed samples, respectively, were obtained with a step size of 0.1 μm. The OIM data points with confidence index below 0.1 were replaced by black dots. The mean grain size (D) and the kernel average misorientation (KAM) were obtained using OIM Analysis 6 software (EDAX Inc., version 6.2.0, Mahwah, NJ, USA). The dislocation density was evaluated by means of KAM as [14]

$$\rho = 1.15 \cdot KAM/(b\,h), \tag{1}$$

where $b = 0.258$ nm is the Burgers vector and $h = 300$ nm is the distance between the measured points in KAM maps.

The hardness measurements were carried out in order to evaluate the strain hardening and annealing softening. The latter was estimated as [18]

$$X = (Hv_\varepsilon - Hv_T)/(Hv_\varepsilon - Hv_0),\tag{2}$$

where Hv_ε, Hv_T, and Hv_0 are the hardness of cold rolled, annealed, and initial samples, respectively. The tensile tests were carried out at room temperature and at an initial strain rate of $2 \times 10^{-3}\,\mathrm{s}^{-1}$ using an Instron 5882 testing machine (Instron, Norwood, MA, USA). The tensile specimens with a gauge length of 16 mm and cross section of $1.5 \times 3\,\mathrm{mm}^2$ were machined with the tensile axis parallel to the rolling axis.

3. Results and Discussion

3.1. Cold Rolling

Typical OIM images of the deformation microstructures developed in the high-Mn steel during cold rolling to various total strains are shown in Figure 1. The deformation twins readily develop during cold rolling (Figure 1a). An increase in the rolling reduction leads to an increase in the dislocation density. Correspondingly, the number and misorientation of dislocation subboundaries increase. The misorientation of some deformation subboundaries increases over a critical value, separating low-angle subboundaries and high-angle boundaries. Therefore, these deformation subboundaries are indicated as high-angle grain boundaries on OIM image (Figure 1b). Further rolling is accompanied by the development of deformation microbands involving large lattice distortions (Figure 1c). The microbanding during cold rolling is consistent with previous studies on high-Mn austenitic TWIP steels, which involved numerous microshear bands after 60% rolling reduction [16,17]. The microband density and their thickness increases with straining. After large rolling reduction, the microstructural analysis by OIM becomes difficult because of a large fraction of frequently developed deformation microbands and related lattice distortions.

Cold rolling is accompanied by a significant increase in the hardness (Figure 2). The hardness increases almost two-fold from about 2000 to approximately 4000 MPa after rolling reduction of 20%. The rate of strain hardening gradually decreases during cold rolling. An increase in the rolling reduction to 40% leads to the hardness of 5000 MPa. Further rolling to large total reduction of 80% results in the hardness of about 6000 MPa. Such strain hardening behavior is typical for large strain cold deformation of various structural steels and alloys, when the hardness increases with straining and gradually approaches a saturation level at sufficiently large strains [19].

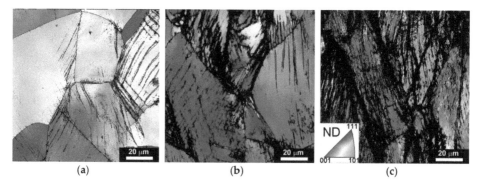

(a)   (b)   (c)

Figure 1. Deformation microstructures in a high-Mn steel subjected to rolling reduction of 20% (a); 40% (b); 60% (c). Low-angle and high-angle boundaries are indicated by thin and thick black lines, respectively. Color orientations are shown for the normal direction (ND).

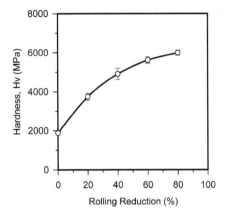

Figure 2. Strain hardening for a high-Mn steel subjected to cold rolling at room temperature.

3.2. Annealing Behavior

The isochronal annealing was carried out to investigate the temperature effect on microstructure evolution in the cold rolled high-Mn steel. The range of annealing temperatures and the annealing time were selected to cover the operation of various restoration mechanisms including recovery and recrystallization and, therefore, to obtain variety of annealed microstructures—i.e., recovered, partially recrystallized, and fully recrystallized [13,14]. The effect of annealing temperature on the hardness and fractional softening of the present high-Mn steel subjected to cold rolling to various total strains is shown in Figure 3. Two temperature intervals are clearly distinguished by their effect on both the hardness and the fractional softening. Annealing at temperatures below about 800 K does not result in any remarkable changes in the hardness irrespective of the previous rolling reductions. The corresponding fractional softening does not exceed 0.2 after annealing within this temperature range. In contrast, the hardness drastically drops after annealing at temperatures above 800 K. This change in the hardness is more pronounced in the samples subjected to larger rolling reductions. Except the sample subjected to rolling reduction of 20%, the fractional softening comprises approximately 0.8 after annealing at temperatures above 800 K. The different softening behavior at temperatures below or above 800 K suggests that different annealing/softening mechanisms operate in these temperature regimes.

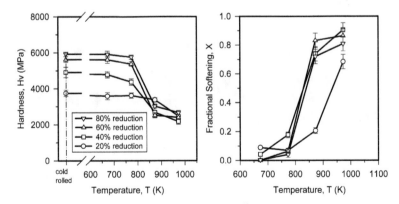

Figure 3. Hardness change and annealing softening for a high-Mn steel subjected to various cold rolling reductions and then annealed for 30 min at 673–973 K.

Typical microstructures that evolved in the steel samples after rolling reductions of 20% and 60% followed by annealing at a temperature of 773 K are shown in Figure 4 as OIM micrographs and *KAM* maps. It is clearly seen that annealing at this temperature is not accompanied by any remarkable changes in the deformation microstructures. It can be concluded, therefore, that static recovery is the only operative softening mechanism in this annealing domain. High-Mn austenitic steels with low SFE have been shown being quite stable against recovery because of large dislocation dissociation spacing [20]. Hence, the larger internal distortions, which were caused by higher dislocation densities in deformation substructures after larger rolling reductions, remain larger after recovery annealing as shown in the *KAM* maps. On the other hand, annealing at temperatures above 800 K results in the recrystallization development (Figure 5). An increase in the cold strain accelerates the recrystallization kinetics. The fraction recrystallized comprises about 0.6 in the sample subjected to rolling reduction of 40% and then annealed at 873 K (Figure 5a), whereas the completely recrystallized microstructures evolve after annealing at the same temperature following the rolling reductions of 60% and 80% (Figure 5b,c). An increase in the rolling reduction promotes the development of fine grained recrystallized microstructure similar to ordinary primary recrystallization behavior [21]. An increase in the annealing temperature increases the size of recrystallized grains (Figure 5d).

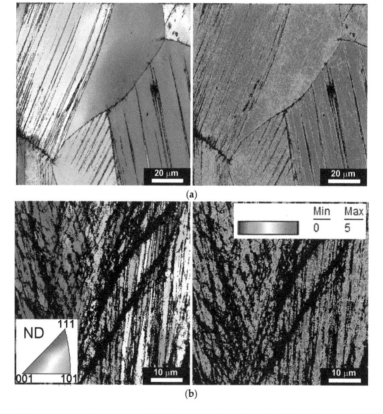

Figure 4. Annealed microstructures of a high-Mn steel subjected to cold rolling reduction of (**a**) 20% or (**b**) 40% and then annealed at 773 K. Left-side images are color orientations corresponding to the normal direction (ND), right-side images are *KAM* maps. Low-angle and high-angle boundaries are indicated by thin and thick black lines, respectively.

The effect of rolling reduction on the recrystallized grain size and the recovered dislocation density in the high-Mn austenitic steel subjected to cold deformation followed by an annealing is shown in Figure 6. The size of recrystallized grains after annealing at 973 K remarkably decreases from 8 to 1.6 μm with an increase in the rolling reduction from 20% to 80%. It is worth noting that annealing temperature does not affect the recrystallized grain size substantially. For instance, following rolling reduction of 80%, annealing at 873 K results in a recrystallized grain size of 1.1 μm. Similar to the recrystallized grain size, the dislocation density in the recovery annealed samples strongly depends on the preceding rolling reduction. The dislocation density varies in the range of 10^{14} m^{-2} to 4×10^{14} m^{-2} in the samples subjected to rolling reductions of 20–40% followed by annealing at 673–873 K. In contrast, much higher dislocation densities of about 10^{15} m^{-2} evolve in the samples subjected to rolling reductions of 60–80% followed by annealing at 673–773 K.

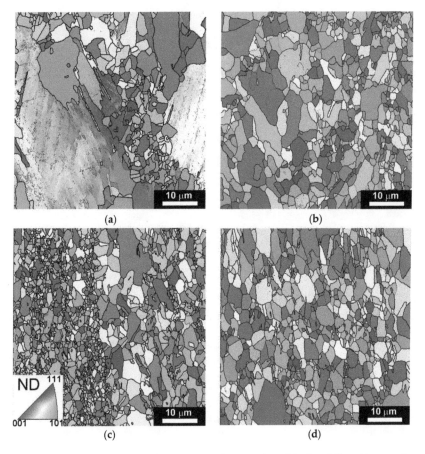

Figure 5. Annealed microstructures of a high-Mn steel subjected to cold rolling followed by annealing; (**a**) rolling reduction of 40%, annealing at 873 K; (**b**) rolling reduction of 60%, annealing at 873 K; (**c**) rolling reduction of 80%, annealing at 873 K; (**d**) rolling reduction of 80%, annealing at 973 K.

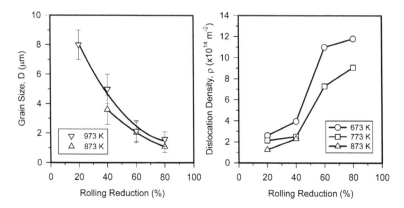

Figure 6. Effect of cold rolling reduction and annealing temperature on the grain size and dislocation density after recrystallization and recovery annealing of a high-Mn steel.

3.3. Mechanical Properties

The engineering stress-elongation curves obtained by tensile tests of the high-Mn austenitic steel after cold rolling and annealing are shown in Figure 7. Recovery annealing at temperatures of 673 and 773 K does not provide significant softening irrespective of the rolling reductions. Therefore, the general shape of the tensile stress–strain curves for the samples annealed at 673 and 773 K depends remarkably on the previous rolling reduction (Figure 7a,b). The steel samples subjected to relatively small rolling reductions of 20% and 40% exhibit increased yield strength of about 700 and 1250 MPa, respectively. Following yielding, the strain hardening quickly decreased to some positive value, which remains almost constant up to necking. An increase in the annealing temperature from 673 to 773 K improves plasticity, although it does not remarkably affect the yield strength and the ultimate tensile strength. The yield strength increases above 1200 MPa after rolling reductions of 60–80% followed by recovery annealing at 673 and 773 K, whereas the total elongation decreases substantially and does not exceed a couple of percent after rolling reduction of 60% and 80% followed by recovery annealing. On the other hand, recrystallization annealing at 873 and 973 K softens the cold rolled steel samples and significantly improves their plasticity (Figure 7c,d). The strength of the recrystallized steel samples commonly decreases with an increase in the temperature and a decrease in the rolling reduction.

The tensile deformation behavior of the present steels samples is closely related to their microstructures. The dislocation substructures and corresponding strain hardening remain almost unchanged during recovery annealing. Therefore, the tensile behavior of the recovery annealed samples depends mainly on the previous cold strain. In contrast, the mechanical properties of the steel samples after recrystallization annealing depend on the mean grain size. In this case, the rolling reduction affects the strength properties of recrystallized steels through the recrystallized grain size that depends on the recrystallization kinetics, which, in turn, depends on the previous cold strain. The relationship between the yield strength of the steel samples after recrystallization annealing and the recrystallized grain size is represented in Figure 8 as a Hall-Petch type plot. The yield strength ($\sigma_{0.2}$) of the recrystallized steel samples obeys an ordinary relationship,

$$\sigma_{0.2} = \sigma_0 + k_y D^{-0.5} \tag{3}$$

where, σ_0 is the yield strength of the same material with unlimited grain size and k_y is the grain boundary strengthening factor [22,23]. The values of $\sigma_0 = 160$ MPa and $k_y = 355$ MPa $\mu m^{0.5}$ are obtained in the present study. Note here, almost the same σ_0 and k_y have been reported in other papers on austenitic steels [13,24–26].

The strengthening of the recovered steel samples can be related to their dislocation density similar to other studies on work hardened materials [27–29]

$$\sigma_{0.2} = \sigma_0 + \alpha\, M\, G\, b\, \rho^{0.5} \tag{4}$$

where α, M, and G are a numerical factor, the Tailor factor ($M = 3$ is frequently used for face centered cubic metallic materials [30,31]) and the shear modulus, respectively. Here, σ_0 means the strength of dislocation-free recrystallized steel and can be taken as that from Equation (3) for the sake of simplicity. The relationship between the dislocation density, which was evaluated by means of *KAM* (Equation (1)), and the yield strength of the recovered steel samples is also represented in Figure 8. It is clearly seen that the dislocation strengthening can be expressed by Equation (4) with $\alpha = 0.6$. The value of α has been reported varying from about 0.2 to 0.5 [14,15,24,30–32]. Relatively large $\alpha = 0.6$ obtained for the present work hardened steel samples may be attributed to an enhanced efficiency of dislocation strengthening in high-Mn austenitic steels with low SFE as well as to somewhat underestimated dislocation density by the KAM values, which are actually associated with excess dislocations of similar Burgers vectors rather than total dislocation density [33]. Then, combining Equations (3) and (4), the yield strength of recovered and recrystallized steel samples including partially recrystallized ones can be expressed as

$$\sigma_{0.2} = \sigma_0 + F_{REX}\, k_y\, D^{-0.5} + (1 - F_{REX})\, \alpha\, M\, G\, b\, \rho^{0.5} \tag{5}$$

where F_{REX} is the fraction recrystallized. The plot in Figure 9 validates the speculation above.

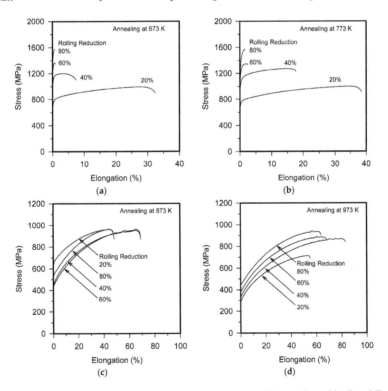

Figure 7. Engineering stress vs. elongation curves for a high-Mn steel subjected to cold rolling followed by annealing at (**a**) 673 K; (**b**) 773 K; (**c**) 873 K; (**d**) 973 K.

The present results suggest that combination of cold rolling with reductions of 20% to 80% and annealing at 673 to 973 K can be used as various advanced thermo-mechanical processing methods to obtain high-Mn austenitic TWIP steel plates with a desired strength and ductility. The required combination of mechanical properties is achieved by the development of appropriate structural state, which may consist of a mixture of various structures including work hardened, recovered, and partially and completely recrystallized ones. An increase in the rolling reduction increases the deformation stored energy and, therefore, promotes the primary recrystallization. It should be noted that the stored energy significantly accelerates the rate of recrystallization nucleation. Hence, the rapid recrystallization nucleation at temperatures close to a critical recrystallization temperature for the steel samples subjected to large strain cold rolling results in beneficial mechanical properties including high strength and large uniform elongation owing to the development of uniform ultrafine-grained microstructures.

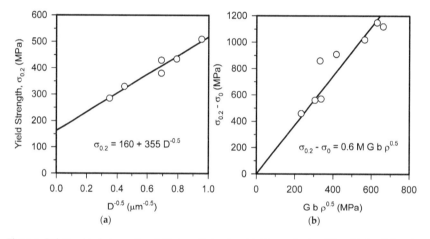

Figure 8. Relationship between the yield strength ($\sigma_{0.2}$) and (**a**) the grain size (*D*) or (**b**) the dislocation density (ρ) for a high-Mn steel subjected to cold rolling followed by annealing.

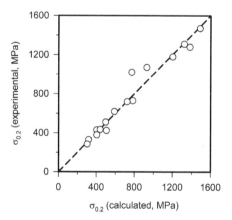

Figure 9. Relationship between the experimental yield strength and that calculated by Equation (5) for a high-Mn steel subjected to cold rolling and then annealed.

Another important finding of the present study is the unique relationship between the microstructural/substructural parameters and the strength of the cold rolled and annealed samples irrespective of processing regimes. The strength of work hardened metallic materials was treated with a modified Hall-Petch type relationship including the term of dislocation strengthening [16,17,34,35]. However, a mutual correlation between the dislocation densities and grain boundary densities in metals and alloys subjected to large strain deformation made the estimation of individual strengthening mechanisms difficult [32,36,37]. In contrast, the present approach of fractional contributions of different strengthening mechanisms, which are originated from specific structural elements, into overall strength allows us to adequately predict the yield strength of steels with a mixture of various work hardened, recovered, and recrystallized microstructures.

4. Conclusions

The microstructure evolution and mechanical properties were studied in an Fe-18Mn-0.6C-1.5Al steel subjected to cold rolling and subsequent isochronal annealing for 30 min. The main results can be summarized as follows.

1. The cold rolling resulted in significant strengthening of the steel. The hardness increased from 1900 to almost 6000 MPa after rolling reduction of 80%.
2. Annealing behavior was characterized by the development of recovery and recrystallization. Recovery took place during annealing at temperatures of 673 and 773 K, leading to fractional softening below 0.2. On the other hand, the static recrystallization readily developed during annealing at 873 and 973 K, leading to fractional softening of about 0.8.
3. The recrystallized grain size depended on annealing temperature and rolling reduction. An increase in the rolling reduction from 20% to 80% led to a decrease in the mean recrystallized grain size from about 8 to 1.6 μm after subsequent annealing at 923 K and to 1.1 μm after annealing at 873 K.
4. The yield strength of the recovered and recrystallized steel samples, as well as partially recrystallized ones, could be expressed by a modified Hall-Petch type relationship taking into account the fractional contributions from grain size strengthening and dislocation strengthening.

Acknowledgments: This study was supported by the Ministry of Science and Education of Russian Federation under the Grant No. 14.575.21.0134 (RFMEFI57517X0134). Authors are grateful to the personal of the Joint Research Center, Technology and Materials, Belgorod State University for their assistance with instrumental analysis.

Author Contributions: Dmitri A. Molodov and Rustam Kaibyshev elaborated the research topic; Alexander Kalinenko and Pavel Kusakin designed and performed experiments; Andrey Belyakov, Dmitri A. Molodov, and Rustam Kaibyshev analyzed the obtained results; all co-authors discussed the paper.

Conflicts of Interest: The authors declare no conflict of interest.

References

1. Grässel, O.; Krüger, L.; Frommeyer, G.; Meyer, L.W. High strength Fe-Mn-(Al, Si) TRIP/TWIP steels development—Properties—Application. *Int. J. Plast.* **2000**, 16, 1391–1409. [CrossRef]
2. Bouaziz, O.; Allain, S.; Scott, C.P.; Cugy, P.; Barbier, D. High manganese austenitic twinning induced plasticity steels: A review of the microstructure properties relationships. *Curr. Opin. Solid State Mater. Sci.* **2011**, 15, 141–168. [CrossRef]
3. Kusakin, P.S.; Kaibyshev, R.O. High-Mn twinning-induced plasticity steels: Microstructure and mechanical properties. *Rev. Adv. Mater. Sci.* **2016**, 44, 326–360.
4. De Cooman, B.C.; Estrin, Y.; Kim, S.K. Twinning-induced plasticity (TWIP) steels. *Acta Mater.* **2018**, 142, 283–362. [CrossRef]
5. Hofmann, H.; Mattissen, D.; Schaumann, T.W. Advanced cold rolled steels for automotive applications. *Steel Res. Int.* **2009**, 80, 22–28. [CrossRef]

6. Nikulin, I.; Sawaguchi, T.; Kushibe, A.; Inoue, Y.; Otsuka, H.; Tsuzaki, K. Effect of strain amplitude on the low-cycle fatigue behavior of a new Fe-15Mn-10Cr-8Ni-4Si seismic damping alloy. *Int. J. Fatigue* **2016**, *88*, 132–141. [CrossRef]

7. Ryan, N.D.; McQueen, H.J. Dynamic softening mechanisms in 304 austenitic stainless steel. *Can. Metall. Q.* **1990**, *29*, 147–162. [CrossRef]

8. Belyakov, A.; Tikhonova, M.; Yanushkevich, Z.; Kaibyshev, R. Regularities of Grain Refinement in an Austenitic Stainless Steel during Multiple Warm Working. *Mater. Sci. Forum* **2013**, *753*, 411–416. [CrossRef]

9. Sakai, T.; Belyakov, A.; Kaibyshev, R.; Miura, H.; Jonas, J.J. Dynamic and post-dynamic recrystallization under hot, cold and severe plastic deformation conditions. *Prog. Mater. Sci.* **2014**, *60*, 130–207. [CrossRef]

10. Gill Sevillano, J.; Van Houtte, P.; Aernoudt, E. Large strain work hardening and textures. *Prog. Mater. Sci.* **1980**, *25*, 69–412. [CrossRef]

11. Frommeyer, G.; Brüx, U.; Neumann, P. Supra-ductile and high-strength manganese-TRIP/TWIP steels for high energy absorption purposes. *ISIJ Int.* **2003**, *43*, 438–446. [CrossRef]

12. Saeed-Akbari, A.; Mosecker, L.; Schwedt, A.; Bleck, W. Characterization and prediction of flow behavior in high-manganese twinning induced plasticity steels: Part I. mechanism maps and work-hardening behavior. *Metall. Mater. Trans. A* **2012**, *43*, 1688–1704. [CrossRef]

13. Kusakin, P.; Tsuzaki, K.; Molodov, D.A.; Kaibyshev, R.; Belyakov, A. Advanced thermomechanical processing for a high-Mn austenitic steel. *Metall. Mater. Trans. A* **2016**, *47*, 5704–5708. [CrossRef]

14. Kusakin, P.; Kalinenko, A.; Tsuzaki, K.; Belyakov, A.; Kaibyshev, R. Influence of cold forging and annealing on microstructure and mechanical properties of a high-Mn TWIP steel. *Kov. Mater.* **2017**, *55*, 161–167. [CrossRef]

15. Torganchuk, V.; Belyakov, A.; Kaibyshev, R. Effect of rolling temperature on microstructure and mechanical properties of 18%Mn TWIP/TRIP steels. *Mater. Sci. Eng. A* **2017**, *708*, 110–117. [CrossRef]

16. Kusakin, P.; Belyakov, A.; Haase, C.; Kaibyshev, R.; Molodov, D.A. Microstructure evolution and strengthening mechanisms of Fe-23Mn-0.3C-1.5Al TWIP steel during cold rolling. *Mater. Sci. Eng. A* **2014**, *617*, 52–60. [CrossRef]

17. Yanushkevich, Z.; Belyakov, A.; Haase, C.; Molodov, D.A.; Kaibyshev, R. Structural/textural changes and strengthening of an advanced high-Mn steel subjected to cold rolling. *Mater. Sci. Eng. A* **2016**, *651*, 763–773. [CrossRef]

18. Sakai, T. Dynamic recrystallization microstructures under hot working conditions. *J. Mater. Process. Technol.* **1995**, *53*, 349–361. [CrossRef]

19. Belyakov, A.; Kimura, Y.; Adachi, Y.; Tsuzaki, K. Microstructure evolution in ferritic stainless steels during large strain deformation. *Mater. Trans.* **2004**, *45*, 2812–2821. [CrossRef]

20. Haase, C.; Barrales-Mora, L.A.; Molodov, D.A.; Gottstein, G. Tailoring the mechanical properties of a twinning-induced plasticity steel by retention of deformation twins during heat treatment. *Metall. Mater. Trans. A* **2013**, *44*, 4445–4449. [CrossRef]

21. Humphreys, F.J.; Hatherly, M. *Recrystallization and Related Annealing Phenomena*, 2nd ed.; Elsevier: Oxford, UK, 2004; pp. 215–268. ISBN 0-08-044164-5.

22. Hall, E.O. The deformation and ageing of mild steel: III discussion of results. *Proc. R. Soc. Lond. Ser. B* **1951**, *64*, 747–753. [CrossRef]

23. Petch, N.J. The cleavage strength of polycrystals. *J. Iron Steel Inst.* **1953**, *174*, 25–28.

24. Yanushkevich, Z.; Belyakov, A.; Kaibyshev, R.; Haase, C.; Molodov, D.A. Effect of cold rolling on recrystallization and tensile behavior of a high-Mn steel. *Mater. Charact.* **2016**, *112*, 180–187. [CrossRef]

25. Shakhova, I.; Dudko, V.; Belyakov, A.; Tsuzaki, K.; Kaibyshev, R. Effect of large strain cold rolling and subsequent annealing on microstructure and mechanical properties of an austenitic stainless steel. *Mater. Sci. Eng. A* **2012**, *545*, 176–186. [CrossRef]

26. Odnobokova, M.; Belyakov, A.; Enikeev, N.; Molodov, D.A.; Kaibyshev, R. Annealing behavior of a 304L stainless steel processed by large strain cold and warm rolling. *Mater. Sci. Eng. A* **2017**, *689*, 370–383. [CrossRef]

27. Mecking, H.; Kocks, U.F. Kinetics of flow and strain-hardening. *Acta Metall.* **1981**, *29*, 1865–1875. [CrossRef]

28. Estrin, Y.; Toth, L.S.; Molinari, A.; Brechet, Y. A dislocation-based model for all hardening stages in large strain deformation. *Acta Mater.* **1998**, *46*, 5509–5522. [CrossRef]

29. Haase, C.; Barrales-Mora, L.A.; Roters, F.; Molodov, D.A.; Gottstein, G. Applying the texture analysis for optimizing thermomechanical treatment of high manganese twinning-induced plasticity steel. *Acta Mater.* **2014**, *80*, 327–340. [CrossRef]
30. Ardell, A.J. Precipitation hardening. *Metall. Trans. A* **1985**, *16*, 2131–2165. [CrossRef]
31. Ma, K.; Smith, T.; Hu, T.; Topping, T.D.; Lavernia, E.J.; Schoenung, J.M. Distinct hardening behavior of ultrafine-grained Al-Zn-Mg-Cu alloy. *Metall. Mater. Trans. A* **2014**, *45*, 4762–4765. [CrossRef]
32. Shakhova, I.; Belyakov, A.; Yanushkevich, Z.; Tsuzaki, K.; Kaibyshev, R. On strengthening of austenitic stainless steel by large strain cold working. *ISIJ Int.* **2016**, *56*, 1289–1296. [CrossRef]
33. Calcagnotto, M.; Ponge, D.; Demir, E.; Raabe, D. Orientation gradients and geometrically necessary dislocations in ultrafine grained dual-phase steels studied by 2D and 3D EBSD. *Mater. Sci. Eng. A* **2010**, *527*, 2738–2746. [CrossRef]
34. Hansen, N. Hall-Petch relation and boundary strengthening. *Scr. Mater.* **2004**, *51*, 801–806. [CrossRef]
35. Morozova, A.; Kaibyshev, R. Grain refinement and strengthening of a Cu-0.1Cr-0.06Zr alloy subjected to equal channel angular pressing. *Philos. Mag.* **2017**, *97*, 2053–2076. [CrossRef]
36. Starink, M.J. Dislocation versus grain boundary strengthening in SPD processed metals: Non-causal relation between grain size and strength of deformed polycrystals. *Mater. Sci. Eng. A* **2017**, *705*, 42–45. [CrossRef]
37. Yanushkevich, Z.; Dobatkin, S.V.; Belyakov, A.; Kaibyshev, R. Hall-Petch relationship for austenitic stainless steels processed by large strain warm rolling. *Acta Mater.* **2017**, *136*, 39–48. [CrossRef]

 *metals*

Article

Effect of Post Weld Heat Treatment on the Microstructure and Mechanical Properties of a Submerged-Arc-Welded 304 Stainless Steel

Tae-Hoon Nam [1,2], Eunsol An [1], Byung Jun Kim [1], Sunmi Shin [1], Won-Seok Ko [3], Nokeun Park [4], Namhyun Kang [2,*] and Jong Bae Jeon [1,*]

[1] Functional components and Materials R&D Group, Korea Institute of Industrial Technology, Yangsan 50635, Korea; skgnsl41@kitech.re.kr (T.-H.N.); sol@kitech.re.kr (E.A.); jun7741@kitech.re.kr (B.J.K.); smshin@kitech.re.kr (S.S.)
[2] School of Materials Science and Engineering, Pusan National University, Busan 46241, Korea
[3] School of Materials Science and Engineering, University of Ulsan, Ulsan 44610, Korea; wonsko@ulsan.ac.kr
[4] School of Materials Science and Engineering, Yeungnam University, Gyeongsan 38541, Korea; nokeun_park@yu.ac.kr
* Correspondence: nhkang@pusan.ac.kr (N.K.); jbjeon@kitech.re.kr (J.B.J.); Tel.: +82-55-367-9407 (N.K.); +82-51-510-3274 (J.B.J.)

Received: 8 November 2017; Accepted: 22 December 2017; Published: 2 January 2018

Abstract: The present study is to investigate the effect of post heat treatment on the microstructures and mechanical properties of a submerged-arc-welded 304 stainless steel. The base material consisted of austenite and long strips of delta-ferrite surrounded by Cr-carbide, and the welds consisted of delta ferrite and austenite matrix. For the heat treatment at 850 °C or lower, Cr-carbides were precipitated in the weld metal resulting in the reduction of elongation. The strength, however, was slightly reduced despite the presence of Cr-carbides and this could possibly be explained by the relaxation of internal stress and the weakening of particle hardening. In the heat treatment at 1050 °C, the dissolution of Cr-carbide and disappearance of delta ferrite resulted in the lower yield strength and higher elongation partially assisted from deformation-induced martensitic transformation. Consequently, superior property in terms of fracture toughness was achieved by the heat treatment at 1050 °C, suggesting that the mechanical properties of the as-weld metal can be enhanced by controlling the post weld heat treatment.

Keywords: austenitic 304 stainless steels; sub-merged arc welding; post-weld heat treatment

1. Introduction

Austenitic stainless steels are widely used in important parts of chemical plants such as pressure vessels, chemical machineries, and pipes because they have excellent corrosion resistance, acid resistance, formability, and weldability. They are also extensively applied in gas turbines, jet propellants, and nuclear power plants because of their excellent toughness in high-temperature and high-pressure environments [1–4].

Submerged Arc Welding (SAW) is widely used for joining ship structures and piping thick-wall stainless steels [5,6]. In SAW, a solid filler wire is automatically inserted to pre-dispersed granular flux and an arc generates between the base metal and the filler wire while being covered with flux. The flux in contact with the arc melts and then the molten slag covers and protects the molten metal. Generally high-efficiency welding is achieved by flowing a large current through a large-diameter wire. It has the advantage of improving welding productivity through high speed solidification during welding and improving weld quality through welding heat control [7].

Due to the heat input from the SAW welding process, there is a degradation problem of microstructural and mechanical properties such as inhomogeneity of the microstructure, concentration of residual stress, brittleness, and deterioration of toughness in the welded material [8]. Post-weld heat treatment (PWHT) is usually introduced to solve such degradation. However, if the delta ferrite in austenitic stainless steel is excessively present in the weld metal during SAW welding, embrittlement of the welded part may occur due to transformation into another brittle phase such as a sigma phase during post heat treatment. In order to solve these problems, it is important to investigate post-heat treatment process to control the content of delta ferrite and the harmful carbides [9]. Proper post-heat treatment can improve the mechanical properties by homogenizing the microstructure of the welded material and controlling the formation of beneficial and harmful phases [10–13].

Although extensive researches have conducted on the effect of PWHT on the welded austenitic stainless steels, industrial field still requires the comprehensive research on the resultant effect of PWHT, especially for the practical and the cost effective process condition of PWHT. The present study thus investigated the effect of PWHT on the change of microstructural and mechanical properties of the welded part of AISI STS 304 steel produced by SAW. The microstructural changes during the heat treatment were analyzed and the resultant effect on mechanical properties were discussed.

2. Materials and Methods

In this study, submerged-arc welding was performed in austenitic stainless steel STS304 using STS308L as a welding electrode. The chemical compositions of the materials used are shown in Table 1. A plate with the width of 100 mm, the length of 1000 mm and the thickness of 25 mm was fabricated by the SAW. In order to evaluate the microstructure and mechanical properties, PWHT was performed from 650–1050 °C with an interval of 200 °C as shown in Figure 1a. After maintaining at these temperatures for 0.5–4 h, the specimens were then quenched in water.

In order to observe the microstructure, the test pieces were grinded using sandpapers and then were micro-polished using a diamond suspension with a 1 μm diameter. According to ASTM E-340, the polished surface was etched using a mixed solution of distilled water, hydrochloric acid and nitric acid with a ratio of 4:3:3. Through the optical microscopy, the microstructures of the base metal, the weld part and the heat affected zone (HAZ) were analyzed. According to ASTM E-8, subsized plate-type tensile specimens were taken from the middle of the welded plate with perpendicular to the welding direction as shown in Figure 1b.

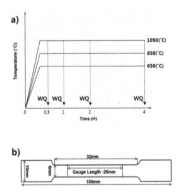

Figure 1. Schematic diagrams showing (a) post-weld heat treatment schedules and (b) ASTM-E8 subsized tensile specimen.

The tensile tests were conducted to measure yield strength, tensile strength and elongation with the strain rate of 10^{-3}/s according to ASTM A370 using universal tensile machine (MTS, Eden Praire,

MN, USA). The yield strength was determined by the 0.2% offset method and the elongation was measured by the travel distance of crosshead. Vickers hardness test was conducted with the load of 2.94 N for 15 s dwell time and the hardness values were measured 10 times in total, and the average value was calculated by excluding the maximum and the minimum value. To measure the content of delta ferrite in the weld metal, the mean value was measured 10 times per specimen using FERITSCOPE-FMP30 (Fischer, Windsor, CT, USA). Electron backscattered diffraction (EBSD) and Electron Probe Microanalysis (EPMA) was performed to observe the phase change after post heat treatment using JEOL FE-SEM 7200F (JEOL, Tokyo, Japan). For thermodynamic calculations to predict the volume fraction of equilibrium phase at given temperatures, ThermoCalc software (Thermo-Calc Software, 2017b, Solna, Sweden) package was adopted with the latest steel database (TCFE9).

Table 1. Chemical composition of the base metal and the filler wire (wt %).

Alloys	C	Mn	Si	Cr	Ni	Mo	Al	Co	Nb	Cu	N
STS 304 (Base)	0.046	1.19	0.42	18.2	8.02	0.15	0.003	0.164	0.01	0.23	0.05
STS 308L (Filler)	0.02	1.98	0.41	19.7	10.79	0.03	-	-	-	0.13	0.05

3. Results and Discussion

Figure 2 shows the microstructure of weld metal, heat-affected zone (HAZ) and base metal after sub-merged arc welding (SAW). The base metal consisted of austenite single phase and hot-rolling strips. These rolling strips correspond to delta ferrite surrounded by Cr-based carbides precipitated during hot rolling (Figure 2d–f). The weld metal was composed of austenite (white) and delta ferrite (black). The delta ferrite is classified into lathy and vermicular type by shape (white circle and rectangle in Figure 2c, respectively), and this shape difference is considered to be caused by welding heat input. In the sequential order of multi-pass SAW, the welded structure initially appeared in a lathy shape, and then the lathy shape was transformed into a twisty vermicular shape induced by the next welding pass. From the results of the Energy Dispersive Spectroscopy (EDS) analysis, it was confirmed that the formation of the primary Cr-carbide did not occur in the weld metal. Scanning electron microscope (SEM) analysis revealed that pores and cracks possibly formed during solidification were hardly found in the weld metal. Therefore, it was confirmed that the welding conditions such as welding power and electrodes were correctly selected.

The heat affected zone of the austenitic stainless steel is generally prone to be Cr-deficient due to the precipitation of Cr-Fe rich carbides at phase boundaries or grain boundaries causing brittleness and welding decay [14]. From the EPMA analysis in Figure 2d, there was no remarkable increase of Cr-carbide in HAZ, compared to the base metal where Cr-carbides initially present around long-elongated delta ferrite. Also the shape of Cr-carbides near HAZ still remained in a long elongated shape, which means that those carbides were likely to be formed during hot-rolling process. It is thus considered that the formation of Cr-carbide by welding heat effect hardly occurred in the present specimen. Figure 3a,b show Vickers hardness measured over base metal, HAZ and welding metal. The hardness value is considered not to vary significantly over those regions despite that the hardness increased at the regions where austenite and carbides existing together. The hardness of the base material and the HAZ were similar with each other and the hardness of the weld metal was slightly lower than those of the base metal and the HAZ. The hardness results provide that microstructural changes such as grain coarsening/refining or Cr-carbides precipitation barely occurs in the HAZ. The as-welded specimens were always fractured in the middle of the weld metal, not in the HAZ during tensile tests. This also indicates that there is almost no formation of brittle phases in the HAZ. Therefore, the analysis of microstructural changes and mechanical properties during post weld heat treatment is focused on the weld metal, not on the HAZ.

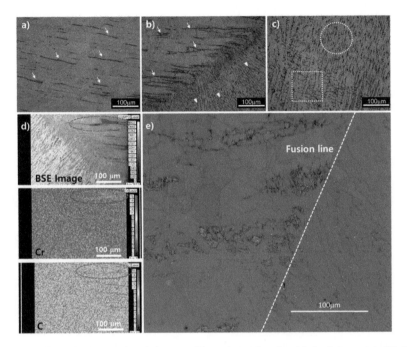

Figure 2. Optical micrographs of the as-weld specimen showing (**a**) the base metal, (**b**) the heat-affected zone, and (**c**) the weld metal (White arrows and triangles are indicating the pre-existing strips and delta ferrite, respectively.); (**d**) Electron Probe Microanalysis (EPMA) images showing the presence of Cr-carbides in the base metal; (**e**) Electron backscattered diffraction (EBSD) phase map on the fusion line including base metal (right) and heat affected zone (HAZ) (left). Red, blue, and cyan color indicate delta-ferrite, austenite, and $Cr_{23}C_6$, respectively.

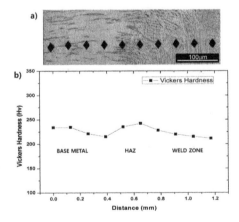

Figure 3. (**a**) Optical micrographs indicating the indenting positions; (**b**) Vicker hardness measured through the base metal, the HAZ, and the weld metal.

Figure 4 shows the microstructure of the base material and the HAZ after heat-treated at 650–850 °C for 0.5–4 h. The size of the austenite grain boundaries of the base material and the HAZ is 40–60 μm on average, which is not significantly changed, compared to the as-weld specimen.

The strips already existing in the base material still remained, without noticeable changes in the shape and the volume fraction even after the heat treatment at 650 °C. In the 850 °C heat treatment, however, the volume of the strips decreased and their shapes were blurred as compared with the as-weld and the heat-treated specimens at 650 °C. It can be inferred that the Cr-based carbide was decomposed into an austenite phase at 850 °C.

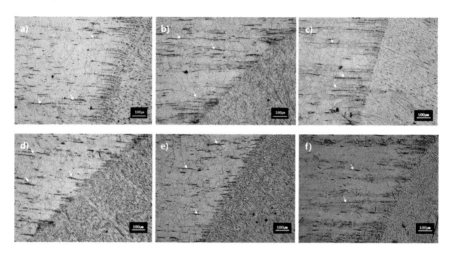

Figure 4. Optical micrographs on the base metal and the HAZ of the heat-treated specimens at (**a**) 650 °C for 0.5 h, (**b**) 650 °C for 1 h, (**c**) 650 °C for 4 h, (**d**) 850 °C for 0.5 h, (**e**) 850 °C for 1 h, and (**f**) 850 °C for 4 h. Arrows indicating pre-existing strips of delta ferrite and Cr-carbide in the HAZ, respectively.

It has been reported that Cr-based carbide is precipitated and a delta phase is transformed into a sigma phase in the temperature range of 600–850 °C in 300-series austenitic stainless steels [11,15]. The results of the thermodynamic calculations in Figure 5 show that at 650 °C, about 1% of the $Cr_{23}C_6$ and 8% of the sigma phase can be precipitated in equilibrium from the base metal composition. This $Cr_{23}C_6$ is known to be precipitated in the austenite grain boundaries where segregated chromium atoms tend to be bound with carbon atoms. Therefore, the formation of $Cr_{23}C_6$ makes chromium concentration in the grain boundaries lowered, i.e., chromium depletion, which results in the less formation of the Cr-based oxides and thus provides harmful effect on corrosion properties and toughness of grain boundaries. On the other hand, sigma phase of a tetragonal structure composed of Fe-Cr-Ni-Mo is generally reported to be precipitated in ferrite/austenite phase boundaries and to be grown into ferrite phase [16,17]. In the present thermodynamic calculations, the sigma phase was expected to have a high fraction of 8%, which could be regarded as the much higher value than the actual experimental value. The reason for this overestimation is that the thermodynamic calculation does not consider kinetics of phase transformation; alpha ferrite was calculated to exist in 650 °C from the present thermodynamic calculation but was not actually present in the base metal where the single austenite phase only existed. In the base metal, the preferred nucleation site of the sigma phase, i.e., ferrite, is hardly existing, and thus the actual volume of the sigma phase could be much lower than the calculated predictions. On the other hand, in the case of the heat treatment at 850 °C, it was predicted that 0.4% of $Cr_{23}C_6$ phase would be precipitated and the sigma phase could not be precipitated. The volume fraction of $Cr_{23}C_6$ decreased with increasing temperature up to 900 °C above which the Cr-carbide would not be precipitated.

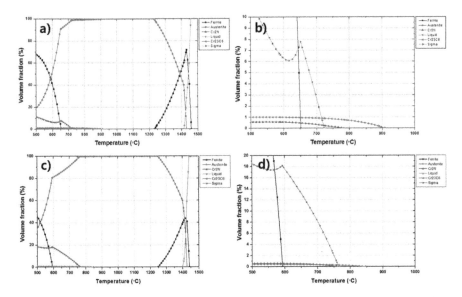

Figure 5. Predicted volume fraction of equilibrium phases at given temperatures in (**a**,**b**) the base metal (STS 304) and (**c**,**d**) the filler metal (STS 308L).

Figure 6 shows the microstructure of the weld metal heat-treated at 650–850 °C for 0.5 and 4 h. Similar to the as-weld microstructure, the austenite matrix contained delta ferrite dendrites. After heat treatment at 650–850 °C, the lathy type delta ferrite changed to vermicular type due to the heat treatment and the size of the individual dendrite of delta ferrite increased at 850 °C in comparison with 650 °C. The phase boundary between austenite and delta ferrite is considered to migrate in favor of reducing the interfacial energy. The initially thin dendrites of delta ferrite could thus be thicker during heat treatment without losing their total volume fraction as confirmed in Figure 6e,f. The fraction of delta ferrite was measured to be slightly decreased during heat treatment at 650–850 °C. Figure 7 shows the volume fractional change of delta ferrite with temperature and time obtained from the magnetic induction method. Because possible phases such as austenite, Cr-carbide, and sigma phase are all paramagnetic, only ferromagnetic ferrite reacts to the magnetic induction, so that the surface volume fraction of ferrite can be precisely measured [18]. At 650–850 °C, the ferrite fraction of the weld metal was reduced by 1% compared to the as-weld specimen. From the thermodynamic calculation results shown in Figure 5c, a small amount of ferrite has been transformed into austenite because the A_{e3} temperature is less than 650 °C. However, as observed from optical photographs in Figure 6, the volume fraction variation of delta ferrite seems to be insignificant but rather the size of individual delta ferrite is likely to become larger as temperature increases from 650 to 850 °C.

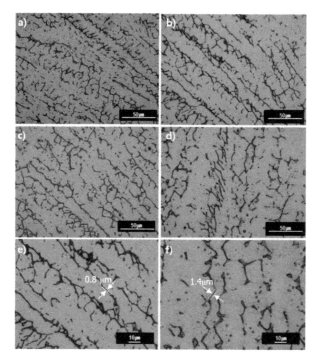

Figure 6. Optical micrographs (OM) on the weld metal of the heat-treated specimens for (**a**) 0.5 h at 650 °C, (**b**) 4 h at 650 °C, (**c**) 0.5 h at 850 °C, and (**d**) 4 h at 850 °C, respectively. To show the thickening of delta ferrite dendrite, markers indicate the thickness of delta ferrite dendrite for the heat-treated specimens for (**e**) 4 h at 650 °C and (**f**) 4 h at 850 °C.

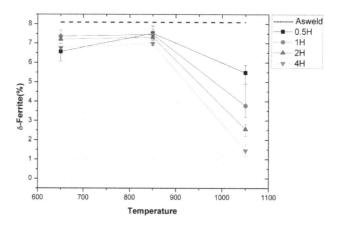

Figure 7. Volume fraction of delta ferrite in the weld metal measured by magnetic induction method.

Figure 8 is the EBSD phase map that shows the presence of $Cr_{23}C_6$ in the weld after annealing at 650 °C for 4 h. The $Cr_{23}C_6$, which did not exist in the as-weld specimen, was found to precipitate finely in the austenite/delta-ferrite interface after 4 h of heat treatment at 650 °C. $Cr_{23}C_6$ carbide is reported to precipitate from austenite/ferrite phase boundary or austenite grain boundary in

300-series austenitic stainless steels [11,15,16]. After precipitation, it grows into austenite matrix rather than ferrite matrix. It is known that the precipitates are finely located in the boundaries with the size distribution from several hundred nm to few μm, depending on the composition and the formation temperature [15]. As can be seen from the thermodynamic calculations shown in Figure 5, the temperature range near 650 °C could have 0.5% volume fraction of $Cr_{23}C_6$, which is the maximum value in whole temperature range. Although it is a small volume fraction, it is presumed that if the precipitates are finely distributed along the phase boundaries, it could deteriorate the ductility and toughness of the material. In the thermodynamic calculation, it was predicted that about 12% of the sigma phase would be precipitated in the weld metal at 650 °C. However, the EBSD results hardly showed the presence of sigma phase. It is believed that the transformation kinetics of sigma phase is relatively slow, so that sigma phase could not be transformed from ferrite with the given temperature and time. Even if sufficient time is provided, it is expected that the sigma phase would not be present as much as the predicted value since the volume fraction of ferrite, which is the preferred nucleation site of the sigma phase, is relatively small in the actual weld metal.

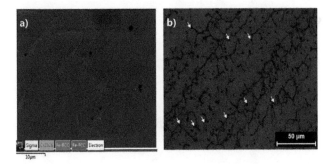

Figure 8. (**a**) EBSD map showing presence of $Cr_{23}C_6$ at phase boundary; (**b**) OM image after color-etching showing the presence of $Cr_{23}C_6$ (dark).

Figure 9a–d show the microstructures of the base material and HAZ of the specimens annealed at 1050 °C for 0.5–4 h. The grain size of the base metal after PWHT in 1050 °C for 4 h became larger (72 μm) than the as-weld specimen (52 μm), which shows the grain growth of austenite with increasing time and temperature. This grain growth is believed to occur because of the acceleration of thermally-assisted migration of the austenite grain boundary. The strip already existing in the base metal was observed to disappear as the heat treatment time became longer. As can be seen from the thermodynamic calculation in Figure 5, the Cr-based carbide was considered to transform into the austenite phase at 1050 °C since the dissolution temperature of the Cr-based carbide is about 900 °C.

Figure 9e–h shows the microstructure of welded specimens after heat treatment at 1050 °C for 0.5–4 h. The delta ferrite, which existed as lathy or vermicular shape in the as-weld specimen, lost its shape into the elliptical or the spheroidized after the heat treatment of 1050 °C. This is due to the migration of austenite/delta-ferrite interface for decreasing the interfacial energy, similar to the mechanism of the shape change from the lathy to the vermicular type observed in the heat treatment at 650–850 °C. The volume fraction of delta ferrite at 1050 °C decreased by heat treatment time as confirmed in Figure 7 where delta-ferrite, which was 8% of volume fraction in the as-weld specimen decreased down to 1.2% after heat treatment at 1050 °C for 4h. From the thermodynamic point of view, the transformation of delta ferrite to austenite was believed to be promoted at 1050 °C, well above the A_{e3} temperature. From the thermodynamic calculations, austenite single phase exists at 1050 °C, without Cr-carbide or delta-ferrite phase, and well agree with the experimental observations. The residual delta-ferrite, which was not transformed into austenite by the rapid cooling after the welding, was gradually austenized by heat treatment at 1050 °C. In addition, at that temperature, residual stress due to welding could be released and carbides could dissolve sufficiently, which could possibly lead to noticeable changes in the mechanical properties.

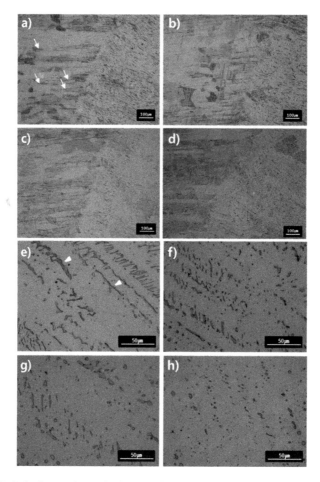

Figure 9. Optical micrographs on the base metal and the HAZ of the heat-treated specimens at 1050 °C for (**a**) 0.5 h, (**b**) 1 h, (**c**) 2 h, and (**d**) 4 h, respectively. Optical micrographs on the weld metal of the heat-treated specimens at 1050 °C for (**e**) 0.5 h, (**f**) 1 h, (**g**) 2 h, and (**h**) 4 h, respectively. Arrows and triangles indicate the pre-existing strips in the base/HAZ and delta ferrite in the weld metal, respectively.

Figure 10 summarized the yield and tensile strength, uniform and total elongation, and fracture toughness of specimens after PWHT. The yield and tensile strengths in all range of heat treatment temperatures and times were lower than those of the as-weld specimen. The yield strength gradually decreased as the temperature increased while there was no significant change depending on the time. The tensile strength of the specimens heat-treated at 650–850 °C was 570–580 MPa and a noticeable difference was hardly found depending on the heat treatment time. In the case of the heat treated specimen at 1050 °C, the tensile strength was markedly lower and continuously decreasing with the time. As shown in Figure 10b, the uniform and total elongation decreased with increasing temperature up to 850 °C, compared with the as-weld specimen. On the other hand, the specimen at

1050 °C showed higher elongation than the as-weld specimen, and prolonged its elongation as the heat treatment time increased.

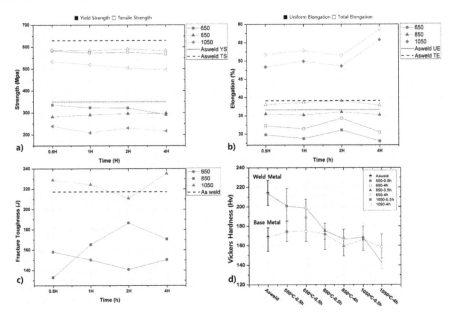

Figure 10. (**a**) Yield and tensile strength of the as-weld and the heat-treated specimens; (**b**) uniform and total elongation of the as-weld and the heat-treated specimens; (**c**) fracture toughness of the as-weld and the heat-treated specimens; (**d**) Vickers hardness of the as-weld and the heat-treated specimens.

These changes of mechanical properties can be understood based on the microstructural evolution during heat treatment. In the case of reduction in strength and ductility shown at 650–850 °C, relaxation of internal stress, decreased volume fraction of delta ferrite and increased size of individual delta ferrite dendrite could be responsible for such degradation. As shown in Figure 8, the carbides were found to form in the austenite/ferrite phase boundaries. It is expected that material strength would increase by compromising ductility with the formation of Cr-carbide, but inversely the strength decreased at the corresponding temperature. This reduction in strength, first of all, could probably be explained by the stress relaxation due to the recovery of the hot-rolled microstructure in the base metal and by the relaxation of the internal stress in the base and the weld metal initially caused by the welding process [19]. Second, as shown in Figure 7, the slight decrease in volume fraction of the delta ferrite in the weld metal could possibly cause the strength reduction in terms of modulus hardening. Third, as can be seen in Figure 6, when the individual size of the delta ferrite increased without losing its volume fraction substantially, the strength can be reduced owing to the weakening of dispersion strengthening or precipitation strengthening [20]. Stochastically, the probability of the existence of dislocation sources is higher in the austenite matrix whose volume fraction and size is much higher than delta ferrite. Given that austenite has generally lower yield strength than ferrite, the operation of dislocation is likely to initiate in the austenite matrix. The dislocation movement in the austenite could then be impeded by stiff delta ferrite arrays and if the individual size of the delta ferrite increased, particle strengthening could be degraded.

On the other hand, in the case of heat treatment above 1050 °C, the yield and tensile strengths were lower but elongations were higher than the as-weld specimen. As shown in Figure 8, the decrease in the volume fraction of delta ferrite at 1050 °C is considered to be the cause of the strength reduction. Since the thickness of delta ferrite was small as about 1–2 μm, it is believed that the delta ferrite is strengthening the weld metal in terms of modulus hardening and particle hardening. Therefore, decreasing the fraction of delta ferrite may have affected the decrease of strength and increase of elongation inversely. In addition, as shown in Figure 9, the austenite grain size of the base and welds increased during the heat treatment at 1050 °C, and thus Hall-Petch effect due to the increase of grain size also has an effect on the reduction in yield and tensile strengths. Furthermore, the stress relaxation due to the heat treatment at 1050 °C could make the base and weld metal softer. Figure 10c shows the fracture toughness obtained by integrating stress-strain curves. When heat treatment temperature was in the range of 650–850 °C, the fracture toughness was lower than that of the as-weld specimen. When the heat treatment temperature was more than 1050 °C, the fracture toughness was in overall similar or higher than the as-weld specimen. In the temperature range studied in the present work, the strengths were reduced as compared with the as-weld specimen, but ductility significantly increased at 1050 °C. This ductility increase is reflected in the increase of the fracture toughness, and the PWHT at a temperature of 1050 °C was found to provide positive effect on improving the toughness of the as-weld material.

Figure 10d shows the hardness variation of the welds with the PWHT temperature and time. After 10 times of indentations, the values were averaged by excluding the maximum and the minimum. Hardness is determined by the magnitude of strengthening under compression, and thus it can be interpreted similarly to the strength. Vickers hardness showed a monotonic decrease with increasing the heat treatment temperature. The hardness change with the heat treatment time was negligible for 1 to 4 h, and temperature was more of an influential factor on the hardness. As described above, reduction of hardness with increasing temperature is also able to be explained by the relaxation of the residual stress, the decrease of delta-ferrite volume fraction and the increase of the individual delta ferrite thickness.

Figure 11 contained the true stress-strain curves together with work hardening rate for analyzing the difference in work hardening behaviors after PWHT. In the as-weld specimen, the work hardening rate decreased rapidly after passing yield point and maintained in the steady state. On the other hand, the work hardening rate of the specimen at 650 °C for 2 h was generally higher than that of the as-weld specimen, and decreased continuously with further straining after having a short steady-state flow regime. The work hardening rate of PWHT 850 °C increased continuously with deformation after reaching the steady state. This is presumably due to the deformation-induced martensitic transformation (DIMT) occurring in both of 304 base metal and 308 weld metal. When the DIMT occurs, the martensite phase is produced with the deformation, and the volume fraction of martensite increases steadily as the deformation continues, leading to the gradual increase of the work hardening rate [18]. The inflection shown in true stress-strain curve is related to the initiation of DIMT. A number of researches reported that the DIMT indeed occurs in 300-series austenitic stainless steels and results in the continuous increase of the work hardening rate during deformation [18,21,22]. At 1050 °C, the work hardening rate gradually increased and an inflection was observed in true stress-strain curve. In this temperature, the initially present Cr-carbide arrays would dissolve and thus make austenite matrix denser with carbon atoms. The austenite is thus more stabilized and the DIMT kinetics could be slower, leading to the increase of elongation. Since the delta ferrite already existing in the weld metal was mostly transformed into austenite and also the grain size of austenite increased at 1050 °C, the yield strength became lower than the as-weld specimen.

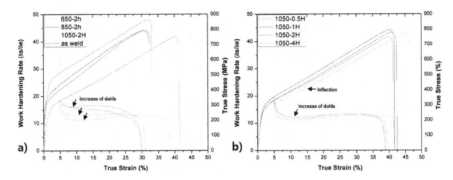

Figure 11. True stress-strain curves and work hardening rates in terms of (**a**) PWHT temperature and (**b**) PWHT time.

4. Conclusions

The effect of PWHT on the microstructure and mechanical properties of submerged-arc-welded 304 austenitic stainless steel was investigated. The results are summarized as follows:

(1) The base metal initially had a microstructure consisting of austenite matrix and strips of delta-ferrite surround by $Cr_{23}C_6$. The weld metal consisted of lathy and vermicular type of delta ferrite and austenite matrix. There was insignificant variation of microstructures and mechanical properties between HAZ and the base metal, implying that heat affection by welding was negligible.

(2) Cr-carbides were precipitated at the delta-ferrite/austenite interface in the weld metal after the heat treatment at 650–850 °C. Despite of the presence of Cr-carbide, the yield strength was lower than the as-weld specimen, which could be possibly explained by the relaxation of internal stress accumulated during the hot-rolling and the welding.

(3) After heat-treatment at 1050 °C, the pre-existing strips in the base metal was dissolved into austenite and the delta-ferrite of the weld metal was transformed into austenite, leading to the microstructure close to the austenite single phase. Although the yield strength was lower than that of the as-weld specimen due to the disappearance of hard phases and the relaxation of residual stress, the elongation became higher than that of the as-weld metal due to the compromising effect of strength and partially to the DIMT occurring during deformation. In terms of fracture toughness, the heat treatment at 1050 °C were suggested to provide superior property than as-weld material due to the trade-off between strength and ductility.

Acknowledgments: This study has been conducted with the support of the Korea Institute of Industrial Technology as "Development of metal powder large-area DED process including pre- and post- processes based on simulations (KITECH EO-17-0041)" and "Development of Manufacturing Technology of Cryogenic Stainless Steel Welded Steel Pipe (KITECH JG-17-0006)".

Author Contributions: Tae-Hoon Nam and Jong Bae Jeon conceived and designed the experiments; Tae-Hoon Nam and Eunsol An performed the experiments; Byung Jun Kim, Sunmi Shin, Nokeun Park, Won-Seok Ko, Namhyun Kang and Jong Bae Jeon analyzed and discussed the data; Tae-Hoon Nam and Jong Bae Jeon wrote the paper.

Conflicts of Interest: The authors declare no conflict of interest.

References

1. Weiss, B.; Stickler, R. Phase instabilities during high temperature exposure of 316 austenitic stainless steel. *Metall. Trans.* **1972**, *3*, 851–866. [CrossRef]

2.   Chen, X.H.; Lu, J.; Lu, L.; Lu, K. Tensile properties of a nanocrystalline 316L austenitic stainless steel. *Scr. Mater.* **2005**, *52*, 1039–1044. [CrossRef]

3.   Menthe, E.; Rie, K.T. Further investigation of the structure and properties of austenitic stainless steel after plasma nitriding. *Surf. Coat. Technol.* **1999**, *116–119*, 199–204. [CrossRef]

4.   Altenberger, I.; Scholtes, B.; Martin, U.; Oettel, H. Cyclic deformation and near surface microstructures of shot peened or deep rolled austenitic stainless steel AISI 304. *Mater. Sci. Eng. A* **1999**, *264*, 1–16. [CrossRef]

5.   Tabatabaeipour, S.M.; Honarvar, F. A comparative evaluation of ultrasonic testing of AISI 316L welds made by shielded metal arc welding and gas tungsten arc welding processes. *J. Mater. Process. Technol.* **2010**, *210*, 1043–1050. [CrossRef]

6.   Li, X.R.; Zhang, Y.M.; Kvidahl, L. Penetration Depth Monitoring and Control in Submerged Arc Welding. *Weld. J.* **2013**, *92*, 48–56.

7.   Gunaraj, V.; Murugan, N. Application of response surface methodology for predicting weld bead quality in submerged arc welding of pipes. *J. Mater. Process. Technol.* **1999**, *88*, 266–275. [CrossRef]

8.   Gunaraj, V.; Murugan, N. Prediction and comparison of the area of the heat-affected zone for the bead-on-plate and bead-on-joint in submerged arc welding of pipes. *J. Mater. Process. Technol.* **1999**, *95*, 246–261. [CrossRef]

9.   Kumar, S.; Shahi, A.S. Effect of heat input on the microstructure and mechanical properties of gas tungsten arc welded AISI 304 stainless steel joints. *Mater. Des.* **2011**, *32*, 3617–3623. [CrossRef]

10.   Luo, J.; Dong, Y.; Li, L.; Wang, X. Microstructure of 2205 duplex stainless steel joint in submerged arc welding by post weld heat treatment. *J. Manuf. Process.* **2014**, *16*, 144–148. [CrossRef]

11.   Hamada, I.; Yamauchi, K. Sensitization behavior of type 308 stainless steel weld metals after postweld heat treatment and low-temperature aging and its relation to microstructure. *Metall. Mater. Trans. A* **2002**, *33*, 1743–1754. [CrossRef]

12.   Jang, D.; Kim, K.; Kim, H.C.; Jeon, J.B.; Nam, D.-G.; Sohn, K.Y.; Kim, B.J. Evaluation of Mechanical Property for Welded Austenitic Stainless Steel 304 by Following Post Weld Heat Treatment. *Korean J. Met. Mater.* **2017**, *55*, 664–670. [CrossRef]

13.   Kang, N.-H. Development of Alloy Design and Welding Technology for Austenitic Stainless Steel. *J. KWJS* **2010**, *28*, 10–14. [CrossRef]

14.   Barbosa Gonçalves, R.; Henrique Dias de Araújo, P.; José Villela Braga, F.; Augusto Hernandez Terrones, L.; Pinheiro da Rocha Paranhos, R. Effect of conventional and alternative solution and stabilizing heat treatment on the microstructure of a 347 stainless steel welded joint. *Weld. Int.* **2017**, *31*, 196–205. [CrossRef]

15.   Tseng, C.C.; Shen, Y.; Thompson, S.W.; Mataya, M.C.; Krauss, G. Fracture and the formation of sigma phase, M23C6, and austenite from delta-ferrite in an AISI 304L stainless steel. *Metall. Mater. Trans. A* **1994**, *25*, 1147–1158. [CrossRef]

16.   Kington, A.V.; Noble, F.W. σ phase embrittlement of a type 310 stainless steel. *Mater. Sci. Eng. A* **1991**, *138*, 259–266. [CrossRef]

17.   Vitek, J.; David, S. The sigma phase transformation in austenitic stainless steels. *Weld. J.* **1986**, *65*, 106s–111s.

18.   Shin, H.C.; Ha, T.K.; Chang, Y.W. Kinetics of deformation induced martensitic transformation in a 304 stainless steel. *Scr. Mater.* **2001**, *45*, 823–829. [CrossRef]

19.   Mújica Roncery, L.; Weber, S.; Theisen, W. Welding of twinning-induced plasticity steels. *Scr. Mater.* **2012**, *66*, 997–1001. [CrossRef]

20.   Zhang, Z.; Chen, D.L. Consideration of Orowan strengthening effect in particulate-reinforced metal matrix nanocomposites: A model for predicting their yield strength. *Scr. Mater.* **2006**, *54*, 1321–1326. [CrossRef]

21.   Huang, C.X.; Yang, G.; Deng, B.; Wu, S.D.; Li, S.X.; Zhang, Z.F. Formation mechanism of nanostructures in austenitic stainless steel during equal channel angular pressing. *Philos. Mag.* **2007**, *87*, 4949–4971. [CrossRef]

22.   Lee, T.-H.; Oh, C.-S.; Kim, S.-J. Effects of nitrogen on deformation-induced martensitic transformation in metastable austenitic Fe–18Cr–10Mn–N steels. *Scr. Mater.* **2008**, *58*, 110–113. [CrossRef]

Article

Evolution of Microstructure and Mechanical Properties of a CoCrFeMnNi High-Entropy Alloy during High-Pressure Torsion at Room and Cryogenic Temperatures

Sergey Zherebtsov [1], Nikita Stepanov [1], Yulia Ivanisenko [2], Dmitry Shaysultanov [1], Nikita Yurchenko [1], Margarita Klimova [1,*] and Gennady Salishchev [1]

[1] Laboratory of Bulk Nanostructured Materials, Belgorod State University, 308015 Belgorod, Russia; zherebtsov@bsu.edu.ru (S.Z.); stepanov@bsu.edu.ru (N.S.); shaysultanov@bsu.edu.ru (D.S.); yurchenko_nikita@bsu.edu.ru (N.Y.); salishchev@bsu.edu.ru (G.S.)

[2] Karlsruhe Institute of Technology, Institute of Nanotechnology, 76021 Karlsruhe, Germany; julia.ivanisenko@kit.edu

* Correspondence: klimova_mv@bsu.edu.ru; Tel.: +7-472-258-5416

Received: 16 January 2018; Accepted: 7 February 2018; Published: 10 February 2018

Abstract: High-pressure torsion (HPT) is applied to a face-centered cubic CoCrFeMnNi high-entropy alloy at 293 and 77 K. Processing by HPT at 293 K produced a nanostructure consisted of (sub)grains of ~50 nm after a rotation for 180°. The microstructure evolution is associated with intensive deformation-induced twinning, and substructure development resulted in a gradual microstructure refinement. Deformation at 77 K produces non-uniform structure composed of twinned and fragmented areas with higher dislocation density then after deformation at room temperature. The yield strength of the alloy increases with the angle of rotation at HPT at room temperature at the cost of reduced ductility. Cryogenic deformation results in higher strength in comparison with the room temperature HPT. The contribution of Hall–Petch hardening and substructure hardening in the strength of the alloy in different conditions is discussed.

Keywords: high-entropy alloys; high-pressure torsion; microstructure evolution; twinning; mechanical properties

---

1. Introduction

The concept of high-entropy alloys (HEAs) as a mixture of more than five metallic elements in equimolar proportions was proposed first by Yeh et al. in 2004 [1]. Due to increased configurational entropy, the HEAs were expected to exist in the form of a single solid solution phase while formation of intermetallic compounds and secondary phases would be suppressed. In reality, the microstructure of alloys of multiple principle elements is usually more complicated and can consist of a mixture of different phases including intermetallic ones [2]. Among many proposed HEA compositions, CoCrFeMnNi alloy indeed has a single phase disordered fcc microstructure [3–5]. Due to such a 'model' structure, this alloy is currently one of the most studied HEAs so far.

It was found that the CoCrFeMnNi alloy demonstrated very high elongation (7080%), but rather low yield strength of 200 MPa at room temperature [6,7]. Strengthening of HEA can be achieved via microstructure refinement. A reduction of a grain size from 150 to 5 μm in the CoCrFeMnNi alloy increased the yield strength by a factor of two while maintained good ductility [6]. Severe plastic deformation of the CoCrFeNiMn alloy via high-pressure torsion (HPT) expectably refined the microstructure to a grain size d~40–50 nm and increased the microhardness of the alloy by a factor of ~3 [8–10]. The ultimate tensile strength of the alloy after HPT was found to be ~2000 MPa [8].

The formation of the nanostructure was attributed to fragmentation of nanotwins [8] and to accelerated atomic diffusivity under the shear strain during HPT [9]. However, the microstructure and mechanical properties evolution of the alloy was insufficiently studied. For instance, the evolution of tensile properties during HPT had not been studied yet. Besides, considerably higher intensity of deformation twinning in the alloy at cryogenic temperature was observed during tension [6,11] or rolling [12]. Increased susceptibility to twinning can significantly accelerate the microstructure refinement, as it was shown earlier [12–14]. However, no information on attempts of HPT of the CoCrFeMnNi HEA under cryogenic conditions was found in the literature.

Therefore, the aim of the present work was to investigate in details the microstructure and mechanical properties evolution of the high-entropy CoCrFeMnNi alloy during high-pressure torsion at room temperature and to evaluate the effect of temperature decrease to 77 K on structure and properties of the alloy.

2. Materials and Methods

The equiatomic alloy with the composition of CoCrFeMnNi was produced by arc melting of the components in high-purity argon inside a water-cooled copper cavity. The purities of the alloying elements were above 99.9 %. To ensure chemical homogeneity, the ingots were flipped over and re-melted at least five times. The produced ingots of the CoCrFeMnNi alloy had dimensions of about $6 \times 15 \times 60$ mm^3. Homogenization annealing was carried out at 1000 °C for 24 h [15,16]. Prior to homogenization, the samples were sealed in vacuumed (10^{-2} Torr) quartz tubes filled with titanium chips to prevent oxidation. After annealing, the tubes were removed from the furnace and the samples were cooled inside the vacuumed tubes down to room temperature. After the homogenization procedure, the alloy was cold rolled with ~80% height reduction and then annealed at 850 °C for 1 h to produce a recrystallized structure. The alloy after such a treatment is hereafter referred to as the initial condition in the current study.

For HPT processing discs measuring 15 mm diameter $\times$ 0.8 mm thick were cut from the recrystallized specimens, then ground and mechanically polished. The samples were subjected to HPT at 293 K (to 90, 180, or 720° of the anvils turn) or 77 K (to 180°) under a pressure of 4.3 GPa in a Bridgman anvil-type unit with a rate of 1 rpm using a custom-built computer-controlled HPT device (W. Klement GmbH, Lang, Austria). The corresponding shear strain level γ was calculated as [17]

$$\gamma = \frac{2\pi Nr}{h} \tag{1}$$

where N is the number of revolutions, r is the radius, and h is the thickness of the specimen.

The microstructure of the alloys was studied using transmission (TEM) and scanning (SEM) electron microscopy and electron back-scattered diffraction (EBSD) analysis. For SEM observations the specimens were mechanically polished in water with different SiC papers and a colloidal silica suspension; the final size of the Al_2O_3 abrasive was 0.04 μm. The samples were examined after polishing without etching. The SEM back-scattered electron (BSE) images of microstructures in the initial condition were obtained using a FEI Quanta 3D microscope. EBSD was conducted in a FEI Nova NanoSEM 450 field emission gun SEM equipped with a Hikari EBSD detector and a TSL OIM™ system version 6.0. EBSD examinations with the step size of 50 nm were carried out in an axial section in three different characteristic areas: (i) in the central part of the disc, i.e., within ±0.5 mm of the center; (ii) at half of a radius, i.e., at ~3.5 mm from the center; (iii) at the edge, i.e., at ~6–7 mm from the center. The points with confidence index (CI) < 0.1 were excluded from the analysis; these points are shown with black color to avoid any artificial interference in the microstructure. In the presented inverse pole figure (IPF) maps, the high angle (>15°) and low angle (2–15°) boundaries are shown respectively as black and white lines.

TEM examination was performed in the mid-thickness shear plane of deformed samples in the vicinity (~1 mm away) of the edge. The samples for TEM analysis were prepared by conventional twin-jet electro-polishing of mechanically pre-thinned to 100 μm foils, in a mixture of 90% CH₃COOH and 10% HClO₄ at the 27 V potential at room temperature. TEM investigations were performed using JEOL JEM-2100 apparatus at accelerating voltage of 200 kV.

The dislocation density was determined using X-ray diffraction (XRD) profiles analysis obtained by a RIGAKU diffractometer with Cu Kα radiation at 45 kV and 35 mA as it was done earlier for the similar alloy [18]. The value of the dislocation density, ρ, was calculated using the equation [19]

$$\rho = \frac{3\sqrt{2\pi}\,\langle\varepsilon_{50}^2\rangle}{Db} \tag{2}$$

where $\langle\varepsilon_{50}^2\rangle$ is microstrains, D is the crystallite size, and b is the Burgers vector. The microstrains, $\langle\varepsilon_{50}^2\rangle$, and the crystallite size, D, values were estimated on the basis of the Williamson–Hall plot [20], using the equation

$$\frac{\beta_s cos\Theta}{\lambda} = \frac{2\langle\varepsilon_{50}^2\rangle sin\Theta}{\lambda} + \frac{K}{D} \tag{3}$$

where β_s is the corrected full width at half maximum (FWHM) of the selected Kα₁ reflection of the studied material, Θ is the Bragg angle of the selected reflection, λ is the Kα₁ wavelength, and K is Scherrer constant. In the present study, the FWHM values and the positions of fcc (111) and (222) reflections were determined and used for further calculations. The instrumental broadening was determined from the FWHM values of the annealed silicon powder.

Mechanical properties of the alloy after HPT were determined through tensile tests of flat dog-bone specimens with gauge measuring 5 mm length × 1 mm width × 0.6 mm thickness. The samples for the tensile tests were cut from the HPT-processed discs using an electric discharge machine so that the gage was located at a distance of ~3.5 mm from the sample center. The specimens were then pulled at a constant crosshead speed of 3 mm/min (strain rate of 0.01 s⁻¹) in a custom built computer controlled tensile stage for miniature samples to fracture. The stress–strain curves were obtained using a high-precision laser extensometer.

3. Results

In the initial condition the CoCrFeNiMn alloy had an fcc single-phase microstructure (Figure 1a). An SEM-BSE image shows a homogeneous microstructure with a grain size of ~15 μm (Figure 1b). Numerous annealing twins were found inside grains. Some pores visible as black dots can also be found in the microstructure.

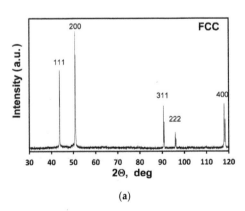

(a) (b)

Figure 1. Microstructure of the initial condition of the CoCrFeNiMn alloy: (a) XRD pattern; (b) SEM-BSE image.

The influence of HPT on the microstructure depended essentially on the distance from the specimen center (Figure 2). In comparison with the initial condition, a rotation of 90° at room temperature resulted in rather small changes in the microstructure (Figure 2a) associated with the formation of low-angle boundaries in some grains and appearance of individual deformation twins. It should be noted that boundaries of both deformation and annealing twins are crystallographycally identical. However, deformation twins are usually much thinner than annealing twins and intersect grains. The thickness of deformation twins in a single phase fcc microstructure is usually several tens of nanometers [12,21], whereas the thickness of annealing twins can reach several micrometers [22]. Some of the deformation twins are indicated with arrows in Figure 2b,c. The analysis of the EBSD data showed that observed (both annealing and deformation) twins belong to (111) <112> system with twin/matrix misorientation of 60° around <111>. An increase in the number of deformation twins and the development of substructure (sometimes appearing as black, poorly distinguished areas mainly in the vicinity of twin or grain boundaries) with the increase of imposed strain can be observed (Figure 2b,c).

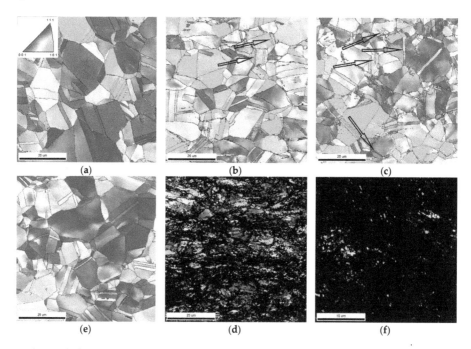

Figure 2. EBSD IPF maps of the CoCrFeNiMn alloy after HPT at 293 K: (**a,d**) central part of the specimen; (**b,e**) half-radius; (**c,f**) edge of the specimens; (**a–c**) 90° and (**d–f**) 180° rotation. The corresponding true shear strain is (**a**) 1.1; (**b**) 8.4; (**c**) 13.5; (**d**) 2.2; (**e**) 16.8; (**f**) 26.9. The color code is inserted in Figure 2a. The deformation twins are indicated with arrows in Figure 2b,c.

The central part of the specimens after rotation for 180° (Figure 2d) had a microstructure very similar to that observed after 90° rotation; i.e., only little changes can be observed in comparison with the initial microstructure. This is possibly due to relatively low true strain in the central part of both specimens ($\varepsilon \leq 2.2$). However, much more pronounced changes were observed at larger strain (Figure 2e,f). The microstructure at the half-radius consisted of relatively large areas of 5–7 μm with developed substructure and small fragments of ~0.5 μm (Figure 2e). Due to a high level of internal stresses which prevents obtaining identifiable Kikuchi patterns, a considerable part of the EBSD map was presented by black dot areas with a low CI. The comparison with the IPF map of the edge part of the specimen after the rotation for 90° (Figure 2c) suggests that the true shear strain of ~15 is required to produce a significant fraction of such highly-deformed areas.

At the edge of the specimen, the average size of the visible fragments decreased to ~0.5 μm (Figure 2f). However, the black dot areas with low CI occupied the majority of the scanned area thereby suggesting a very high level of internal stresses and considerable microstructure refinement. This finding is in good agreement with an increase in the true shear strain to 26.9. Detailed TEM analysis of the microstructure of those severely deformed areas is given below. Note that EBSD analysis of the sample after the rotation for 720° has produced IPF maps almost completely occupied with black dots both at the center, half-radius, and edge of the specimen that most likely associated with a considerable increase in homogeneity of the microstructure refinement. The corresponding images are not shown as they do not present any meaningful information.

The effect of a decrease in deformation temperature to 77 K was studied after the rotation for 180° (Figure 3). Microstructure changes in comparison with the specimen deformed at room temperature (Figure 2d-f) depended considerably on the examined part of the disk and the imposed strain. Microstructure of the central part of the specimen deformed at 77 K (Figure 3a) was evidentially more refined than that at room temperature (Figure 2d) and consisted of rather large areas of ~8 μm which contained individual twins and relatively poor developed substructure. Low quality of the EBSD map in between these areas indicated quite a high level of internal stresses, most likely due to strain localization. Small fragments < 1 μm in diameter were observed in severely deformed areas. Higher strain level obtained at the half of radius (ε = 16.8, Figure 3b) resulted in a microstructure which is quite similar to that at room temperature (Figure 2e). At the edge of specimens the microstructure after HPT at 77 K (Figure 3c) had an obvious metallographic texture with fragments elongated along the shear direction. The size of the majority of the fragments was quite small (0.2–0.4 μm). However, big fragments > 2 μm with poorly developed substructure were also observed in the microstructure. In addition, the fraction of visible fragments was greater than that after room temperature HPT (Figure 2f) that can suggest lower level of internal stresses in the former case.

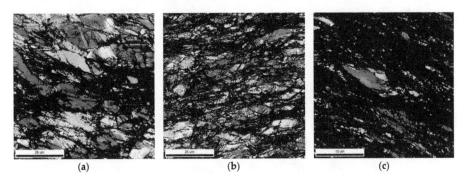

(a) **(b)** **(c)**

Figure 3. Microstructure of the CoCrFeNiMn alloy after HPT at temperature of 77 K to 180°: (a) central part; (b) half-radius; (c) edge. The corresponding true shear strain is (a) 2.2; (b) 16.8; (c) 26.9.

TEM micrographs of the CoCrFeNiMn alloy microstructure after HPT at room temperature is shown in Figure 4. After HPT rotation for 90° at 293 K (Figure 4a,b) the microstructure at the edge of specimens consisted of crossing deformation twins of 50–100 nm width (corresponding diffraction pattern inserted in Figure 4a) and high dislocation density. Twins either spaced 0.5–1 μm apart (Figure 4a) or cluster together with the formation of twin bundles (Figure 4b). After the rotation for 180° at room temperature the microstructure at the edge of specimens was considerably refined (Figure 4c,d). Although the microstructure can be mainly described as a cellular one with a very high dislocation density (Figure 4c), very small grains of ~50 nm can also be recognized at larger magnification. Twin boundaries were not detected in the microstructure. However, a chain of small elongated (sub)grains in some places (indicated with dotted lines in Figure 4d) suggested that these grains can originated from twins. Rotation for 720° resulted in the formation of a homogeneous ultra-fine microstructure consisted of irregular dislocation pile-ups of different shapes and sizes and very small grains with a size of ~20–60 nm (Figure 4e,f). The diffraction pattern consisted of rings with many diffracted beams indicating the presence of many small (sub)grains with mainly high misorientation angle boundaries within the selected field of view (insert on Figure 4e).

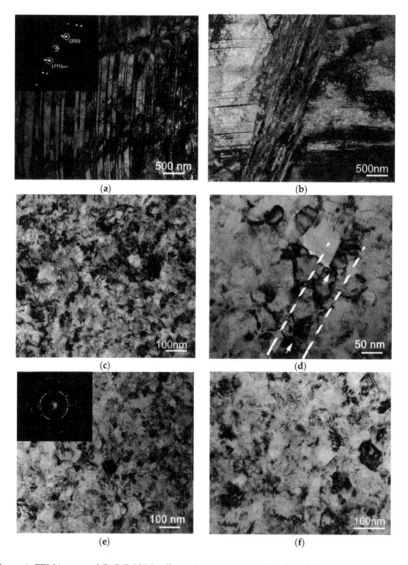

(a)

(b)

(c)

(d)

(e)

(f)

Figure 4. TEM images of CoCrFeNiMn alloy microstructure at the edge of specimens after HPT at temperature of 293 K to: (**a**,**b**) 90° (true strain of ≈12); (**c**,**d**) 180° (true strain ≈25); (**e**,**f**) 720° (true strain ≈50). Elongated (sub)grains and transverse subboundaries are indicated with dash lines and arrows, respectively in Figure 4d.

In contrast to the room-temperature deformation (Figure 4c,d), the microstructure observed after HPT at 77 K was found to be highly inhomogeneous (Figure 5). It consisted of areas with twin bundles and structure with high dislocation density and subgrain orientation (Figure 5a). The selected area diffraction pattern (insert in Figure 5a) was obtained from a crystallite with zone axis $[0\bar{6}2]$; within these crystallite areas, with both twin and subgrain structure can be observed. The reflexes characterizing the twin orientation are indexed on the diffraction pattern. The matrix reflexes show azimuthal spread, thereby suggesting the presence of small misorientations inside the crystallite. An example of such microstructure is shown in the Figure 5b. The microstructure revealed sites of different shapes and sizes with low-angle misorientation between them; meanwhile, no grains were observed in the microstructure (Figure 5b).

| (a) | (b) |

Figure 5. TEM image of the CoCrFeNiMn alloy microstructure at the edge of specimens after HPT at temperature of 77 K to 180° (true strain ~25). (**a**) low magnification; (**b**) high magnification.

According to the XRD analysis, the dislocation density after cryogenic deformation was 2.9×10^{15} m^{-2}. In comparison, after similar processing at room temperature the dislocation density was 9.14×10^{14} m^{-2}, i.e., three-fold increase in dislocation density due to decrease of HPT temperature occurred. In turn, the imposed strain at room temperature had weakly affected the dislocation density. For example, the dislocation density values after rotations for 90° and 720° were 1.48×10^{15} m^{-2} and 8.86×10^{14} m^{-2}, respectively, i.e., they were rather similar to each other and to dislocation density after rotation for 180° at room temperature.

Tensile stress–strain curves of the alloy are shown in Figure 6. The resulting mechanical properties including yield strength (YS), ultimate tensile strength (UTS), uniform elongation (UE), and total elongation (TE), are summarized in Table 1. Already after a rotation for 90° the alloy had a high ultimate tensile strength approaching 1 GPa but rather limited ductility of ~12%. Increasing HPT rotation from 90° to 180° resulted in a moderate rise in strength and a pronounced decrease in ductility to ~5%. At further increase in strain to two full rotations (720°) the strength of the alloy increased considerably (approximately by a factor of 2) while the ductility decreased moderately. A decrease in temperature of HPT from 293 to 77 K considerably increased the strength of the alloy but not affected ductility much. All flow curves exhibited a peak flow stress at the initial stages of deformation followed by a moderate decrease in flow stress. The uniform elongation of the alloy was rather small in all cases which is quite typical of severely deformed materials [23]. However, after strain localization (at the maximum flow stress), when the plastic deformation had become concentrated in the neck, the alloy had showed quite pronounced additional elongation.

Table 1. Tensile mechanical properties [1] of the CoCrFeNiMn alloy after HPT at 293 and 77 K.

HPT Condition		YS, MPa	UTS, MPa	UE, %	TE, %
Temperature, K	Rotation, °				
	90	860	981	1.4	11.9
293	180	1134	1197	0.5	6.4
	720	1834	2069	1.4	7.4
77	180	1442	1596	1.3	10.4

[1] YS—yield stress, UTS—ultimate tensile stress, UE—uniform elongation, TE—total elongation.

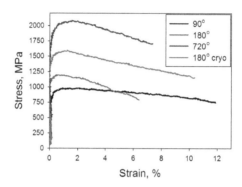

Figure 6. Engineering stress–strain curves obtained during tensile tests of the CoCrFeNiMn alloy after HPT at 293 and 77 K.

4. Discussion

The results of the present work clearly demonstrate that the microstructure development in the CoCrFeNiMn alloy during HPT strongly depended on both the imposed strain (i.e., the angle of rotation and the distance from the center of the discs) and deformation temperature. At room temperature, the formation of ultra-fine grained (UFG) structure on the periphery of the HPT discs occurred quite fast: 50 nm-sized (sub)grains were observed after a 180° turn (Figure 4c,d). Fast kinetics of microstructure evolution in the alloy during HPT has been already reported [9]. However, it should be noted that the microstructure developed in the beginning of HPT-straining was highly heterogeneous (Figure 2d–f). In the center of the discs, initial grains with weakly developed deformation twins/substructure had survived even after rotation for 180° (Figure 2d). This is an obvious result of a strong gradient of the imposed strain along the radius of the discs and is typical of the HPT process [17]. Meanwhile, according to the present results, homogenous UFG microstructure can be produced after the rotation for 720°.

Formation of UFG microstructure after HPT at room temperature is well documented for various metallic materials [17] including high-entropy alloys [8–10,24–26]. The microstructure evolution during HPT is usually associated with the formation of low-angle grain boundaries at low strains and transformation of some of these low-angle subgrain boundaries into high-angle grain boundaries at higher strains giving rise to a considerable microstructure refinement. However, in the materials that deform by both twinning and slip the kinetics of microstructure refinement can be enhanced due to intensive twin formation [12–14]. Thin twin laths which are separated from the matrix by high-angle boundaries readily transform into a chain of grains during deformation. This process is particularly relevant during initial stages of deformation when the number of twins and the density of high-angle twin boundaries are the highest [13]. It is already established that mechanical twinning can occur in the CoCrFeNiMn alloy during plastic deformation at room temperature, however, the extent and the role of twinning in microstructure development remains quite debatable [11,12,27,28].

The results of the present work indicate that twinning during HPT develops at the edge of specimens till rather small strains (rotation for ~90°, Figure 4a,b). The occurrence of twinning during HPT observed in the present study is in agreement with the results of previous investigations [8,9]. At later stages of strain (rotation for 180°), an equiaxed structure composed of a mixture of dislocation cells and grains was observed (Figure 4c). In general the process of the equiaxed fine-grained structure formation can be associated with: (i) deformation-induced twinning; (ii) fragmentation of twin laths by secondary twinning and/or transverse subboundaries (indicated with arrows in Figure 4d); (iii) formation of low-angle boundaries in those parts of microstructure which were not involved into twinning; and (iv) further increase of their misorientation to the high-angle range due to interaction with lattice dislocation like it was observed in different metallic materials—including titanium, copper based alloys, austenitic steel and some others [17,29–32]. Finally, these processes are likely to be responsible for the microstructure refinement at large strain (Figure 4e,f). This behavior is similar to that observed during room-temperature rolling of the same alloy [12]. Therefore, fast kinetics of the microstructure refinement in the CoCrFeNiMn alloy during HPT can be attributed to extensive deformation twinning which readily occurs at the early stages of deformation.

The effect of twinning on microstructure refinement should be sensitive to deformation temperature. Since the critical resolved shear stress (CRSS) for twinning in various metals and alloys depends weakly on temperature compared to that for slip [33], a decrease in deformation temperature from 293 to 77 K was believed to have resulted in more intensive twinning and increased kinetics of microstructure refinement [12,13]. However, a decrease in deformation temperature in case of HPT of the CoCrFeNiMn alloy leaded to the formation of a highly inhomogeneous microstructure composed of a mixture of twinned and fragmented areas (Figure 5). This is possibly a result of considerable increase in strength (approximately by a factor of 2 [6]) due to the lower temperature which increased strain localization and could provoke slippage of anvils at HPT and, thus, reduction of the actually imposed strain. However, the dislocation density estimated using XRD (Table 2) after cryo-deformation was considerably higher than that after room temperature HPT. This finding does not agree with the results of the dislocation density measurements during rolling at 77 and 293 K [12]. During rolling, active twinning at 77 K promoted microstructure refinement. Possibly, dislocation mechanisms play more important role in cryo-HPT of the CoCrFeNiMn alloy than during HPT at room temperature, and therefore, the formation of homogeneous UFG structure occurs slower under cryogenic conditions.

The microstructure analysis is in reasonable agreement with the results of mechanical properties. The observed microstructure refinement during HPT at room temperature was accompanied by a considerable rise in strength and some decrease in ductility (Figure 6). It should be noted, however, that the difference in the microstructure in terms of (sub)grain size and dislocation density (Table 2) was negligible between the specimens processed for 180° or 720° turns. Meanwhile, the alloy after rotation for 720° was found to be approximately 30% stronger than that after strain to 180°. Also, strength of the cryo-deformed alloy was substantially higher than that of the specimen after room temperature HPT despite highly inhomogeneous, coarser microstructure.

The contributions of the most relevant hardening mechanisms in strength of the deformed alloy can be expressed as

$$YS = \sigma_0 + \sigma_\rho + \sigma_{H-P} \qquad (4)$$

where σ_0 denotes the friction stress, σ_ρ is the substructure hardening and σ_{H-P} is the Hall–Petch hardening. The substructure hardening σ_ρ can be expressed as

$$\sigma_\rho = M\alpha Gb\sqrt{\rho} \qquad (5)$$

where M is the average Taylor factor, α is a constant, G is the shear modulus, b is the Burgers vector, and ρ is the dislocation density. The Hall–Petch contribution to the strength is typically of the form

$$\sigma_{H-P} = K_y \, d^{\frac{-1}{2}} \qquad (6)$$

in which K_y is the Hall–Petch coefficient and d is the grain size.

The dislocation density was found to be quite similar for the specimens processed for 180° or 720° at room temperature (Table 2). This result is in agreement with the equality of measured size of cells or subgrains in both structures since it is well established that $d_{sub} \sim \sqrt{\rho}$ [34], where d_{sub}—is the average size of subgrains. Therefore, it can be assumed that the only relevant factor which changes during HPT of the alloy from 180° to 720° strain is the grain size. Here the "grain" term is used to denote the crystallites bordered by high ($\geq 15°$) angle boundaries. The evolution of grain size during deformation was evaluated using equation (1); the input parameters for calculation were: $K_y = 0.494$ MPa m$^{1/2}$ and $\sigma_0 = 125$ MPa (both parameters were taken from [6]), $M = 3$, $\alpha = 0.2$, $G = 81$ GPa [35], $b = 2.54 \times 10^{-10}$ m [28]. Estimated dislocation densities and yield stress are tabulated in Table 2.

Table 2. Input parameters and results of calculation for the evaluation of contributions of the Hall–Petch and substructure hardening mechanisms.

Strain, ° (Temperature, K)	Dislocation Density, m^{-2}	Calculated Grain Size, nm	σ_{H-P}, MPa	σ_ρ, MPa	Predicted YS, MPa
90 (293)	1.48×10^{15}	2500	311	475	911
180 (293)	9.14×10^{14}	580	646	373	1144
720 (293)	8.86×10^{14}	140	1317	367	1809
180 (77)	2.9×10^{15}	625	625	664	1414

The best fit of experimental results was obtained when the grain sizes take on values shown in Table 2. Although microstructural investigation did not obviously support these findings, the obtained data seems quite reasonable. At the initial stages of deformation the fraction of the high-angle boundaries is rather low and consists of initial grain boundaries and boundaries of deformation twins. Therefore, the effective grain size should be approximately equal to the average space between twin boundaries. This similarity can indeed be observed in the microstructure.

During further deformation, new grains developed in place of subgrains within initial grains. This is a result of gradual transformation of geometrically necessary boundaries, which separate microvolumes with different combinations of slip systems [36], from low-angle subboundaries into high-angle grain boundaries. The fraction of high-angle grain boundaries gradually increased, thereby decreasing the effective grain size. At the final stages of HPT all sub-boundaries are expected to be transformed into boundaries; this situation was nearly reached after the rotation for 720° when the calculated grain size was only three times larger than the measured size of subgrains. Taking into account some decrease in strain along the radius (the yield stress was measured exactly at the middle of radius, while TEM investigations were performed closer to the edge of the discs) than the point where the microstructure was analyzed) this approximation can be considered quite reasonable.

The results of the calculation show that the Hall–Petch hardening contribution increased with strain while the substructure hardening maintained at approximately the same level (Table 2, Figure 7). Thus, the Hall–Petch hardening contribution becomes much more important than the substructure hardening in samples rotated for angles larger than 180°. This result is in good agreement with data reported earlier for the nanocrystalline HEA [37].

It is worth noting that, according to the calculations, the effective grain size in the CoCrFeNiMn alloy after HPT at 293 and 77 K is approximately equal (Table 2). Higher strength of the sample deformed in cryogenic condition can be ascribed to the contribution of substructure (dislocation) hardening. After the rotation for 180° at cryogenic temperature, the contribution of the substructure hardening is comparable to that of the Hall–Petch hardening (Figure 7).

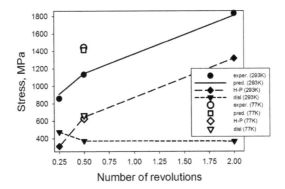

Figure 7. The contributions of different hardening mechanisms in strength of CoCrFeNiMn alloy during HPT at 293 K and 77 K. Exper.—experimental yield stress; pred.—predicted yield stress, obtained from. Equation (4); H-P—Hall-Petch hardening; disl—substructure hardening. Temperatures of HPT are indicated in brackets.

5. Conclusions

Microstructure evolution and mechanical properties of CoCrFeMnNi high-entropy alloy during high-pressure torsion (HPT) at 293 and 77 K was studied. Following conclusions were made:

(1) HPT processing at room temperature results in the formation of in inhomogeneous severely deformed microstructure with (sub)grains of ~50 nm already after the rotation for 180°. The microstructure evolution was associated with intensive deformation-induced twinning, and substructure development resulted in a gradual microstructure refinement.

(2) HPT at 77 K produced more heterogeneous structure in comparison with the room-temperature deformation. The dislocation density was much higher after cryogenic deformation.

(3) Tensile strength of the alloy after HPT at 293 K was found to be strongly dependent on HPT strain. The ultimate tensile strength increased from 981 to 2069 MPa when the rotation angle at HPT increased from 90° to 720°. In all examined conditions, the alloy exhibited limited ductility.

(4) A decrease of HPT temperature from 293 to 77 K resulted in higher tensile strength. A rotation for 180° resulted in the ultimate tensile strength of 1193 MPa and 1596 MPa after processing at room and cryogenic temperature, respectively.

(5) An increase of yield strength of the alloy with an increase of the angle of rotation can mostly be ascribed to a contribution of Hall–Petch strengthening. In turn, higher strength of the alloy after HPT at 77 K was attributed to substructure (dislocation) hardening.

Acknowledgments: This study was supported by Russian Foundation for Basic Research (grant no. 16-38-60061). Authors are thankful to J. Beach for performing tensile tests. The authors are also grateful to the personnel of the Joint Research Center, "Technology and Materials", Belgorod State National Research University, for their assistance with the instrumental analysis.

Author Contributions: Sergey Zherebtsov, Yuliya Ivanisenko, Nikita Stepanov, and Gennady Salishchev conceived and designed the experiments. Dmitry Shaysultanov, Nikita Yurchenko, Margarita Klimova and Yuliya Ivanisenko performed the experiments. Gennady Salishchev, Nikita Stepanov and Sergey Zherebtsov analyzed the data and wrote the paper.

Conflicts of Interest: The authors declare no conflict of interest.

References

1. Yeh, J.-W.; Chen, S.-K.; Lin, S.-J.; Gan, J.-Y.; Chin, T.-S.; Shun, T.-T.; Tsau, C.-H.; Chang, S.-Y. Nanostructured High-Entropy Alloys with Multiple Principal Elements: Novel Alloy Design Concepts and Outcomes. *Adv. Eng. Mater.* **2004**, *6*, 299–303. [CrossRef]

2. Miracle, D.B.; Senkov, O.N. A critical review of high entropy alloys and related concepts. *Acta Mater.* **2017**, *122*, 448–511. [CrossRef]

3. Cantor, B.; Chang, I.T.H.; Knight, P.; Vincent, A.J.B. Microstructural development in equiatomic multicomponent alloys. *Mater. Sci. Eng. A* **2004**, *375*, 213–218. [CrossRef]

4. Otto, F.; Yang, Y.; Bei, H.; George, E.P.P. Relative effects of enthalpy and entropy on the phase stability of equiatomic high-entropy alloys. *Acta Mater.* **2013**, *61*, 2628–2638. [CrossRef]

5. Laurent-Brocq, M.; Akhatova, A.; Perrière, L.; Chebini, S.; Sauvage, X.; Leroy, E.; Champion, Y. Insights into the phase diagram of the CrMnFeCoNi high entropy alloy. *Acta Mater.* **2015**, *88*, 355–365. [CrossRef]

6. Otto, F.; Dlouhý, A.; Somsen, C.; Bei, H.; Eggeler, G.; George, E.P. The influences of temperature and microstructure on the tensile properties of a CoCrFeMnNi high-entropy alloy. *Acta Mater.* **2013**, *61*. [CrossRef]

7. Gali, A.; George, E.P. Tensile properties of high- and medium-entropy alloys. *Intermetallics* **2013**, *39*, 74–78. [CrossRef]

8. Schuh, B.; Mendez-Martin, F.; Völker, B.; George, E.P.P.; Clemens, H.; Pippan, R.; Hohenwarter, A.; Völker, B.; George, E.P.P.; Clemens, H.; et al. Mechanical properties, microstructure and thermal stability of a nanocrystalline CoCrFeMnNi high-entropy alloy after severe plastic deformation. *Acta Mater.* **2015**, *96*, 258–268. [CrossRef]

9. Lee, D.-H.; Choi, I.-C.; Seok, M.-Y.; He, J.; Lu, Z.; Suh, J.-Y.; Kawasaki, M.; Langdon, T.G.; Jang, J. Nanomechanical behavior and structural stability of a nanocrystalline CoCrFeNiMn high-entropy alloy processed by high-pressure torsion. *J. Mater. Res.* **2015**, *30*, 2804–2815. [CrossRef]

10. Heczel, A.; Kawasaki, M.; Lábár, J.L.; Jang, J.; Langdon, T.G.; Gubicza, J. Defect structure and hardness in nanocrystalline CoCrFeMnNi High-Entropy Alloy processed by High-Pressure Torsion. *J. Alloy. Compd.* **2017**, *711*, 143–154. [CrossRef]

11. Laplanche, G.; Kostka, A.; Horst, O.M.M.; Eggeler, G.; George, E.P.P. Microstructure evolution and critical stress for twinning in the CrMnFeCoNi high-entropy alloy. *Acta Mater.* **2016**, *118*, 152–163. [CrossRef]

12. Stepanov, N.; Tikhonovsky, M.; Yurchenko, N.; Zyabkin, D.; Klimova, M.; Zherebtsov, S.; Efimov, A.; Salishchev, G. Effect of cryo-deformation on structure and properties of CoCrFeNiMn high-entropy alloy. *Intermetallics* **2015**, *59*, 8–17. [CrossRef]

13. Zherebtsov, S.V.; Dyakonov, G.S.; Salem, A.A.; Sokolenko, V.I.; Salishchev, G.A.; Semiatin, S.L. Formation of nanostructures in commercial-purity titanium via cryorolling. *Acta Mater.* **2013**, *61*, 1167–1178. [CrossRef]

14. Zherebtsov, S.V.; Dyakonov, G.S.; Salishchev, G.A.; Salem, A.A.; Semiatin, S.L. The Influence of Grain Size on Twinning and Microstructure Refinement during Cold Rolling of Commercial-Purity Titanium. *Metall. Mater. Trans. A* **2016**, *47*, 5101–5113. [CrossRef]

15. Salishchev, G.A.; Tikhonovsky, M.A.; Shaysultanov, D.G.; Stepanov, N.D.; Kuznetsov, A.V.; Kolodiy, I.V.; Tortika, A.S.; Senkov, O.N. Effect of Mn and V on structure and mechanical properties of high-entropy alloys based on CoCrFeNi system. *J. Alloy. Compd.* **2014**, *591*, 11–21. [CrossRef]

16. Stepanov, N.D.; Shaysultanov, D.G.; Salishchev, G.A.; Tikhonovsky, M.A.; Oleynik, E.E.; Tortika, A.S.; Senkov, O.N. Effect of V content on microstructure and mechanical properties of the CoCrFeMnNiVx high entropy alloys. *J. Alloy. Compd.* **2015**, *628*, 170–185. [CrossRef]

17. Zhilyaev, A.P.; Langdon, T.G. Using high-pressure torsion for metal processing: Fundamentals and applications. *Prog. Mater. Sci.* **2008**, *53*, 893–979. [CrossRef]

18. Stepanov, N.D.; Shaysultanov, D.G.; Chernichenko, R.S.; Yurchenko, N.Y.; Zherebtsov, S.V.; Tikhonovsky, M.A.; Salishchev, G.A. Effect of thermomechanical processing on microstructure and mechanical properties of the carbon-containing CoCrFeNiMn high entropy alloy. *J. Alloy. Compd.* **2017**, *693*, 394–405. [CrossRef]

19. Smallman, R.E.; Westmacott, K.H. Stacking faults in face-centred cubic metals and alloys. *Philos. Mag.* **1957**, *2*, 669–683. [CrossRef]

20. Williamson, G.K.; Hall, W.H. X-ray line broadening from filed aluminium and wolfram. *Acta Metall.* **1953**, *1*, 22–31. [CrossRef]

21. Klimova, M.; Zherebtsov, S.; Stepanov, N.; Salishchev, G.; Haase, C.; Molodov, D.A. Microstructure and texture evolution of a high manganese TWIP steel during cryo-rolling. *Mater. Charact.* **2017**, *132*, 20–30. [CrossRef]

22. Odnobokova, M.; Tikhonova, M.; Belyakov, A.; Kaibyshev, R. Development of S3n CSL boundaries in austenitic stainless steels subjected to large strain deformation and annealing. *J. Mater. Sci.* **2017**, *52*, 4210–4223. [CrossRef]

23. Zherebtsov, S.; Kudryavtsev, E.; Kostjuchenko, S.; Malysheva, S.; Salishchev, G. Strength and ductility-related properties of ultrafine grained two-phase titanium alloy produced by warm multiaxial forging. *Mater. Sci. Eng. A* **2012**, *536*, 190–196. [CrossRef]

24. Tang, Q.H.; Huang, Y.Y.; Huang, Y.Y.; Liao, X.Z.; Langdon, T.G.; Dai, P.Q. Hardening of an $Al_{0.3}CoCrFeNi$ high entropy alloy via high-pressure torsion and thermal annealing. *Mater. Lett.* **2015**, *151*, 126–129. [CrossRef]

25. Shahmir, H.; He, J.; Lu, Z.; Kawasaki, M.; Langdon, T.G. Effect of annealing on mechanical properties of a nanocrystalline CoCrFeNiMn high-entropy alloy processed by high-pressure torsion. *Mater. Sci. Eng. A* **2016**, *676*, 294–303. [CrossRef]

26. Moon, J.; Qi, Y.; Tabachnikova, E.; Estrin, Y.; Choi, W.-M.; Joo, S.-H.; Lee, B.-J.; Podolskiy, A.; Tikhonovsky, M.; Kim, H.S. Deformation-induced phase transformation of $Co_{20}Cr_{26}Fe_{20}Mn_{20}Ni_{14}$ high-entropy alloy during high-pressure torsion at 77 K. *Mater. Lett.* **2017**, *202*, 86–88. [CrossRef]

27. Joo, S.-H.; Kato, H.; Jang, M.J.; Moon, J.; Tsai, C.W.; Yeh, J.W.; Kim, H.S. Tensile deformation behavior and deformation twinning of an equimolar CoCrFeMnNi high-entropy alloy. *Mater. Sci. Eng. A* **2017**, *689*, 122–133. [CrossRef]

28. Jang, M.J.; Ahn, D.-H.; Moon, J.; Bae, J.W.; Yim, D.; Yeh, J.-W.; Estrin, Y.; Kim, H.S. Constitutive modeling of deformation behavior of high-entropy alloys with face-centered cubic crystal structure. *Mater. Res. Lett.* **2017**, *5*, 350–356. [CrossRef]

29. Zherebtsov, S.; Lojkowski, W.; Mazur, A.; Salishchev, G. Structure and properties of hydrostatically extruded commercially pure titanium. *Mater. Sci. Eng. A* **2010**, *527*, 5596–5603. [CrossRef]

30. Salishchev, G.; Mironov, S.; Zherebtsov, S.; Belyakov, A. Changes in misorientations of grain boundaries in titanium during deformation. *Mater. Charact.* **2010**, *61*, 732–739. [CrossRef]

31. Sakai, T.; Belyakov, A.; Kaibyshev, R.; Miura, H.; Jonas, J.J. Dynamic and post-dynamic recrystallization under hot, cold and severe plastic deformation conditions. *Prog. Mater. Sci.* **2014**, *60*, 130–207. [CrossRef]

32. Templeman, Y.; Ben Hamu, G.; Meshi, L. Friction stir welded AM50 and AZ31 Mg alloys: Microstructural evolution and improved corrosion resistance. *Mater. Charact.* **2017**, *126*, 86–95. [CrossRef]

33. Meyers, M.A.; Vöhringer, O.; Lubarda, V.A. The onset of twinning in metals: A constitutive description. *Acta Mater.* **2001**, *49*, 4025–4039. [CrossRef]

34. Hull, D.; Bacon, D.J. *Introduction to Dislocations*; Butterworth-Heinemann: Oxford, UK, 2011; ISBN 9780080966724.

35. Laplanche, G.; Gadaud, P.; Horst, O.; Otto, F.; Eggeler, G.; George, E.P.P. Temperature dependencies of the elastic moduli and thermal expansion coefficient of an equiatomic, single-phase CoCrFeMnNi high-entropy alloy. *J. Alloy. Compd.* **2015**, *623*, 348–353. [CrossRef]

36. Hughes, D.; Hansen, N. Microstructure and strength of nickel at large strains. *Acta Mater.* **2000**, *48*, 2985–3004. [CrossRef]

37. Fu, Z.; Chen, W.; Wen, H.; Zhang, D.; Chen, Z.; Zheng, B.; Zhou, Y.; Lavernia, E.J. Microstructure and strengthening mechanisms in an FCC structured single-phase nanocrystalline $Co_{25}Ni_{25}Fe_{25}Al_{7.5}Cu_{17.5}$ high-entropy alloy. *Acta Mater.* **2016**, *107*, 59–71. [CrossRef]

Article

Characterization on the Microstructure Evolution and Toughness of TIG Weld Metal of 25Cr2Ni2MoV Steel after Post Weld Heat Treatment

Xia Liu [1,2], Zhipeng Cai [1,*], Sida Yang [3], Kai Feng [3,*] and Zhuguo Li [3]

1   Department of Mechanical Engineering, Tsinghua University, Beijing 100084, China; liuxia@shanghai-electric.com
2   Shanghai Electric Power Generation Equipment Co., Ltd., Shanghai 200240, China
3   Shanghai Key Laboratory of Materials Laser Processing and Modification, School of Materials Science and Engineering, Shanghai Jiao Tong University, Shanghai 200240, China; yangsida@sjtu.edu.cn (S.Y.); lizg@sjtu.edu.cn (Z.L.)
*   Correspondence: czpdme@mail.tsinghua.edu.cn (Z.C.); fengkai@sjtu.edu.cn (K.F.); Tel.: +86-21-5474-5878 (K.F.)

Received: 16 January 2018; Accepted: 3 March 2018; Published: 6 March 2018

Abstract: The microstructure and toughness of tungsten inert gas (TIG) backing weld parts in low-pressure steam turbine welded rotors contribute significantly to the total toughness of the weld metal. In this study, the microstructure evolution and toughness of TIG weld metal of 25Cr2Ni2MoV steel low-pressure steam turbine welded rotor under different post-weld heat treatment (PWHT) conditions are investigated. The fractography and microstructure of weld metal after PWHT are characterized by optical microscope, SEM, and TEM, respectively. The Charpy impact test is carried out to evaluate the toughness of the weld. The optical microscope and SEM results indicate that the as-welded sample is composed of granular bainite, acicular ferrite and blocky martensite/austenite (M-A) constituent. After PWHT at 580 °C, the blocky M-A decomposes into ferrite and carbides. Both the number and size of precipitated carbides increase with holding time. The impact test results show that the toughness decreases dramatically after PWHT and further decreases with holding time at 580 °C. The precipitated carbides are identified as $M_{23}C_6$ carbides by TEM, which leads to the dramatic decrease in the toughness of TIG weld metal of 25Cr2Ni2MoV steel.

Keywords: welded rotor; weld metal; impact toughness; PWHT; microstructure evolution

1. Introduction

Steam turbines play an important role in modern power plants. As a result of concerns related to the environmental impact from pollutants and greenhouse gas, the power industry has had to improve thermal efficiency by developing ultra-supercritical (USC) combustion technology, which has led to elevated steam temperatures and continuously increasing demand being placed on their components' performance [1]. The harsher working conditions exert extra stress on all moving parts, demanding higher reliability with much-enhanced mechanical properties for all components [2,3]. Steam turbine rotors are one of the most critical and highly stressed parts in steam turbines, experiencing centrifugal force, torsional force and bending stress. Thus, the material used to manufacture steam turbine rotors needs to have excellent performance, such as high ductility, deep hardenability, high strength, high fatigue strength and creep resistance [4–6]. NiCrMoV refractory steel has been proved to be a suitable material for meeting these requirements of steam turbine rotor materials [7–9]. However, it is difficult to manufacture high-quality steam turbine rotors by forging directly, because of their heavy section and large dimensions. Thus, welded rotors have been widely used, with advantages including lighter weight, higher rigidity, easier manufacturing process, shorter production cycle, lower cost, and excellent performance [10,11].

In practice, for large-scale rotors, multi-layer and multi-pass welding technologies are utilized due to their advantages in terms of normalizing the pre-layer or pre-pass microstructure, thus increasing the ductility and improving the welding quality. The narrow gap-welding method is frequently chosen in the manufacturing of large-scale rotors, in order to join several forged parts with less filler wire consumption and weld deformation. Despite the weld quality of narrow-gap tungsten inert-gas welding (NG-TIG) is better than that of narrow-gap submerged-arc welding (NG-SAW), NG-SAW is generally employed to manufacture heavy section rotors, due to its higher efficiency and lower cost [12–15]. In the manufacturing process of large-scale rotors, NG-TIG is firstly employed for backing weld, and then multi-layer and multi-pass NG-SAW is used to manufacture the welded rotor.

After the welding process, post-weld heat treatment (PWHT) for the welded joints is invariably carried out, and is indispensable for eliminating welding stress and improving comprehensive mechanical properties [16–20]. Tempering is one of the methods for PWHT that is able to release the welding stress and stabilize the microstructure of the welded rotor steel [21–24]. Generally, the welded rotor is heated to a higher temperature with a certain heating speed and then the welded rotor is insulated for a period of time followed by a cooling process with a lower cooling rate in the furnace or air because of the large size and high thickness of the welded rotor [25,26].

For a qualified weld metal, the creep property and fatigue resistance are essential performances, while good impact toughness at room temperature should be also taken into account to ensure reliability during the testing and startup or shut-down of plants. However, the weld metal of NiCrMoV refractory steel is susceptible to tempered embrittlement as a consequence of tempering within a specified temperature range, leading to a dramatic decrease in impact toughness [27]. Salemi et al. [8] investigated the effect of tempering temperature on the mechanical properties and fracture morphology of NiCrMoV steel. They found that the impact energy was improved by increasing the tempering temperature without any evidence of tempered martensite embrittlement. Wang et al. [28] studied the microstructure and impact toughness of 15Cr2Ni3MoW steel after tempering at different temperatures. The results showed that a large amount of stabilized martensite/austenite (M-A) islands appeared after the tempering treatment at 350 °C, while the M-A islands decomposed into precipitated carbide distributing along the grain boundary, which is responsible for the tempering brittleness. As the multi-layer and multi-pass NG-SAW part constitutes the majority of weld metal, most research has placed an emphasis on the microstructure and mechanical properties of NG-SAW parts. Li et al. [29,30] investigated toughness weak point of bainite weld metal of NiCrMoV refractory steel. They found that the carbon-rich areas containing much more M-A blocks in the incomplete phase change zone of layers between two adjacent beads were responsible for the decrease of toughness. Furthermore, the influence of the M-A blocks on the weld toughness was closely related to the distribution, pattern and dimension. Zhang et al. [31] concluded that the precipitation and aggregation of carbides was disadvantageous to the toughness. However, little study has been focused on the microstructure and toughness of TIG backing weld parts, which is an indispensable part and contributes to the toughness of whole weld metal significantly. In this study, the toughness of the TIG weld metal of 25Cr2Ni2MoV steel is evaluated by Charpy impact test. The fractography and microstructure evolution under different PWHT conditions are characterized by optical microscope (OM), scanning electron microscopy (SEM) and transmission electron microscopy (TEM) to reveal the fracture mechanism.

2. Materials and Methods

The base metal used in this study is 25Cr2Ni2MoV steel and the filler metal is 2.5% Ni low-alloy steel. The chemical compositions of the base metal (BM) and the filler metals (FM) are listed in Table 1. The 25Cr2Ni2MoV steel is welded by TIG and the welding parameters are listed in Table 2. The dimensions and position of samples used in this study are shown in Figure 1. All the samples were carefully taken from the middle of the weld joint. After the welding process, the samples were immediately subjected to PWHT in the furnace with different heat treatment parameters, including hold times of 5 h, 10 h and 20 h at temperature of 580 °C. The detailed PWHT process is shown in Figure 2.

Table 1. The composition of base metal and filler metals (BM: base metal; FM: filler metal).

Elements	C	Si	Mn	P	S	Cr	Mo	Ni	V	Cu	Fe
BM	0.24	0.05	0.20	0.004	0.003	2.36	0.76	2.23	0.07	<0.17	Balance
FM	0.094	0.23	1.42	<0.010	0.002	0.56	0.52	2.20	<0.01	0.059	Balance

Table 2. The detailed welding process parameters.

Parameters	Wire Diameter (mm)	Current (A)	Voltage (V)	Speed (mm/min)	Gas
-	1.0	270	11.2	70	Ar 99.999% 20/30

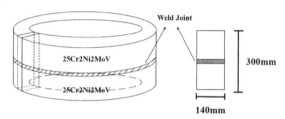

Figure 1. The schematic figure showing the dimensions and position of samples used for study.

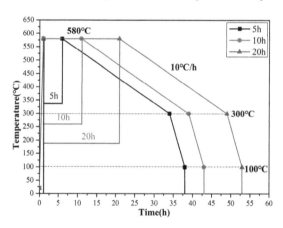

Figure 2. The schematic diagram of thermal cycles of PWHT for weld metal with different hold times of 5 h, 10 h and 20 h at 580 °C.

The Charpy impact test was conducted at room temperature to evaluate the impact toughness of the weld metal treated by different PWHT process according to the ASTM E23-2008 standard procedure. Standard Charpy V-notch (CVN) samples with dimension: length × width × thickness = 55 mm × 10 mm × 10 mm, were cut from the weld metal and Charpy impact tests were performed by pendulum impact tester of PTM2200-D1 impact testing system (SUNS Technology Stock Co. Ltd., Shenzhen, China). To ensure the reliability of the results, each kind of sample was tested five times. After impact tests, the fractography for the impact fractures of different samples were observed by scanning electron microscope (SEM).

In order to analyze the microstructure evolution and mechanism of impact fracture of weld metal with different PWHT, OM, SEM and TEM were conducted to observe the microstructure. The weld metal samples without PWHT, as well as those treated by different PWHTs, were mounted and mechanically polished according to standard metallographic procedures for microstructure observation. The samples were etched by 4% Nitric acid and alcohol solution, then the microstructure of these specimens was analyzed by OM and SEM. The fine microstructure of weld metal was further investigated by transmission electron microscope (TEM) on FEI Tecnai G2 F20 FEG-TEM (FEI, Hillsboro, OR, USA) equipped with an energy-dispersive X-ray spectrometer (EDS) at 200 kV. The TEM samples were prepared by mechanical grinding and polishing followed by thinning to an electron-transparent thickness by low-energy Ar ion milling and subsequent ion polishing using Gatan PIPS, model 691 (Gatan, Inc., Pleasanton, CA, USA).

3. Results and Discussion

3.1. Microstructure Analysis

Figure 3 shows the representative metallographic images of the as-welded sample and the samples with PWHT with different hold times. SEM was also carried out to further observe the microstructure of the weld metal, and the results are shown in Figure 4. Figures 3a and 4a,b show the microstructure of the as-welded sample. The microstructure of as-welded sample is mainly composed of granular bainite, (GB) and a certain amount of acicular ferrite (AF). In addition, a large number of blocky M-A (martensite/austenite) are homogeneously distributed in the bainite and ferrite matrix. With further observation, it can be seen that blocky M-A with size of less than 10 μm is not decomposed, and no carbide precipitation is observed. Whereas, after PWHT, the blocky M-A decomposes into ferrite and carbide, and the decomposition continues with the increase of the holding time. In addition, the size of blocky M-A also increases with the increase of holding time. As shown in Figures 3b and 4c,d, after PWHT with a holding time of 5 h, the blocky M-A decomposes slightly. After PWHT with holding time of 10 h, the blocky M-A blocks decompose with small granular carbides distributing around the blocky M-A, as can be seen from Figures 3c and 4e,f. For PWHT with holding time of 20 h, the blocky M-A dramatically decomposes and a large amount of thick carbides precipitate around the boundary. Furthermore, some of the carbides have become spheroidizing.

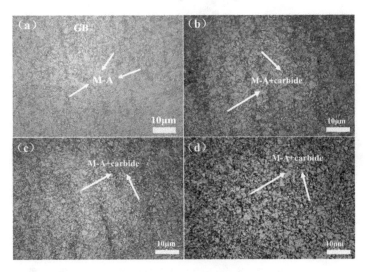

Figure 3. Optical microscopy (OM) images of the as-welded sample (**a**) and the samples after PWHT with different hold time at 580 °C: (**b**) 5 h, (**c**) 10 h, and (**d**) 20 h.

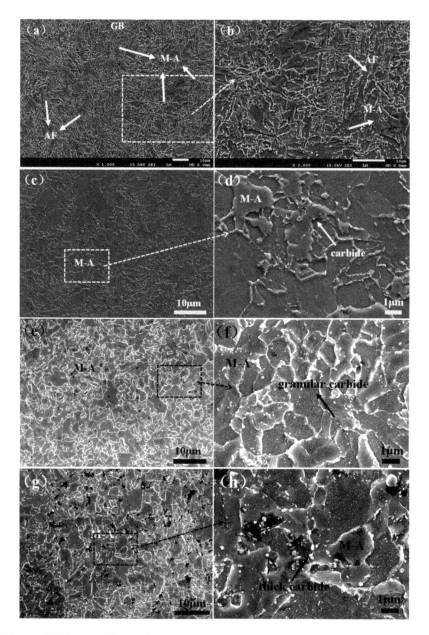

Figure 4. SEM images of the as-welded sample (**a,b**) and the samples after PWHT with different hold times at 580 °C: (**c,d**) 5 h, (**e,f**) 10 h, and (**g,h**) 20 h.

3.2. Charpy Impact Results

The Charpy impact test was carried out at room temperature to evaluate the toughness of the weld metal with different hold times at 580 °C in PWHT. The absorbed energies in the Charpy impact test versus hold times, as well as the deviation from the average value, are shown in Figure 5. It can be seen that the as-welded metal had the highest absorbed energy, at around 177.4 J. The absorbed energies of the weld metal dramatically decrease after PWHT and that gradually decreases with the hold time from 5 h to 20 h. It can be expected that the absorbed energies of the weld metal will continue to decrease with increasing of the hold time in PWHT to a certain extent. Specifically, the absorbed energy of weld metal with a holding time of 5 h is 69.0 J. When the hold time is 10 h, the absorbed energy is about 54.4 J. The weld metal with holding time of 20 h has the lowest absorbed energy of 38.5 J, which is more than four times lower than the as-welded one. The dramatic decrease of the toughness of the weld metal can be attributed to the microstructure evolution caused by the different PWHT processing. Associating the microstructure evolution of weld metal after PWHT with the Charpy impact test results, it is concluded that the precipitation of carbides along the boundaries of the blocky M-A has a great influence on the toughness of the sample.

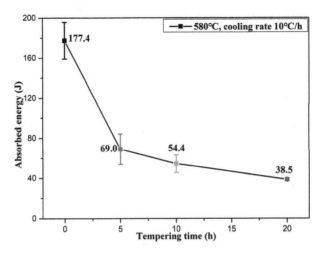

Figure 5. The absorbed energies in Charpy impact test of the as-welded metal and the samples with PWHT with different hold time of 5 h, 10 h and 20 h at 580 °C.

3.3. Fractography Characterization

The morphology and cross-section of the fracture surface of the samples with different holding times were characterized by SEM to investigate the mechanism of fracture. Figure 6a,b presents the SEM fractograph of the as-welded sample. The presence of dimples is obvious at the fracture surface of this sample, indicating a ductile fracture mechanism. However, in general, the fracture mechanism transforms from ductile fracture to the combined mechanism, with both brittle fracture and ductile fracture after PWHT. Specifically, the fractography of weld metal after 5 h holding time shows characteristics of quasi cleavage fracture with numbers of cleavage facets as shown in Figure 6c. Figure 6d shows the unbroken blocky M-A blocks located along the fracture surface, and the size of cleavage fracture surface is also in agreement with that of the blocky M-A blocks, indicating that the boundary of blocky M-A blocks provides a crack propagation path when the sample is impacted. When the holding time is 10 h, the fracture morphology of the sample shows the characteristics of quasi cleavage fracture containing a large number of cleavage facets. However, compared to the sample with 5 h holding time, there are a number of dimples around the cleavage facets, which reveals that the initiation of cleavage facets is possibly the micro crack around the dimples. As shown in Figure 6f, the crack propagation path is still along the boundary of blocky M-A blocks, which is similar to the mechanism of sample with 5 h holding time. When the holding time increases to 20 h, the fracture morphology of the sample is still that of a quasi cleavage fracture with lots of cleavage facets. More dimples can easily be found on the fracture surface when compared to the sample with 10 h holding time. However, carbide particles can be clearly seen in the dimples, as shown in

Figure 6g. From this result, it can be inferred that the initiation of micro cracks, which grow into cleavage facets, is attributed to carbide particles precipitated around the blocky M-A. The cross-section of the fracture surface (Figure 6h) shows that the crack propagation path is also along the boundary of blocky M-A. In association with the OM and SEM microstructure characterization, it can be concluded that the carbide particle precipitated around the boundary of blocky M-A grows larger and the number becomes more with the increase of the holding time. Consequently, there is more potential initiation of micro cracks in the weld metal, which is responsible for the decrease of toughness after PWHT.

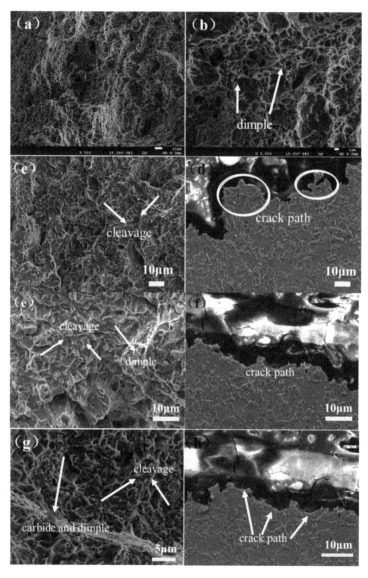

Figure 6. The SEM fractograph of Charpy samples with different PWHT hold times at 580 °C: (**a,b**) as-welded sample; (**c,d**) 5 h, (**e,f**) 10 h, and (**g,h**) 20 h. The typical evidence of the brittle fracture and ductile fracture are pointed out by arrows.

3.4. TEM Characterization

In order to further investigate the fine microstructure and the precipitated carbides around blocky M-A, TEM was conducted, and the results are presented as Figures 7–9. As shown in Figure 7, there is no carbide precipitated from the blocky M-A blocks in the as-welded sample. The selected area electron diffraction (SAED) pattern as shown in Figure 7b reveals that the substrate is mainly composed of ferrite. Figure 8 displays the typical TEM images of the samples after PWHT with a hold time of 20 h. It can be obviously seen that a large amount of needle-like and blocky carbides with dimensions of about 100–200 nm are precipitated from the M-A blocks and ferrite substrate. According to the SAED patterns of precipitated carbides and HR-TEM result (in Figure 9), the precipitated carbides are identified as M23C6 carbides. Associating the OM, SEM and TEM microstructure characterization with the dramatic decrease in toughness of TIG weld metal of 25Cr2Ni2MoV steel, it is concluded that the precipitation of M23C6 carbides leads to the significant decrease in the toughness.

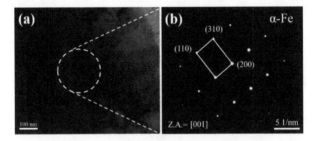

Figure 7. (a) Bright field image of the as-welded sample and (b) corresponding SAED patterns.

Figure 8. The typical TEM images of the samples after PWHT with hold times of 20 h (a,b), and corresponding selected area electron diffraction (SAED) patterns of precipitated carbide (c).

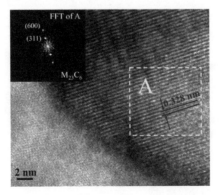

Figure 9. HR-TEM images of the boundary between the precipitated carbides and the substrate, with the corresponding FFT results shown in the inset.

4. Conclusions

The microstructure evolution, toughness and the fracture morphology of the weld metal of 25Cr2Ni2MoV steel treated with different PWHT holding times have been studied. The results obtained can be summarized as follows:

1. The microstructure of the as-welded sample is composed of granular bainite and acicular ferrite. In addition, a large number of blocky M-A is homogeneously distributed in the bainite and ferrite matrix. The as-welded sample exhibits the highest toughness of 177.4 J.

2. After PWHT at 580 °C, the blocky M-A decomposes into ferrite and carbides, and the carbide particles are distributed along the boundary of the blocky M-A. As the holding time in PWHT is increased, more blocky M-A decomposes, and both the number and size of the precipitated carbides grow.

3. As the precipitation of carbide particles distributes along the boundary of blocky M-A, the toughness of the weld metal dramatically decreases after PWHT. Both the number and size of precipitated carbide particles grow with the increase in holding time during PWHT. Since the initiation of micro cracks is attributed to the precipitated carbide particles and the crack propagation path is along the boundary of the blocky M-A, the sample with the longer holding time during PWHT has the lower toughness. Therefore, a shorter holding time during PWHT is suggested in order to achieve higher toughness for TIG-welded 25Cr2Ni2MoV steel.

4. The precipitated carbide particles are identified as M23C6 carbides by TEM characterization, which is responsible for the dramatic decrease in the toughness of the TIG weld metal of 25Cr2Ni2MoV steel.

Acknowledgments: Financial support from "Chen Guang" project Shanghai Municipal Education Commission and Shanghai Education Development Foundation (Grant Number 13CG07), "Chenxing" young scholar project of Shanghai Jiao Tong University (Grant Number 14X100010017) is acknowledged.

Author Contributions: Xia Liu performed most of experiments and wrote this manuscript; Zhipeng Cai and Kai Feng conceived and designed the experiments; Sida Yang performed the Charpy impact experiments; Zhuguo Li analyzed the data.

Conflicts of Interest: The authors declare no conflict of interest.

References

1. Liu, X.J.; Kong, X.B.; Hou, G.L.; Wang, J.H. Modeling of a 1000 MW power plant ultra super-critical boiler system using fuzzy-neural network methods. *Energy Convers. Manag.* **2013**, *65*, 518–527. [CrossRef]

2. Kosman, W. Thermal analysis of cooled supercritical steam turbine components. *Energy* **2010**, *35*, 1181–1187. [CrossRef]

3. Huo, W.; Li, J.; Yan, X. Effects of coolant flow rates on cooling performance of the intermediate pressure stages for an ultra-supercritical steam turbine. *Appl. Therm. Eng.* **2014**, *62*, 723–731. [CrossRef]

4. Chen, C.Y.; Yen, H.W.; Kao, F.H.; Li, W.C.; Huang, C.Y.; Yang, J.R.; Wang, S.H. Precipitation hardening of high-strength low-alloy steels by nanometer-sized carbides. *Mater. Sci. Eng. A* **2009**, *499*, 162–166. [CrossRef]

5. Wang, B.; Liu, Z.Y.; Zhou, X.G.; Wang, G.D.; Misra, R.D.K. Precipitation behavior of nanoscale cementite in hypoeutectoid steels during ultra fast cooling (UFC) and their strengthening effects. *Mater. Sci. Eng. A* **2013**, *575*, 189–198.

6. Liu, P.; Lu, F.; Liu, X.; Ji, H.; Gao, Y. Study on fatigue property and microstructure characteristics of welded nuclear power rotor with heavy section. *J. Alloys Compd.* **2014**, *584*, 430–437. [CrossRef]

7. Hong, S.; Lee, J.; Lee, B.J.; Kim, H.S.; Kim, S.K.; Chin, K.G.; Lee, S. Effects of intergranular carbide precipitation on delayed fracture behavior in three Twinning Induced Plasticity (TWIP) steels. *Mater. Sci. Eng. A* **2013**, *587*, 85–99. [CrossRef]

8. Salemi, A.; Abdollah-zadeh, A. The effect of tempering temperature on the mechanical properties and fracture morphology of a NiCrMoV steel. *Mater. Charact.* **2008**, *59*, 484–487. [CrossRef]

9. Tanaka, Y.; Azuma, T.; Yaegashi, N. Isothermal aging test results (up to 100,000 h) of NiCrMoV steels for low pressure steam turbine. *Int. J. Pressure Vessels Pip.* **1994**, *59*, 71–81. [CrossRef]

10. Zhu, M.L.; Wang, D.Q.; Xuan, F.Z. Effect of long-term aging on microstructure and local behavior in the heat-affected zone of a Ni-Cr-Mo-V steel welded joint. *Mater. Charact.* **2014**, *87*, 45–61. [CrossRef]

11. Zhu, M.L.; Xuan, F.Z. Correlation between microstructure, hardness and strength in HAZ of dissimilar welds of rotor steels. *Mater. Sci. Eng. A* **2010**, *527*, 4035–4042. [CrossRef]

12. Bhamji, I.; Preuss, M.; Threadgill, P.L.; Moat, R.J.; Addison, A.C.; Peel, M.J. Linear friction welding of AISI 316L stainless steel. *Mater. Sci. Eng. A* **2010**, *528*, 680–690. [CrossRef]

13. Hokamoto, K.; Nakata, K.; Mori, A.; Tsuda, S.; Tsumura, T.; Inoue, A. Dissimilar material welding of rapidly solidified foil and stainless steel plate using underwater explosive welding technique. *J. Alloys Compd.* **2009**, *472*, 507–511. [CrossRef]

14. Torkamany, M.J.; Tahamtan, S.; Sabbaghzadeh, J. Dissimilar welding of carbon steel to 5754 aluminum alloy by Nd:YAG pulsed laser. *Mater. Des.* **2010**, *31*, 458–465. [CrossRef]

15. Beidokhti, B.; Kokabi, A.H.; Dolati, A. A comprehensive study on the microstructure of high strength low alloy pipeline welds. *J. Alloys Compd.* **2014**, *597*, 142–147. [CrossRef]

16. Villaret, V.; Deschaux-Beaume, F.; Bordreuil, C.; Rouquette, S.; Chovet, C. Influence of filler wire composition on weld microstructures of a 444 ferritic stainless steel grade. *J. Mater. Process. Technol.* **2013**, *213*, 1538–1547. [CrossRef]

17. Gunaraj, V.; Murugan, N. Prediction of heat-affected zone characteristics in submerged arc welding of structural steel pipes. *Weld. J.* **2002**, *81*, 45–53.

18. Avazkonandeh-Gharavol, M.H.; Haddad-Sabzevar, M.; Haerian, A. Effect of chromium content on the microstructure and mechanical properties of multipass MMA, low alloy steel weld metal. *J. Mater. Sci.* **2009**, *44*, 186–197. [CrossRef]

19. Shekhter, A.; Kim, S.; Carr, D.G.; Croker, A.B.L.; Ringer, S.P. Assessment of temper embrittlement in an ex-service 1Cr-1Mo-0.25V power generating rotor by Charpy V-Notch testing, KIc fracture toughness and small punch test. *Int. J. Pressure Vessels Pip.* **2002**, *79*, 8–10. [CrossRef]

20. Wu, Q.; Lu, F.; Cui, H.; Liu, X.; Wang, P.; Tang, X. Role of butter layer in low-cycle fatigue behavior of modified 9Cr and CrMoV dissimilar rotor welded joint. *Mater. Des.* **2014**, *59*, 165–175. [CrossRef]

21. Wu, Q.; Lu, F.; Cui, H.; Ding, Y.; liu, X.; Gao, Y. Microstructure characteristics and temperature-dependent high cycle fatigue behavior of advanced 9% Cr/CrMoV dissimilarly welded joint. *Mater. Sci. Eng. A* **2014**, *615*, 98–106. [CrossRef]

22. Farabi, N.; Chen, D.L.; Li, J.; Zhou, Y.; Dong, S.J. Microstructure and mechanical properties of laser welded DP600 steel joints. *Mater. Sci. Eng. A* **2010**, *527*, 1215–1222. [CrossRef]

23. Fattahi, M.; Nabhani, N.; Vaezi, M.R.; Rahimi, E. Improvement of impact toughness of AWS E6010 weld metal by adding TiO_2 nanoparticles to the electrode coating. *Mater. Sci. Eng. A* **2011**, *528*, 8031–8039. [CrossRef]

24. Gaffard, V.; Gourgues-Lorenzon, A.F.; Besson, J. High temperature creep flow and damage properties of 9Cr1MoNbV steels: Base metal and weldment. *Nucl. Eng. Des.* **2005**, *235*, 2547–2562. [CrossRef]

25. Karthikeyan, T.; Thomas Paul, V.; Saroja, S.; Moitra, A.; Sasikala, G.; Vijayalakshmi, M. Grain refinement to improve impact toughness in 9Cr-1Mo steel through a double austenitization treatment. *J. Nucl. Mater.* **2011**, *419*, 256–262. [CrossRef]

26. Zhao, M.C.; Huang, X.F.; Atrens, A. Role of second phase cementite and martensite particles on strength and strain hardening in a plain C-Mn steel. *Mater. Sci. Eng. A* **2012**, *549*, 222–227. [CrossRef]

27. Klueh, R.; Hashimoto, N.; Maziasz, P. Development of new nano-particle-strengthened martensitic steels. *Scr. Mater.* **2005**, *53*, 275–280. [CrossRef]

28. Wang, B.; Yi, D.; Liu, H.; Wu, B.; Yuan, J. Effect of tempering temperature on microstructure and mechanical properties of 15Cr2Ni3MoW steel. *Trans. Mater. Heat Treat.* **2008**, *29*, 73–77.

29. Li, Y.; Wang, L.; Wu, J.; Cai, Z.; Pan, J.; Liu, X.; Qiao, S.; Ding, Y.; Shen, H. Determination of toughness weak points of bainite weld metal of NiCrMoV refractory steel. *J. Mech. Eng.* **2013**, *49*, 83–88. [CrossRef]

30. Li, Y.; Cai, Z.; Pan, J.; Liu, X.; Wang, P.; Huo, X.; Shen, H. Research on toughness weak points of joints of NiCrMoV regractory steel for manufacturing steam turbine rotor. *Trans. China Weld. Inst.* **2014**, *35*, 73–76.
31. Zhang, B.; Cai, Z.; Wu, J.; Pan, J.; Liu, X.; Qian, S.; Xu, X.; Huo, X.; Shen, H. Influence of tempering parameter on aged toughness of welding joint of Ni-Cr-Mo-V turbine rotor steel. *Electr. Weld. Mach.* **2013**, *43*, 27–32.

Article

Precipitation, Recrystallization, and Evolution of Annealing Twins in a Cu-Cr-Zr Alloy

Xiaobo Chen [1,2], Feng Jiang [1,2,3,4,*], Jingyu Jiang [1,2], Pian Xu [2,3], Mengmeng Tong [1,2] and Zhongqin Tang [2,3]

1   School of Materials Science and Engineering, Central South University, Changsha 410083, China; rewillcsu@hotmail.com (X.C.); csu_msejiang@hotmail.com (J.J.); csu_tong@hotmail.com (M.T.)
2   Key Laboratory of Ministry of Education for Non-ferrous Materials Science and Engineering, Central South University, Changsha 410083, China; xupian@hotmail.com (P.X.); csu_tzq816@hotmail.com (Z.T.)
3   Light Alloy Research Institute, Central South University, Changsha 410083, China
4   Science and Technology on High Strength Structural Materials Laboratory, Changsha 410083, China
*   Correspondence: jfeng2@csu.edu.cn; Tel.: +86-731-8887-7693

Received: 14 February 2018; Accepted: 27 March 2018; Published: 1 April 2018

Abstract: In this paper, the precipitation, recrystallization, and evolution of twins in Cu-Cr-Zr alloy strips were investigated. Tensile specimens were aged at three different temperatures for various times so as to bring the strips into every possible aging condition. The results show that the appropriate aging parameter for the 70% reduced cold-rolled alloy strips is 723 K for 240 min, with a tensile strength of 536 MPa and an electrical conductivity of 85.3% International Annealed Copper Standards (IACS) at the peak aged condition. The formation of fcc (face-centered cubic) ordered Cr-rich precipitates (β') is an important factor influencing the significant improvement of properties near the peak aged condition. In terms of crystallographic orientation relationships, there are basically two types of β' precipitates in the alloy. Beyond the Cr-rich precipitates ($\beta'(\text{I})$) formed during the early aging stages, which mimic a cube-on-cube orientation relationship (OR) with the matrix, another Cr-rich precipitate ($\beta'(\text{II})$) is observed in the peak aged condition. $\beta'(\text{II})$ is coherent with the matrix, with the following ORs: $[111]_{\beta'(\text{II})}//[100]_{Cu}$, $\{02\text{-}2\}_{\beta'(\text{II})}//\{02\text{-}2\}_{Cu}$ and $[011]_{\beta'(\text{II})}//[211]_{Cu}$, $\{200\}_{\beta'(\text{II})}//\{\text{-}111\}_{Cu}$. These precipitates have a strong dislocation and grain boundary pinning effect, which hinder the dislocation movement and crystal boundary migration, and eventually delay recrystallization and enhance the recrystallization resistance of the peak aged strips. During the subsequent annealing process, the transition phase β' gradually loses the coherence mismatch and grows into a larger equilibrium phase of chromium with a bcc (body-centered cubic) structure (β), resulting in the reduction of the pinning effect to dislocations and sub-grains, so that recrystallization occurs. Annealing twins are formed during the recrystallization process to release the deformation energy and to reduce the drive force for interface migration, eventually hindering grain growth.

Keywords: Cu-Cr-Zr; precipitation; orientation relationship; recrystallization; annealing twins

---

1. Introduction

Copper and its alloys are widely used in automobile, electronic, and electric power industries due to their high strength, high thermal and electrical conductivity, and good ductility, as well as because they can be easily shaped [1–6]. To satisfy applications in the frame materials in large-scale integrity circuits, besides the requirement of high strength and high electrical conductivity, excellent recrystallization resistance is also needed. Cu-Cr-Zr might be able to satisfy these requirements. Extensive literature is available on the microstructures, physical properties, and mechanical properties of Cu-Cr-Zr alloys [6–13]. From the literature, cold working and aging treatment are considered to be the two most important materials-processing techniques for Cu-Cr-Zr alloys. Fiber and banded structures can be formed during the cold work process, resulting in an increase of strength and hardness. The strength and electrical conductivity can be further improved via aging treatment through the formation of second phase precipitation. The precipitation of binary Cu-Cr alloys is well studied. The main strengthening phase formed during the aging process is the equilibrium phase of chromium (β) with a body-centered cubic structure (bcc, a = 0.2895 nm). No transition phase is observed at the early aging stage [14,15]. There is a general agreement about the fact that a transition phase β′ is formed during the ternary Cu-Cr-Zr alloy aging process [16,17]. Nevertheless, the structure and orientation of the precipitated phase has not yet formed a consistent theory. Tang et al. [16] reported that, for Cu-Cr-Zr-Mg alloys, the peak hardness was associated with the fine scale of an ordered compound, possibly of the Heusler type, with the suggested composition of CrCu$_2$ (Zr,Mg) and a face-centered cubic (fcc) crystal structure with a large unit cell containing 8 Cu, 4 Cr, and 4 Zr or 4 Mg. This unit cell can be regarded as containing 8 bcc sub-cells with each housing a Cu atom at its center and Cr, Zr, or Mg atoms alternatively occupying the corners. This Heusler type (CrCu$_2$ (Zr,Mg)) phase was also reported by Liu et al. [18], Qi et al. [19], and Su et al. [20]. Using multi-angle electron diffraction, Batra et al. [21] proved the existence of a transition fcc structure Cr-rich phase with a cube-on-cube OR during the early aging stage. This kind of fcc structure was also observed by Xia et al. [22] and Chen et al. [23]. Moreover, other precipitation phases have also been observed, like Cu$_4$Zr [16] or Cu$_{51}$Zr$_{14}$ [24], and some Zr and Fe segregations along Cu/Cr interfaces [25] have also been reported. As discussed above, the transition phases have not yet been elucidated. As a result, it is of great importance to elucidate the structure of the precipitates, especially the structure of the phases formed during peak aged conditions.

A high operating temperature can be achieved for the application of frame materials in large-scale integrity circuits, making the recrystallization resistance of great importance for the successful application in the frame materials. However, only very few reports are available to discuss the recrystallization resistance behavior as well as the mechanism of Cu-Cr-Zr alloys. Liu et al. [18] studied the aging process of Cu-Cr-Zr-Mg alloys prepared via the rapid solidification method. Recrystallization occurred and the precipitates kept coarsening when aged at 723 K for 240 h. Su et al. [26] reported that when the temperature exceeded 773 K, recrystallization occurred in a 45% cold-rolled Cu-Cr-Zr-Mg alloy, while the precipitates coarsened at the same time. These reports clearly suggest that the recrystallization resistance properties of Cu-Cr-Zr alloys are related to the coarsening of the precipitates. This is necessary to elucidate the evolution of the precipitation as well as the microstructure in order to facilitate the preparation of highly recrystallization-resistant Cu-Cr-Zr alloys.

In this work, we studied the evolution of precipitation, recrystallization, and twins during the heat treatment process in a Cu-Cr-Zr alloy. We identified the crystal structure and the orientation relationships of the strengthening phases at the peak aged condition. In addition, the corresponding precipitation strengthening effect and recrystallization resistance mechanism were discussed.

2. Experiments

The chemical composition of the hot-rolled Cu-Cr-Zr alloy employed in this work is as follows (wt %): 0.80-Cr, 0.20-Zr, and Cu balance (Bal.). The initial hot-rolled sheet was 50 mm in thickness. After solution heat treatment at 1253 K for 60 min and after quenching with water, the sheet was cold-rolled to 1.5 mm with a reduction ratio of about 70%. The cold-rolled sheet was isothermally aged at 673 K, 723 K, and 773 K for various lengths of time. Tensile strength and electrical conductivity measurements were carried out to determine the best aging condition. Afterwards, the best aged sample was treated by stabilization annealing from 573 K to 973 K with a temperature interval of 50 K for 60 min.

Tensile tests were performed on a CSS-44100 electronic universal testing machine (Sinotest Equipment Co. Ltd., Changchun, China) and carried out at room temperature with a tensile speed of 2 mm/min. Hardness (HB) was tested with an HBE-3000 type digital Brinell hardness (Guangzhou material Testing Machine Factory, Guangzhou, China) with a 25-kg load and a 30-s loading time. The electrical conductivity was measured via an eddy current conductivity meter (Xiamen Xin Bote science and Technology Co. Ltd., Xiamen, China) under a work frequency of 60 kHz. The resistivity was calculated and transformed into electrical conductivity according to International Annealed Copper Standards (IACS).

Specimens for optical microscopy (OM) observations were polished and etched in a solution of 10% potassium dichromate, 5% sulfuric acid, and 85% distilled water, and performed on a LEICA EC3 optical microscope (Nanjing Jiangnan Novel Optics Co. Ltd., Nanjing, China) equipped with a digital camera. Samples for transmission electron microscopy (TEM) observations were prepared via double jet electropolishing techniques. The electrolyte consisted of 30% nitric acid in methanol and the solution was maintained at a temperature between 243 K and 253 K. The TEM images and selected area electron diffraction (SAED) were taken by an FEI Tecnai G2 20 transmission electron microscope (FEI, Hillsboro, OR, USA) operating at 200 kV. TEM samples from all conditions were prepared from the center of the tensile or hardness test samples in a direction parallel to the rolling direction. The aged or annealed samples were cooled in air. To investigate areas of the partly recrystallized samples, the techniques of electron backscatter diffraction (EBSD) in a scanning electron microscopy (SEM) (FEI, Hillsboro, OR, USA) were applied. The local orientation measurements were investigated by a Sirion 200 Field-Emission Scanning Electron Microscope (FEI, Hillsboro, OR, USA). The EBSD measurements were carried out at an accelerating voltage of 20 kV and a scan step of 0.2 μm. The results were analyzed via orientation image microscopy (OIM) analysis software.

3. Results

3.1. Properties

Figure 1 shows the influence of the aging time on both the tensile strength (Figure 1a) and electrical conductivity (Figure 1b) of the studied alloy. It can be observed that tensile strength and electrical conductivity increase rapidly during the early stage of aging as the second phase particles rapidly form in the matrix. It can also be seen that the time required to reach peak strength decreases with the increasing aging temperature (Figure 1a). On the other hand, it can also be observed that, when the aging temperature is increased to 773 K, the tensile strength in the peak aged condition is lower than that of aging at 723 K. As shown in Figure 1b, the electrical conductivity can reach to above 80% IACS, when aged at 723 K or 773 K. An excellent combination of those properties, such as a tensile strength of 536 MPa and an electrical conductivity of 85.3% IACS, is obtained in the Cu-Cr-Zr alloy aged at 723 K for 240 min (minimum). The most suitable aging condition is considered to be one where good combinations of strength and conductivity can be obtained after aging. As a result, the optimized aging treatment for this alloy is 723 K for 240 min.

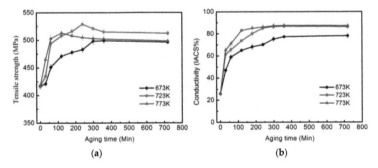

Figure 1. (a) Tensile strength and (b) electrical conductivity of the Cu-Cr-Zr alloy as a function of aging time at various temperatures.

Figure 2 shows the hardness of the peak aged samples after annealing at different temperatures for 60 min. The hardness in the peak aged condition is 163 HB. When the annealing temperature is lower than 773 K, the hardness of the alloy can be kept at a high value. The hardness of the alloy begins to decrease rapidly when the annealing temperature is above 773 K. After annealing at 823 K, the decrease in hardness is significant, and the hardness is 143 HB, about 87.4% of the unannealed sample. The hardness of the alloy continues to decrease as the temperature increases. When the annealing temperature increases to 873 K, the hardness is 131 HB, just about 80.4% of the unannealed sample.

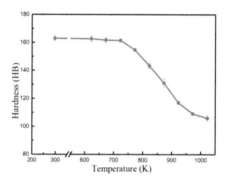

Figure 2. The hardness of the Cu-Cr-Zr alloy as a function of the peak age samples after annealing at different temperatures for 60 min.

3.2. Microstructure

Figure 3 shows OM micrographs of differently treated Cu-Cr-Zr alloys. Typical cold rolling deformation characteristics are still retained in the peak aged samples, as shown in Figure 3a. The grains are elongated and wavy along the rolling direction. Figure 3b,c show the structure of the peak aged samples after annealing at different temperatures. Clear recrystallized grains are observed in the samples after annealing at 873 K, as shown in Figure 3b. After annealing at 973 K, a full recrystallized structure is observed, as shown in Figure 3c. The grain size is uniform, with an average grain size of about 10 μm, and no obvious grain growth is observed.

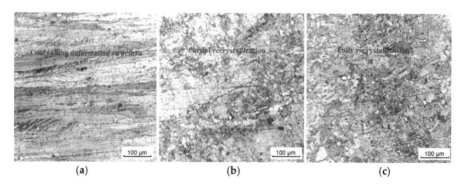

(**a**)   (**b**)   (**c**)

Figure 3. Optical microscopy (OM) micrographs of Cu-Cr-Zr alloys with 240-min-aged initial states at (**a**) 723 K, and 60-min-annealed states at different temperatures of (**b**) 873 K and (**c**) 973 K.

Figure 4 shows the bright-field (BF) image of Cu-Cr-Zr alloys in different heat treatment conditions. The grains are deformed to different degrees after heavy cold rolling, resulting in the formation of different deformation zones. A large number of dislocations are intertwined to form dislocation walls or dislocation transition zones. There are generally two types of boundaries that form and evolve via heavy cold rolling. These boundaries have been termed geometrically necessary boundaries (GNBs) and incidental dislocation boundaries (IDBs), where GNBs form between regions of different strain patterns to accommodate the accompanying difference in lattice rotation, and IDBs form via the random trapping of dislocations [27]. After the aging treatment, the dislocations are rearranged to form small dislocation cells, and the dislocation cells aggregate in the dislocation cell walls, resulting in the decrease of the dislocation density inside the dislocation cells. The IDBs in the local shear zone start to evolve into small angle sub-grains with a size of 0.2–0.5 μm, as shown in Figure 4a. After aging at 723 K for 240 min, when the alloy is annealed at a temperature higher than 773 K, the sub-grains are clearly changed. After annealing at 823 K, well-developed sub-grains with a size of 0.2–0.5 μm can be observed, accompanied by the weakening of GNBs and IDBs in the shear deformation zone (Figure 4b). After annealing at 873 K, recrystallized grains are observed and the size is slightly larger than that of the sub-grains in Figure 4b. When the annealing temperature increases to 973 K, the grain size in the alloy significantly grows and grain boundaries become straight and smooth (Figure 4d).

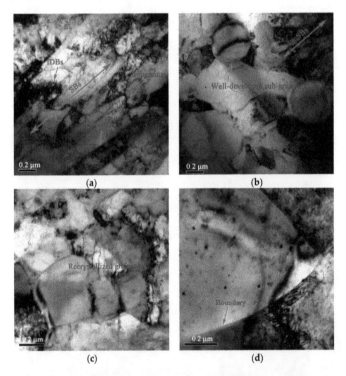

Figure 4. Bright-field (BF) images of Cu-Cr-Zr alloys with 240-min-aged initial states at (**a**) 723 K, and 60-min-annealed states at different temperatures of (**b**) 823 K, (**c**) 873 K, and (**d**) 973 K.

Figure 5 shows the BF and corresponding SAED patterns of the precipitates formed during aging. When aged at 723 K for 30 min, very fine precipitates are observed with a weak contrast. According to high-magnification near-double beam observation, the precipitated phase entails spherical particles with a size of about 2 nm, and there is a double leaf petal-like contrast, which suggests that the precipitate phase is coherent with the matrix, as shown in Figure 5a. Figure 5b shows the structure of the alloy after aging at 723 K for 240 min. A large number of precipitates are observed, significantly enhancing the contrast. The size of the precipitates is slightly larger, about 4 nm. A non-contrast line perpendicular to the manipulated vector is also observed, indicating that there is a coherent relationship between the precipitated particles and the matrix. From the SAED of the sample aged for 30 min, a few superlattice spots can be observed in Figure 4c,d (marked by a triangle arrow). This is probably due to the enrichment of solute atoms on alternate {011} planes [17]. Therefore, the precipitate is likely to have an ordered lattice structure. Diffraction spots from the precipitated particles are shown in Figure 4c,d, which indicates that the discontinuous precipitates are fcc Cr-rich phases with a cube-on-cube OR (designated as $\beta'(I)$). The result is also in accordance with previous studies on Cu-Cr-Zr alloys [17,21–23]. The precipitate lattice parameter is calculated to be about 0.4180 nm. Figure 5e,f show the corresponding SAED of the sample aged for 240 min (peak aged). Superlattice reflections are also observed and the results indicate that the produced precipitates here are also ordered (marked by a triangle arrow). The SAED shows that these precipitates also have an fcc structure with a lattice constant of 0.4150 nm (designated as $\beta'(II)$). The orientation relationship exhibited here is determined to be: $[211]_{Cu}//[011]_{\beta'(II)}$, $\{-111\}_{Cu}//\{200\}_{\beta'(II)}$ and $[100]_{Cu}//[111]_{\beta'(II)}$, $\{02\text{-}2\}_{Cu}//\{02\text{-}2\}_{\beta'(II)}$. The above results show that $\beta'(II)$ has a lattice constant close to as well as an ordered structure with $\beta'(I)$. As a result, it can be concluded that $\beta'(II)$ and $\beta'(I)$ are the same crystal structure with different orientation relationships to the matrix. The same precipitates formed during the aging process show a variety of interphase relationships confirmed in Cu-Ni-Si [28], Mg-Zn-Al [29], and Al-Mg-Si [30] alloys. Although the fcc precipitates have been reported, the existence of two orientation relationships for the transition phase β' is proposed and confirmed in this work for the first time, for Cu-Cr-Zr alloy after a cold-solution deformation-aging treatment.

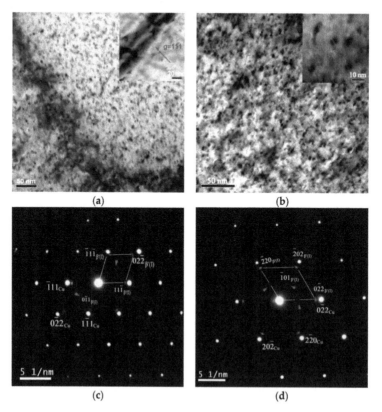

(a)

(b)

(c)

(d)

Figure 5. *Cont.*

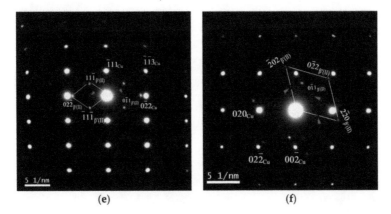

Figure 5. BF micrographs and selected area electron diffraction (SAED) patterns of the studied alloy aged at 723 K for various durations: (**a**) 30 min and (**b**) 240 min. SAED patterns were taken from samples aged for 30 min, from beams along the zone axis of: (**c**) $[110]_{Cu}$ and (**d**) $[111]_{Cu}$; and aged for 240 min, from beams along the zone axis of: (**e**) $[211]_{Cu}$ and (**f**) $[100]_{Cu}$.

Figure 6 shows the BF micrographs of the precipitates of the peak aged samples after annealing at different temperatures. As shown in Figure 6a, a large number of β′ precipitates still exist in the alloy. The β′ precipitates at the sub-grain boundary still exhibit a double petal-like contrast, indicating coherence with the matrix. The β phase, having a larger size, can be clearly observed in the grain boundary, as shown in Figure 6b. This indicates that during the formation of the crystal boundary, the transition phase β′ gradually loses coherence and grows into a larger equilibrium phase β. When the annealing temperature further increases to 873 K, the precipitates all evolve to coarse β particles, as shown in Figure 6c. The size of the precipitates significantly increases, and precipitation-free zones form at the grain boundaries, as shown in Figure 6d.

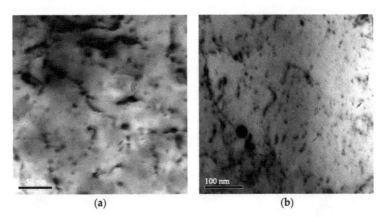

Figure 6. *Cont.*

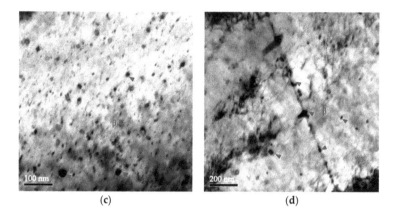

(c)   **(d)**

Figure 6. BF micrographs of Cu-Cr-Zr alloys with 60-min-annealed states at different temperatures: (**a**,**b**) 823 K and (**c**,**d**) 873 K.

Figure 7 shows the micrographs of twins of the peak aged samples annealed at different temperatures. The growth of the recrystallized grains is accompanied by the formation of annealing twins. Figure 7a,b show the BF images of annealing twins annealed at 873 K. The twins formed inside the grains have two parallel twin boundaries which traverse the whole grain. The width of the twins is about 30 nm and only a few dislocations are observed inside the twins, as shown in Figure 7a. Besides the parallel twins, annealing twins are also formed at the grain boundaries of the recrystallized grains, as shown in Figure 7b. A high dislocation density is observed in these twins, whereas only a few dislocations were observed around these twins. When the annealing temperature increases to 973 K, the twins grow significantly and the width can increase up to 200 nm, as shown in Figure 7c. Some of the twins can grow in an intergranular way, and their growth is hindered when they meet coarse precipitates, as shown in Figure 7d. The growth of the twins not only forms parallel twins, but some stepped structures are also observed, as shown in Figure 7e. The SAED result in Figure 7c shows that the pattern belongs to typical twins in fcc metals [14]. The twin plane is {111} and the direction is <112>, respectively. The zone axis between the matrix and the twin has a rotational symmetry of 180° and can be considered as $[101]_M//[-10-1]_T$ (M-Matrix, T-Twin), as shown in Figure 7f.

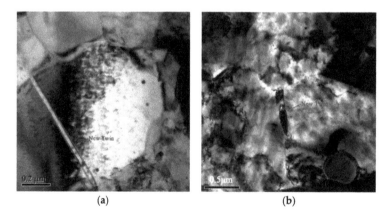

(a)   **(b)**

Figure 7. *Cont.*

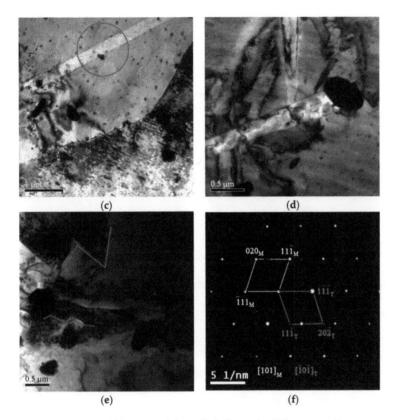

Figure 7. The BF and SAED of the twins of the studied alloy under different annealing temperatures: (a,b) 873 K and (c–e) 973 K; (f) the SAED of twins for the circular selection region in (c).

4. Discussion

4.1. The Influence of the Precipitation on Strength and Conductivity

TEM investigations of the precipitation reveal the presence of two types of β′ phases with two crystal orientation relationships. Beyond the β′(I) Cr-rich precipitates formed at the early stages of aging, which mimic a cube-on-cube OR with the matrix, other fcc-ordered Cr-rich precipitates (β′(II)) were observed in the peak aged condition. β′(II) is coherent with the matrix, with the following orientation relationship with the Cu matrix: $[111]_{β′(II)}//[100]_{Cu}$, $\{02\text{-}2\}_{β′(II)}//\{02\text{-}2\}_{Cu}$ and $[011]_{β′(II)}//[211]_{Cu}$, $\{200\}_{β′(II)}//\{\text{-}111\}_{Cu}$. $[110]_{Cu}$ and $[111]_{Cu}$ are the best observation axes for the β′(I) phase, as shown in Figure 5c,d, and $[100]_{Cu}$ and $[211]_{Cu}$ are the best observation axes for the β′(II) phase, as shown in Figure 5e,f. Attempts [10,25] to identify the composition of precipitates in this alloy have been only partially successful. Hatakeyama et al. [25] reported that Cr clusters enriched with Zr were observed at the prime aging stage, and further aging caused clusters to grow into Cr precipitates. Chbihi et al. [10] found that nano-scaled precipitates contained a large amount of Cu. Following this, during the coarsening stage, precipitates transformed into bcc and progressively increased their Cr contents. As a result, based on our experimental results and previous reports, we assume that the transition phases were mainly composed of Cu and Cr, as well as a small amount of Zr. Moreover, only β′(I) precipitates formed after the initial aging at 723 K for 30 min. Prolonged aging leads to the formation of β′(II) precipitates. After the completion of over-aging, only β precipitates were detected in the matrix. This sequence is similar to that of Al-Mg-Si alloys [30]. Thus, the possible precipitation sequence of this alloy can be understood to be as follows: supersaturated solid solution → β'(I) → β'(II) → β, which is different from Cu-Cr binary alloys, in which the sequence is a supersaturated solid solution → β [15]. As a result, we can conclude that the trace addition of Zr modified the precipitation mechanism and enhanced the age-hardening effects of the binary Cu-Cr alloy. Watanabe et al. [31] proposed that Zr addition probably forms Cu₅Zr precipitates, resulting in an increase in strength. Although some Zr-rich phases have been observed, such as Cu_4Zr [16], $Cu_{51}Zr_{14}$ [24], and Cu_5Zr [31], the formation of Zr-rich precipitates is not likely to be the main cause of the precipitation mechanism change nor the enhancement of the age-hardening effects in this study, because the reported Zr-rich precipitates are coarse phases and their content is low. It is also reported that Zr addition can increase the nucleation rate of Cr-rich precipitates [32]. The nucleation barriers for fcc and bcc Cr-rich clusters are 3×10^{-22} J and 4×10^{-20} J, respectively [10]. This fcc Cr-rich cluster can easily lose coherence due to high distortion and low stability, and can rapidly evolve into the equilibrium β phase in binary alloys. However, in Cu-Cr-Zr alloys, Zr segregates to form the atmosphere of a nucleated cluster [25], which increases the stability of the Cr-rich clusters and promotes the formation of the transition phase β', inhibiting the transformation from β' to β. However, to the authors' knowledge, the causes of this phenomenon remain unclear. The formation of a transition phase with an ordered structure is commonly observed in many systems where cascades of metastable phases are observed (e.g., AlMgSi [33–35] and CuNiSi [28,36]). For Al-Mg-Si alloys, whose precipitation sequence during aging is: α(ssss) → solute clusters →Guinier Preston (GP) zones → β" → β' → β, the β" transition phase, which has an ordered structure, shows the best strengthening effect [37].

The strength of the Cu-Cr-Zr alloy approaches its peak and the conductivity reaches a high and stable value after aging at 723 K for 240 min. A significant improvement in the strength of the Cu-Cr-Zr alloy is obtained after the aging treatment due to the precipitation, which drains the dissolved solute atoms from the copper matrix by forming nano-scale precipitates. After aging at 723 K for 240 min, the alloy is still in the recovery stage, and recrystallization does not occur. A large number of nano-sized coherent β' phases nucleate in the dislocation zone (Figure 8a) through the interaction with the dislocations, producing a good strengthening effect and simultaneously pinning the dislocations and dislocation cells (Figure 8b); this finally results in the enhancement of the alloy's recrystallization resistance capability. Zhang et al. [31] reported that the energy of the precipitate/matrix interface was increased by Zr in the ternary Cu-Cr-Zr alloy and composite, making it more difficult for dislocations to cut the coherent precipitates. Holzwarth et al. [38] found that the experimental data agrees much better with the assumption of an Orowan mechanism than with the shearing strengthening model, especially in the peak-hardened and slightly over-aged state of the Cu-Cr-Zr alloy. Consequently, the enhancement effect of the dispersed nanoparticles can be expressed by the following Orowan-Ashby equation in Equation (1) [39]:

$$\tau_0^{cs} = \frac{0.81 \cdot G \cdot b \cdot \ln(d/b)}{2 \cdot \pi \cdot \sqrt{1-v} \cdot (\lambda - d)} \tag{1}$$

where G is the shear modulus of the matrix, b is the modulus of the Burgers vector of the matrix, d is the average radius of the particles, v is the Poisson's ratio, f is the volume fraction of the nano-sized precipitates, and λ is the spacing between the particles. The latter can be expressed as Equation (2) [36]:

$$\lambda = 0.5 \cdot d \cdot \sqrt{\frac{3\pi}{2f}} \tag{2}$$

The incremental increase of tensile strength caused by the Orowan mechanism can be expressed as:

$$\Delta\sigma_{csmax} = M\Delta\tau_{csmax} \tag{3}$$

where M is the Taylor factor. Table 1 gives the parameters used in the yield strength determination where the Orowan bypass mechanism was considered. The tensile strength is increased by 136 MPa, as calculated by the Orowan mechanism. The tensile strength is increased by 119 MPa after peak aging, and the calculation results are in good agreement with the experimental result in this work, which indicates the strength of the Cu-Cr-Zr alloy after the cold work meets the Orowan mechanism. The result is also in accordance with previous studies on Cu-Cr-Zr alloys [38].

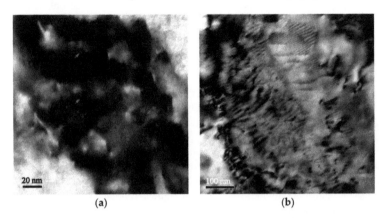

(a)   (b)

Figure 8. Bright-field transmission electron microscopy (TEM) micrographs and SAED of the studied alloy aged at 723 K for 240 min: (**a**) dislocations and second phase precipitates; and (**b**) pinned dislocations.

Table 1. The related parameters in the yield strength calculations, data from [36,38,40].

Parameter	Description	Value	Units	References
G	Shear modulus of copper	45.91	GPa	[38]
b	Burgers vector of copper	0.2556	nm	[40]
v	Poisson's ratio	0.34	-	[36]
M	Taylor factor	3.06	-	[36]
f	Volume fraction	1.09%	-	This work
R	Radius of the particles	4	nm	This work

The fundamental reason for the electrical resistivity of a metal is that the lattice produces the scattering of electrons. The total resistivity of copper alloys is mainly caused by the intrinsic resistivity pure copper, solid solute atoms, and precipitated crystal defects (grain boundaries, dislocations, etc.) according to the theory of conductivity. Different types of crystal defects have different effects on the resistivity, from large to small, followed by solid solute atoms, phase precipitates, vacancies, dislocations, and other defects. After the aging treatment, a large number of solute atoms precipitate from the supersaturated solid solution, resulting in a significant decrease in the electron scattering effect. The size of the β' phases are in the range of 2–5 nm, whereas the mean free path of electrons of pure copper is about 42 nm; the effect of precipitation on the conductivity of the alloy is therefore very small. In addition, a large number of tangled dislocations in the cold deformation state evolve into dislocation cells, sub-grains, and other sub-structures during the aging process, resulting in a significant reduction in the dislocation density of the dislocation cells and sub-grains (Figure 4b). As a result, one can conclude that the aging treatment can greatly increase the conductivity of the cold-worked Cur-Cr-Zr alloy, as shown in Figure 1b.

4.2. The Effect of the Precipitation and Recrystallization Mechanisms on Recrystallization Resistance

The recrystallization resistance of Cu-Cr-Zr alloys is highly related to the precipitate evolution and growth of precipitates. A huge number of finely distributed β' particles can pin the dislocations as well as hinder the grain boundary migration, thereby suppressing the nucleation and growth of the recrystallized grains. When annealed at 823 K, the precipitates at the frontier of sub-grains start to coarsen and lose their coherence to transfer to the equilibrium phase β, leading to the coarsening as well as to the reduction of the number density of precipitates, as shown in Figure 6a,b. Due to the significant difference in the size of the secondary phases, a concentration gradient of the solute atoms exists among the secondary phases. The solute atom concentration around the fine secondary particles is higher than that of the coarse ones, which will cause the solute atoms around the fine secondary phases to migrate to the coarse ones, and which results in the re-dissolution of the fine secondary particles as well as in the growth of the coarse particles. During the process of secondary particle growth and number density reduction, the precipitates at the sub-grain boundary preferentially coarsen and lose their coherence (Figure 6b), which greatly reduces the grain boundary pinning effect. Recrystallization nucleation begins in the Cu-Zr-Cr alloy, and the hardness decreases. When annealed at 873 K, the β′ phase is fully replaced by β, and the density of the precipitates further decreases, as shown in Figure 6c,d; this, in turn, further reduces the capacity for the recrystallization process to be hindered.

Shear zones which have a certain angle with respect to the roll direction are formed after the Cu-Cr-Zr alloy is severely deformed via cold rolling. The shear deformation zones have a higher dislocation density than other regions, and the nucleation and growth of recrystallization occur more easily in the shear deformation region, as shown in Figures 3 and 4. The annealing temperature is a critical external factor for recrystallization, and the difference in the microstructure of different regions is a critical internal factor for inducing a difference in recrystallization behavior. Because of the lower deformation energy in the uniform deformation zones, more energy is required for the nucleation of recrystallization and, as a result, a higher temperature is needed for recrystallization. Unlike uniform deformation zones, higher dislocation densities and deformation are present in the shear deformation zone due to the severe shear deformation. Dislocation cell structures in the shear deformation zone rapidly form the new sub-grains during annealing, and recrystallization first occurs in the shear deformation zones, as shown in Figure 3c.

The growth of the recrystallized grains is accompanied by the formation and growth of annealing twins (marked by a triangle arrow in Figure 7a). It is shown that the growth of the recrystallized grains is restricted by the adjacent grains (Figure 7a). To release the internal deformation energy, twins with parallel interfaces form via the shear mode in order to lower the driving force for recrystallized grains to grow, as shown in Figure 7a. When multiple grains come together, dislocation storage zones are produced by the twins in the intersection of the grain boundary in order to lower the driving force for interface migration and hinder grain combination and growth, as shown in Figure 7b. Thus, the formation of twins can inhibit recrystallization, therefore enhancing the recrystallization resistance of the alloy. Theoretically, without the pinning of the secondary particles, the growth of twins can be infinite in the longitudinal direction. The secondary particles can block the growth of the twins and eventually hinder the recrystallized grains from growing. Due to the formation of twins, as shown in Figure 9a, the recrystallized grain size is uniform and the misorientation of the grains is about 60°, as shown in Figure 9b. This structure is a relatively stable state and reduces the hardness lost after full recrystallization.

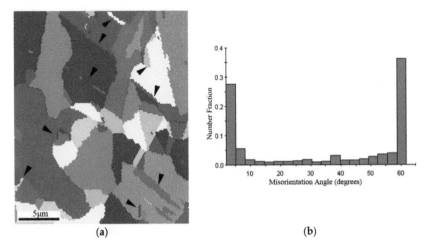

Figure 9. Typical electron backscattered diffraction (EBSD) substructures of recrystallization zone in a Cu-Cr-Zr alloy annealed at 823 K: (a) EBSD orientation imaging map; and (b) the distribution of grain boundary misorientation angles.

5. Conclusions

In this work, precipitation, recrystallization, and the evolution of twins during the heat treatment process in a Cu-Cr-Zr were studied by mechanical testing and microstructure observations. The main results are as follows:

1. A Cu-Cr-Zr alloy with a good combination of strength and conductivity can be obtained. The peak aged condition, with a tensile strength of 536 MPa and electron conductivity of 85.3% IACS, can be achieved after the alloy undergoes a solid solution treatment at 1253 K, before being cold-rolled with a reduction ratio of 70% and aged at 723 K for 240 min. The increasing strength and the conductivity are mainly related to the formation of the transition phase β'. A new type of fcc Cr-rich precipitate $\beta'(\text{II})$ is observed in the peak aged condition, which has an ordered structure and is coherent with the matrix, with the following OR: $[011]_{\beta'(\text{II})}//[211]_{Cu}$, $\{200\}_{\beta'(\text{II})}//\{-111\}_{Cu}$ and $[111]_{\beta'(\text{II})}//[100]_{Cu}$, $\{02\text{-}2\}_{\beta'(\text{II})}//\{02\text{-}2\}_{Cu}$.

2. The Cu-Cr-Zr alloy has a good recrystallization resistance. The dispersed distribution of the β' phase shows a good thermal stability during annealing, and can pin the dislocations and grain boundaries to hinder their migration, inhibiting recrystallization and thus enhancing the recrystallization resistance. When the β' particles are replaced by the coarsened β particles, and when there is a reduction in the density of precipitates, the pinning effect on the dislocations and grain boundaries becomes weak, and recrystallization starts to occur in the Cu-Cr-Zr alloy.

3. Recrystallization nucleation and growth first occurs in the shear deformation zone due to its high deformation energy during the recrystallization process. The growth of the recrystallized grains, as well as the decrease in hardness during annealing, is hindered by the formation of annealing twins due to the release of deformation energy and the reduction of the driving force for interface migration.

Acknowledgments: The authors are grateful to the experimental assistance by Song Bo, and the valuable discussion with Wenzheng Zhang and Wei Wang. This work was financial supported by the international cooperation major Project of Ministry of Science and Technology of the People's Republic of China (2006DFA53240) and the Doctor Innovation Fund founded by Hunan Province Education Department (CX2011B109), as well as the Open-End Fund for the Valuable and Precision Instruments of Central South University (CSUZC20140010).

Author Contributions: Xiaobo Chen and Feng Jiang conceived and designed the experiments; Jingyu Jiang, Pian Xu, Mengmeng Tong, and Zhongqin Tang contributed to the experimental research work. Xiaobo Chen and Feng Jiang analyzed the data; Xiaobo Chen and Feng Jiang discussed the results and analysis with the other authors. Xiaobo Chen and Feng Jiang wrote the paper.

Conflicts of Interest: The authors declare no conflict of interest.

References

1. Igor, A.; Hans-Achim, K.; Mozhgan, G.; Mansour, M.; Lothar, W. Ultrafine-grained precipitation hardened copper alloys by swaging or accumulative roll bonding. *Metals* **2015**, *5*, 763–776.

2. Deng, L.P.; Han, K.; Wang, B.S.; Yang, X.F.; Liu, Q. Thermal stability of Cu–Nb microcomposite wires. *Acta Mater.* **2015**, *101*, 181–188. [CrossRef]

3. Lei, R.S.; Wang, M.P.; Xu, S.Q.; Wang, H.P.; Chen, G.R. Microstructure, hardness evolution, and thermal stability mechanism of mechanical alloyed Cu-Nb alloy during heat treatment. *Metals* **2016**, *6*, 194. [CrossRef]

4. Zhang, X.; Beach, J.A.; Wang, M.; Bellon, P.; Averback, R.S. Precipitation kinetics of dilute Cu-W alloys during low-temperature ion irradiation. *Acta Mater.* **2016**, *120*, 46–55. [CrossRef]

5. Zhang, D.D.; Bai, F.; Wang, Y.; Wang, J.G.; Wang, W.Q. Grain Refinement and Mechanical Properties of Cu-Cr-Zr Alloys with Different Nano-Sized TiCp Addition. *Materials* **2017**, *10*, 919. [CrossRef] [PubMed]

6. Alessandro, F.A.; Carlo, A.B.; Ausonio, T. Synthesis and structural analysis of Copper-Zirconium oxide. *Metals* **2016**, *6*, 195. [CrossRef]

7. Correia, J.B.; Davies, H.A.; Sellars, C.M. Strengthening in rapidly solidified age hardened Cu-Cr and Cu-Cr-Zr alloys. *Acta Mater.* **1997**, *45*, 177–190. [CrossRef]

8. Morozova, A.; Borodin, E.; Bratov, V.; Zherebtsov, S.; Belyakov, A.; Kaibyshev, R. Grain Refinement Kinetics in a Low Alloyed Cu-Cr-Zr Alloy Subjected to Large Strain Deformation. *Materials* **2017**, *10*, 1394. [CrossRef] [PubMed]

9. Hauf, U.; Kauffmann, A.; Kauffmann-Weiss, S.; Feilbach, A.; Boening, M.; Mueller, F.E.H.; Hinrichsen, V.; Heilmaier, M. Microstructure Formation and Resistivity Change in CuCr during Rapid Solidification. *Metals* **2017**, *7*, 478. [CrossRef]

10. Chbihi, A.; Sauvage, X.; Blavette, D. Atomic scale investigation of Cr precipitation in copper. *Acta Mater.* **2012**, *60*, 4575–4585. [CrossRef]

11. Dobatkin, S.V.; Gubicza, J.; Shangina, D.V.; Bochvar, N.R.; Tabachkova, N.Y. High strength and good electrical conductivity in Cu–Cr alloys processed by severe plastic deformation. *Mater. Lett.* **2015**, *153*, 5–9. [CrossRef]

12. Zhang, Y.; Volinsky, A.A.; Hai, T.; Chai, Z.; Liu, P.; Tian, B.H.; Liu, Y. Aging behavior and precipitates analysis of the Cu–Cr–Zr–Ce alloy. *Mater. Sci. Eng. A* **2016**, *650*, 248–253. [CrossRef]

13. Lai, R.L.; He, D.Q.; He, G.A.; Lin, J.Y.; Sun, Y.Q. Study of the microstructure evolution and properties response of a Friction-Stir-Welded Copper-Chromium-Zirconium alloy. *Metals* **2017**, *7*, 381. [CrossRef]

14. Luo, C.P.; Dahmen, U. Morphology and crystallography of Cr precipitates in a Cu-0.33 wt % Cr alloy. *Acta Mater.* **1994**, *42*, 1923–1932. [CrossRef]

15. Fujii, T.; Nakazawa, H.; Kato, M.; Dahmen, U. Crystallography and morphology of nanosized Cr particles in a Cu-0.2% Cr alloy. *Acta Mater.* **2000**, *48*, 1033–1045. [CrossRef]

16. Tang, N.Y.; Taplin, N.M.; Dunlop, G.L. Precipitation and aging in high-conductivity Cu-Cr alloys with additions of zirconium and magnesium. *Mater. Sci. Technol.* **1985**, *1*, 270–275. [CrossRef]

17. Batra, I.S.; Dey, G.K.; Kulkarni, U.D.; Banerjee, S. Microstructure and properties of a Cu–Cr–Zr alloy. *J. Nucl Mater.* **2001**, *299*, 91–100. [CrossRef]

18. Liu, P.; Kang, B.X.; Cao, X.G.; Huang, J.L.; Yen, B.; Gu, H.C. Aging precipitation and recrystallization of rapidly solidified Cu–Cr–Zr–Mg alloy. *Mater. Sci. Eng. A* **1999**, *265*, 262–267. [CrossRef]

19. Qi, W.X.; Tu, J.P.; Liu, F.; Yang, Y.Z.; Wang, N.Y.; Lu, H.M.; Zhang, X.B.; Guo, S.Y.; Liu, M.S. Microstructure and tribological behavior of a peak aged Cu–Cr–Zr alloy. *Mater. Sci. Eng. A* **2003**, *343*, 89–96. [CrossRef]

20. Su, J.H.; Dong, Q.M.; Liu, P.; Li, H.J.; Kang, B.X. Research on aging precipitation in a Cu–Cr–Zr–Mg alloy. *Mater. Sci. Eng. A* **2005**, *392*, 422–426. [CrossRef]

21. Batra, I.S.; Dey, G.K.; Kulkarni, U.D.; Banerjee, S. Precipitation in a Cu-Cr-Zr alloy. *Mater. Sci. Eng. A* **2003**, *356*, 32–36. [CrossRef]

22. Xia, C.D.; Zhang, W.; Kang, Z.Y.; Jia, Y.L.; Wu, Y.F.; Zhang, R.; Xu, G.Y.; Wang, M.P. High strength and high electrical conductivity Cu–Cr system alloys manufactured by hot rolling–quenching process and thermo-mechanical treatments. *Mater. Sci. Eng. A* **2012**, *538*, 295–301. [CrossRef]

23. Cheng, J.Y.; Shen, B.; Yu, F.X. Precipitation in a Cu–Cr–Zr–Mg alloy during aging. *Mater. Charact.* **2013**, *81*, 68–75. [CrossRef]

24. Huang, F.X.; Ma, J.S.; Ning, H.L.; Geng, Z.T.; Liu, C.; Guo, S.M.; Yu, X.T.; Yu, W.T.; Li, H.; Lou, H.F. Analysis of phases in a Cu–Cr–Zr alloy. *Scr. Mater.* **2003**, *48*, 97–102.

25. Hatakeyama, M.; Toyama, T.; Yang, J.; Nagai, Y.; Hasegawa, M.; Ohkubo, T.; Eldrup, M.; Singh, B.N. 3D-AP and positron annihilation study of precipitation behavior in Cu–Cr–Zr alloy. *J. Nucl. Mater.* **2009**, *386*, 852–855. [CrossRef]

26. Su, J.H.; Liu, P.; Dong, Q.M.; Li, H.J.; Ren, F.Z. Recrystallization and precipitation behavior of Cu-Cr-Zr Alloy. *J. Mater. Eng. Perform.* **2007**, *16*, 490–493. [CrossRef]

27. Hughes, D.A.; Hansen, N.; Bammann, D.J. Geometrically necessary boundaries, incidental dislocation boundaries and geometrically necessary dislocations. *Scr. Mater.* **2003**, *48*, 147–153. [CrossRef]

28. Hu, T.; Chen, J.H.; Liu, J.R.; Liu, Z.R.; Wu, C.L. The crystallographic and morphological evolution of the strengthening precipitates in Cu–Ni–Si alloys. *Acta Mater.* **2013**, *61*, 1210–1219. [CrossRef]

29. Shi, Z.Z.; Zhang, W.Z. A transmission electron microscopy investigation of crystallography of τ-Mg$_{32}$ (Al, Zn)$_{49}$ precipitates in a Mg–Zn–Al alloy. *Scr. Mater.* **2011**, *64*, 201–204. [CrossRef]

30. Edwards, G.A.; Stiller, K.; Dunlop, G.L.; Couper, M.J. The precipitation sequence in Al–Mg–Si alloys. *Acta Mater.* **1998**, *46*, 3893–3904. [CrossRef]

31. Watanabe, C.; Monzen, R.; Tazaki, K. Mechanical properties of Cu–Cr system alloys with and without Zr and Ag. *J. Mater. Sci.* **2008**, *43*, 813–819. [CrossRef]

32. Zhang, D.L.; Mihara, K.; Tsubokawa, S.; Suzuki, H.G. Precipitation characteristics of Cu–15Cr–0.15Zr in situ composite. *Mater. Sci. Technol.* **2000**, *16*, 357–363. [CrossRef]

33. Matsuda, K.; Uetani, Y.; Sato, T.; Ikeno, S. Metastable phases in an Al-Mg-Si alloy containing copper. *Metall. Mater. Trans. A* **2001**, *32*, 1293–1299. [CrossRef]

34. Marioara, C.D.; Nordmark, H.; Andersen, S.J.; Holmestad, R. Post-β″phases and their influence on microstructure and hardness in 6xxx Al-Mg-Si alloys. *J. Mater. Sci.* **2006**, *41*, 471–478. [CrossRef]

35. Ninive, P.H.; Strandlie, A.; Gulbrandsen-Dahl, S.; Lefebvre, W.; Marioara, C.D.; Andersen, S.J.; Friis, J.; Holmestad, R.; Lovvik, O.M. Detailed atomistic insight into the β″phase in Al-Mg-Si alloys. *Acta Mater.* **2014**, *69*, 126–134. [CrossRef]

36. Lei, Q.; Xiao, Z.; Hu, W.; Derby, B.; Li, Z. Phase transformation behaviors and properties of a high strength Cu-Ni-Si alloy. *Mater. Sci. Eng. A* **2017**, *697*, 37–47. [CrossRef]

37. Cuniberti, A.; Tolley, A.; Riglos, M.V.C.; Giovachini, R. Influence of natural aging on the precipitation hardening of an Al-Mg-Si alloy. *Mater. Sci. Eng. A* **2010**, *527*, 5307–5311. [CrossRef]

38. Holzwarth, U.; Stamm, H. The precipitation behaviour of ITER-grade Cu-Cr-Zr alloy after simulating the thermal cycle of hot isostatic pressing. *J. Nucl. Mater.* **2000**, *279*, 31–45. [CrossRef]

39. Mabuchi, M.; Higashi, K. Strengthening mechanism of Mg-Si alloy. *Acta Mater.* **1996**, *44*, 4611–4618. [CrossRef]

40. Gladman, T. Precipitation-hardening of metals. *Mater. Sci. Technol.* **1999**, *15*, 30–36. [CrossRef]

Article

Deformation Behavior of High-Mn TWIP Steels Processed by Warm-to-Hot Working

Vladimir Torganchuk [1], Aleksandr M. Glezer [2],*, Andrey Belyakov [1] and Rustam Kaibyshev [1]

[1] Belgorod State University, Belgorod 308015, Russia; torganchuk@bsu.edu.ru (V.T.);
belyakov@bsu.edu.ru (A.B.); rustam_kaibyshev@bsu.edu.ru (R.K.)

[2] National University of Science & Technology (MISIS), Moscow 119049, Russia

* Correspondence: a.glezer@mail.ru; Tel.: +7-495-777-9350

Received: 11 May 2018; Accepted: 1 June 2018; Published: 3 June 2018

Abstract: The deformation behavior of 18%Mn TWIP steels (upon tensile tests) subjected to warm-to-hot rolling was analyzed in terms of Ludwigson-type relationship, i.e., $\sigma = K_1 \cdot \varepsilon^{n1} + \exp(K_2 - n_2 \cdot \varepsilon)$. Parameters of K_i and n_i depend on material and processing conditions and can be expressed by unique functions of inverse temperature. A decrease in the rolling temperature from 1373 K to 773 K results in a decrease in K_1 concurrently with n_1. Correspondingly, true stress approached a level of about 1750 MPa during tensile tests, irrespective of the previous warm-to-hot rolling conditions. On the other hand, an increase in both K_2 and n_2 with a decrease in the rolling temperature corresponds to an almost threefold increase in the yield strength and threefold shortening of the stage of transient plastic flow, which governs the duration of strain hardening and, therefore, manages plasticity. The change in deformation behavior with variation in the rolling temperature is associated with the effect of the processing conditions on the dislocation substructure, which, in turn, depends on the development of dynamic recovery and recrystallization during warm-to-hot rolling.

Keywords: high-Mn steel; deformation twinning; dynamic recrystallization; grain refinement; work hardening

---

1. Introduction

High-manganese austenitic steels with low stacking fault energy (SFE) are currently considered as promising materials for various structural/engineering applications because of their outstanding mechanical properties [1–3]. Owing to their low SFE, these steels are highly susceptible to deformation twinning, which results in the twinning-induced plasticity (TWIP) effect. Austenitic TWIP steels are characterized by pronounced strain hardening, which retards the strain localization and cracking during plastic deformation and, therefore, provides a beneficial combination of high strength with high ductility [1]. Deformation twinning, therefore, is the most crucial deformation mechanism governing the mechanical properties of high-Mn TWIP austenitic steels [4]. The deformation twins appear as bundles of closely spaced twins with thickness of tens nanometers, crossing over the original grains [5,6]. The deformation twins prevent the dislocation motion and promote an increase in dislocation density, leading to strain hardening. Frequent deformation twinning develops in steels with SFE in the range of approx. 20 mJ/m^2 to 50 mJ/m^2, which can be adjusted by manganese and carbon content [4].

The exact values of mechanical properties of austenitic steels, e.g., yield strength, ultimate tensile strength, total elongation, etc., depends on processing conditions. Hot rolling is frequently used as a processing technology for various structural steels and alloys. Final mechanical properties of processed steels and alloys depend on their microstructures that develop during hot working. Metallic materials with low SFE like high-Mn austenitic steels experience discontinuous dynamic recrystallization (DRX) during hot plastic deformation [7]. The developed microstructures depend on the deformation temperature and/or strain rate. Namely, the DRX grain size decreases with a decrease in temperature and/or an increase in strain rate and can be expressed by a power law function of temperature compensated strain rate (Z) [8]. A decrease in the deformation temperature to warm deformation conditions results in a change in the DRX mechanism from discontinuous to continuous, leading to a decrease in the grain size exponent in the relationship between the grain size and the deformation conditions, although this relationship remains qualitatively the same as that for hot working conditions [9]. The grain refinement with an increase in Z is accompanied with an increase in the dislocation density in DRX microstructures, irrespective of the DRX mechanisms [10]. Thus, the yield strength of the warm to hot worked semi-products can be evaluated by using various structural parameters. This approach has been successfully applied for strength evaluation of a range of structural steels and alloys, including high-Mn TWIP steels subjected to various thermo-mechanical treatments [5,6,10–12]. On the other hand, the effect of processing conditions on the deformation behavior of high-Mn TWIP steels has not been qualitatively evaluated, although, particularly for these steels, the deformation behavior is one of the most important properties, which manages the practical applications of the steels. It should be noted that the stress-strain behavior of austenitic steels with low SFE cannot be described by any well elaborated models like Hollomon or Swift equations, especially, at relatively small strains because of exceptional strain hardening [13]. Ludwigson modified the Hollomon-type relationship with an additional term to compensate the large difference between experimental and predicted flow stresses at small strains for such metals and alloys [14]. In spite of certain achievements in the application of the Ludwigson-type equation for the stress-strain behavior prediction, the selection of suitable parameters in this equation is still arbitrary in many ways. The aim of the present study, therefore, is to obtain the relationships between the processing conditions, the developed microstructures, and the stress-strain equation parameters in order to predict the tensile deformation behavior of advanced high-Mn TWIP steels processed by warm-to-hot rolling.

2. Materials and Methods

Two steels, Fe-18%Mn-0.4%C and Fe-18%Mn-0.6%C, have been selected in the present study as typical representatives of high-Mn TWIP steels. The steel melts were annealed at 1423 K, followed by hot rolling with about 60% reduction. The steels were characterized by uniform microstructures consisting of equiaxed grains with average sizes of 60 μm (18Mn-0.4C) and 50 μm (18Mn-0.6C). The steels were subjected to plate rolling at various temperatures from 773 K to 1373 K to a total rolling reduction of 60%. After each 10% rolling reductions, the samples were re-heated to the designated rolling temperature. Structural investigations were carried out using a Quanta 600 scanning electron microscope (SEM), equipped with an electron backscattering diffraction (EBSD) analyzer incorporating orientation imaging microscopy (OIM). The SEM samples were electro-polished at a voltage of 20 V at room temperature using an electrolyte containing 10% perchloric acid and 90% acetic acid. The OIM images were subjected to a clean-up procedure, setting the minimal confidence index of 0.1. The tensile tests were carried out using Instron 5882 testing machine with tensile specimens with a gauge length of 12 mm and a cross section of 3×1.5 mm^2 at an initial strain rate of 10^{-3} s^{-1}. The tensile axis was parallel to the rolling axis.

3. Results and Discussion

3.1. Developed Microstructures

Typical deformation microstructures that develop in the high-Mn steels during warm to hot rolling are shown in Figure 1. The mechanisms of microstructure evolution operating in austenitic steels during warm to hot working and the developed microstructures have been considered in detail in previous studies [10]. The deformation microstructures in the present high-Mn steels subjected to warm to hot rolling at temperatures of 773–1323 K can be briefly characterized here as follows. The temperature range above 1073 K corresponds to hot deformation conditions. Therefore, the deformation microstructures evolved during deformation in this temperature range result from the development of discontinuous DRX. The uniform microstructures consisting of almost equiaxed grains with numerous annealing twins are clearly seen in the samples hot rolled at temperatures above 1073 K (Figure 1a,b). The transverse DRX grain size decreases from 50–80 μm to 5–10 μm with a decrease in the rolling temperature from 1323 K to 1073 K.

In contrast, DRX hardly develops during warm rolling at temperatures below 1073 K. The deformation microstructures composed of flattened original grains evolve during warm rolling (Figure 1c,d). It is worth noting that the transverse grain size in the deformation microstructures developed during warm rolling does not remarkably depend on the rolling temperature. Relatively low deformation temperature suppresses discontinuous DRX. In this case, the structural changes are controlled by dynamic recovery. Under conditions of warm working, continuous DRX, which is assisted by dynamic recovery, can be expected after sufficiently large strains [7]. The present steels, however, are characterized by low SFE of 20–30 mJ/m^2 [4]. Such a low SFE makes the dislocation rearrangements during plastic deformation difficult and slows down the recovery kinetics. Therefore, 60% rolling reduction as applied in the present study is not enough for continuous DRX development in high-Mn steels. The final grain size, therefore, seems to be dependent on the original grain size and the total rolling reduction.

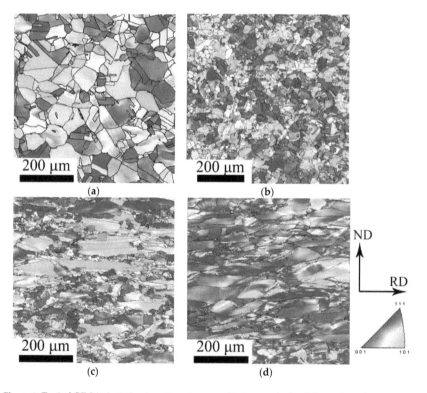

Figure 1. Typical OIM (orientation imaging microscopy) micrographs for deformation microstructures evolved in the Fe-18%Mn-0.4%C steel during hot-to-warm rolling at 1273 K (**a**), 1173 K (**b**), 1073 K (**c**) and 973 K (**d**). Colored orientations are shown for the transverse direction (TD).

3.2. Mechanical Properties

The stress-elongation curves obtained by tensile tests of the high-Mn steels processed by warm-to-hot rolling at different temperatures in the range of 773–1373 K are shown in Figure 2. A decrease in the rolling temperature results commonly in an increase in the strength and a decrease in the plasticity. The effect of the rolling temperature on the tensile tests properties is more pronounced for the warm working domain, i.e., rolling at temperatures below 1073 K, than that for hot working conditions, i.e., rolling at temperatures above 1073 K. A decrease in the temperature from 1373 K to 1073 K results in an increase in the yield strength ($\sigma_{0.2}$) by about 130 MPa while the ultimate tensile strength (UTS) does not change remarkably. In contrast, further decrease in the rolling temperature from 1073 K to 773 K leads to almost twofold increase in $\sigma_{0.2}$, which approaches about 900 MPa, and increases UTS by about 200 MPa. Correspondingly, the strengthening by warm to hot rolling is accompanied by a degradation of plasticity. It is interesting to note that total elongation gradually decreases with a decrease in the rolling temperature for Fe-18%Mn-0.6%C steel, where as that in Fe-18%Mn-0.4%C steel exhibits a kind of bimodal temperature dependence. The total elongation in the Fe-18%Mn-0.4%C steel tends to saturate at a level of 60–65% as the rolling temperature increases above 1073 K, following a rapid increase from 30% to 55% with an increase in the rolling temperature from 773 K to 1073 K.

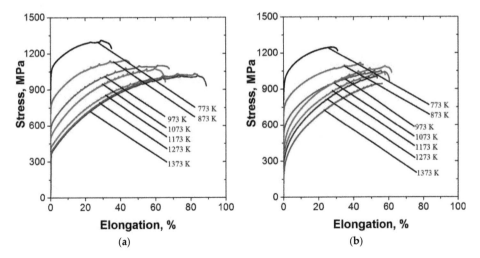

Figure 2. Engineering stress vs elongation curves of the Fe-18%Mn-0.6%C (**a**) and Fe-18%Mn-0.4%C (**b**) steels subjected to rolling at the indicated temperatures.

An apparent saturation for the total elongation of the Fe-18%Mn-0.4%C steel with an increase in the rolling temperature above 1073 K can be associated with a variation in the deformation mechanisms operating during tensile tests. The steel with lower carbon content has somewhat lower SFE [4] and, thus, may involve the strain-induced martensite upon tensile tests at room temperature. This difference in deformation mechanisms has been considered as a reason for the difference in plasticity [10]. The Fe-18%Mn-0.4%C steel subjected to hot rolling at temperatures above 1073 K exhibits the maximal plasticity, which can be obtained in the case of partial ε-martensitic transformation, whereas the Fe-18%Mn-0.6%C steel demonstrates increasing plasticity, which is improved by deformation twinning, with an increase in the rolling temperature. On the other hand, the strength properties, which depend on the grain size and dislocation density, are certainly affected by the rolling temperature, even in the range of hot working.

The Fe-18%Mn-0.6%C steel exhibits higher strength and elongation than the Fe-18%Mn-0.4%C steel after rolling warm to hot rolling in the studied temperature range. This additional strengthening of the Fe-18%Mn-0.6%C steel can be attributed to the difference in carbon content, which has been considered as the contributor to the yield strength of high-Mn TWIP steels [15]. The difference in 0.2 wt % carbon should result in about 85 MPa difference in the yield strength [15].

3.3. Tensile Behavior

Generally, the strength and plasticity during tensile tests depends on strain hardening, which, in turn, depends on the operating deformation mechanisms [1]. The plastic deformation of austenitic steels with low SFE at an ambient temperature is commonly expressed by the Ludwigson relation [14]:

$$\sigma = K_1 \cdot \varepsilon^{n1} + \exp(K_2 - n_2 \cdot \varepsilon), \tag{1}$$

where the first term with the strength factor of K_1 and strain hardening exponent n_1 represents Hollomon equation and the second term has been introduced by Ludwigson to incorporate the transient deformation stage, which differentiates the deformation behavior of fcc-metals/alloys with low-to medium SFE from other materials at relatively small strains. The stress of $\sigma = \exp(K_2)$ is close to the stress of plastic deformation onset and an inverse value of n_2 corresponds to the transient stage duration.

The parameters of K_1, K_2, n_1, n_2 providing the best correspondence between Equation (1) and experimental stress-strain curves are listed in Table 1. Note here, similar values for parameters of Ludwigson relation have been reported in other studies on low SFE austenitic steels [14,16–18]. The larger values of K_2 and K_1 for the present 0.6%C steel as compared to those for the 0.4%C steel reflect the higher stress levels of the former at early deformations and at large tensile strains, respectively. On the other hand, the n_1 values for both steels are close, suggesting similar strain hardening at large tensile strains, irrespective of some differences in the carbon content and SFE. The n_2 values are also close for both steels, indicating the same effect of the rolling temperature on the transient deformation stage during subsequent tensile tests.

Table 1. Parameters of the Ludwigson equation for the Fe-18%Mn-0.4%C and Fe-18%Mn-0.6%C steels processed by warm-to-hot rolling.

Steel	Rolling Temperature (K)	K_1 (MPa)	n_1	K_2	n_2
Fe-18%Mn-0.4%C	773	2248	0.24	6.0	37.8
Fe-18%Mn-0.4%C	873	2260	0.35	6.2	30.8
Fe-18%Mn-0.4%C	973	2342	0.46	6.0	20.3
Fe-18%Mn-0.4%C	1073	2410	0.48	5.8	25.2
Fe-18%Mn-0.4%C	1173	2556	0.58	5.7	12.8
Fe-18%Mn-0.4%C	1273	2575	0.63	5.5	11.4
Fe-18%Mn-0.4%C	1373	2598	0.67	5.5	12.3
Fe-18%Mn-0.6%C	773	2368	0.25	6.2	33.5
Fe-18%Mn-0.6%C	873	2410	0.38	6.3	24.5
Fe-18%Mn-0.6%C	973	2554	0.49	6.2	20.7
Fe-18%Mn-0.6%C	1073	2600	0.55	6.1	20.7
Fe-18%Mn-0.6%C	1173	2608	0.64	5.9	17.5
Fe-18%Mn-0.6%C	1273	2608	0.68	5.9	16.0
Fe-18%Mn-0.6%C	1373	2617	0.73	5.8	12.5

The monotonous changes of obtained parameters with rolling temperature suggest unique relationships between all parameters and deformation conditions. The effects of processing temperature on the parameters of Equation (1) are represented in Figure 3. Except for K_1, all parameters can be expressed by unique linear functions of the inverse rolling temperatures (Figure 3). The bimodal temperature dependencies obtained for K_1 with inflection points at 1073 K in Figure 3 are associated with the transition from warm to hot rolling conditions at this temperature, which reflects clearly on the deformation microstructures (see Figure 1). Using the indicated (Figure 3) linear relationships between the parameters of Ludwigson relation and the inverse rolling temperatures, the true stress vs strain curves calculated by Equation (1) are shown in Figure 4, along with the experimental curves obtained by tensile tests. Figure 4a,c show a general view of the stress-strain curves to validate the first term of Equation (1), whereas Figure 4b,d are plotted in double logarithmic scale to display the deformation behaviors at relatively small strains, which are described by the second term of Equation (1). The clear correspondence between the calculated and experimental plots testifies to the proposed treatments above.

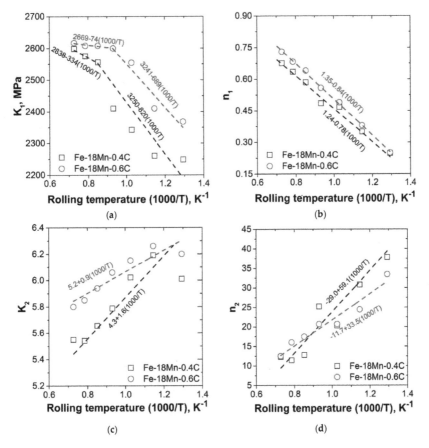

Figure 3. Effect of the rolling temperatures on the parameters of Ludwigson equation, K_1 (a), n_1 (b), K_2 (c), and n_2 (d).

The tensile deformation behavior of the steel samples should indeed be closely related to the steel microstructures, which were evolved by previous thermo-mechanical treatment. In turn, the developed microstructures depend on the processing conditions, i.e., rolling temperature, as the main processing variable in the present study. Generally, the deformation microstructures including the mean grain size and dislocation density that develop in metallic materials during warm-to-hot working can be expressed by power law functions of Zener-Hollomon parameter (temperature-compensated strain rate); $Z = \varepsilon \cdot \exp(Q/RT)$, where Q and R are the activation energy and universal gas constant, respectively [7,12]. Such microstructural changes are associated with thermally activated mechanisms of microstructure evolution in metallic materials. Therefore, the unique linear relationships in Figure 3 between the parameters of the flow stress predicting equation and the inverse rolling temperature suggest exponential relationships between the flow stress and the microstructures developed by warm-to-hot rolling. The second (exponential) term in Equation (1) predicts the flow stresses at relatively small strains (transient deformation), when the deformation behavior is associated with the dislocation ability to planar glide [14]. Thus, the stress-strain relationship of the high-Mn TWIP steels depends on their microstructures, namely, dislocation densities, evolved by previous thermo-mechanical treatments. Similar conclusions about a dominant role of dislocation density in the yield strength [10] and the work-hardening rate [19] were drawn in other studies on TWIP steels.

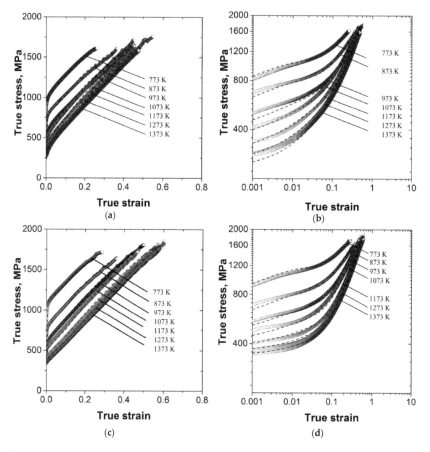

Figure 4. True tensile stress vs strain plots for Fe-18%Mn-0.4%C steel (a,b) and Fe-18%Mn-0.6%C steel (c,d) subjected to warm-to-hot rolling at indicated temperatures. The stress-strain curves obtained by tensile tests are shown by thick gray-scaled lines and those calculated by Equation (1) are shown by dashed lines.

It is worth noting in Figure 4 that maximal true stresses during tensile tests comprise about 1750 MPa for all steel samples, irrespective of the previous rolling conditions. Such gradual change in the altitude and slope of the true stress-strain curves can be represented by a gradual decrease in K_1 concurrently with n_1. An apparent saturation for the true stresses can be attributed to the strain-hardening ability owing to dislocation accumulation.

The strain, at which the transient deformation stage decays (ε_L) can be evaluated from the following relation [14]:

$$\exp(K_2 - n_2 \cdot \varepsilon_L)/(K_1 \varepsilon_L^{n1}) = r, \tag{2}$$

setting an arbitrary small value for r. According to the original Ludwigson treatment [14], $r = 0.02$ is selected in the present study. The values of ε_L calculated by Equation (2) for the present steels subjected to warm-to-hot rolling are shown in Figure 5 as functions of the rolling temperature. Formally, this strain (ε_L) limiting the transient deformation duration can be considered as a critical point below which, the plastic flow cannot be adequately described by Hollomon-type relation, i.e., the first term in Equation (1). The flow stresses during the transient deformation can be calculated taking into account the second term in Equation (1). The strain of ε_L can be roughly related to a strain when cross slip and dislocation rearrangements, which are closely connected with dynamic recovery, impair the strain hardening [14]. Therefore, an increase in ε_L should promote plasticity, including both uniform and total elongations.

It is clearly seen in Figure 5 that ε_L increases from about 0.08 to 0.24 with an increase in the rolling temperature from 773 K to 1373 K; this suggests an improvement in plasticity with increase in rolling temperature. A decrease in SFE promotes planar slip and, thus, should increase ε_L. Indeed, the 0.4%C steel is characterized by a larger ε_L than the 0.6%C steel after hot rolling at temperatures above 1300 K (Figure 5), although the hot rolled steel with higher carbon content exhibits larger total elongation. This relatively low plasticity of the Fe-18%Mn-0.4%C steel is associated with an ε-martensitic transformation [10]. Commonly, transformation-induced plasticity (TRIP) steels demonstrate lower plasticity than TWIP steels [20,21]. After rolling at temperatures below 1300 K, the values of ε_L for the Fe-18%Mn-0.4%C steel are smaller than those for the Fe-18%Mn-0.6%C steel processed under the same conditions (Figure 5). This can be attributed to the effect of warm-to-hot rolling at the evolved dislocation density. The latter has been shown to increase as rolling temperature decreases [10]. Therefore, the transient deformation stage upon the tensile tests is shortened because of the previous plastic deformation during warm-to-hot rolling, which partially consumed the dislocation ability to planar glide.

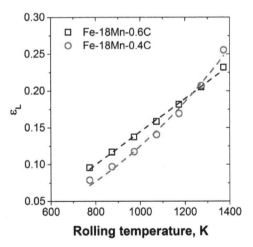

Figure 5. Effect of the rolling temperature on the strain for transient deformation during tensile tests of the Fe-18%Mn-0.4%C and Fe-18%-0.6%C steels.

4. Conclusions

The deformation behavior during tensile tests of Fe-18%Mn-0.4%C and Fe-18%-0.6%C steels subjected to warm to hot rolling was studied. The main results can be summarized as follows.

1. The hot rolling at temperatures above 1073 K was accompanied by the development of discontinuous dynamic recrystallization, leading to a decrease in the transverse grain size with a decrease in the rolling temperature. On the other hand, microstructure evolution during warm rolling at temperatures below 1073 K was controlled by the rate of dynamic recovery, which slowed down with a decrease in the rolling temperature.

2. The true stress-strain curves obtained by tensile tests at ambient temperature can be correctly represented by the Ludwigson-type relationship—$\sigma = K_1 \cdot \varepsilon^{n1} + \exp(K_2 - n_2 \cdot \varepsilon)$— where parameters of K_i and n_i depended on material and processing conditions and can be expressed by unique functions of inverse temperature of previous warm-to-hot rolling. A decrease in the rolling temperature from 1373 K to 773 K resulted in a decrease in K_1 (from approx. 2600 MPa to 2300 MPa) concurrently with n_1 (from approx. 0.7 to 0.25). Correspondingly, the true stress approached a level of about 1750 MPa during tensile tests, irrespective of the previous warm-to-hot rolling conditions. On the other hand, an increase in both K_2 and n_2 with decrease in the rolling temperature corresponded to an almost threefold increase in the yield strength and analogous degradation of plasticity.

3. The stage of transient plastic flow providing initial strain hardening and, therefore, controlling the total plasticity increases from about 0.08 to 0.24 with an increase in the rolling temperature from 773 K to 1373 K. The Fe-18%Mn-0.4%C steel is characterized by smaller values of ε_L than the Fe-18%Mn-0.6%C steel subjected to warm-to-hot rolling at the same temperatures below 1300 K, although the former should possess lower stacking fault energy. The shortening of the transient deformation stage upon tensile tests of the steels subjected to warm-to-hot rolling can be attributed to previous deformation, which partially consumed the dislocation ability to planar glide.

Author Contributions: Conceptualization, A.M.G. and R.K.; Methodology, V.T. and A.B.; Investigation, V.T.; Writing-Original Draft Preparation, A.B.; Writing-Review & Editing, A.M.G. and R.K.; Visualization, V.T.; Supervision, R.K.

Funding: This work was performed partly under the State Order of the Ministry of Education and Science of the Russian Federation No. 20'7/113 (2097).

Acknowledgments: The authors are grateful to the personnel of the Joint Research Centre of "Technology and Materials", Belgorod State University, for their assistance with instrumental analysis.

Conflicts of Interest: The authors declare no conflict of interest.

References

1. Bouaziz, O.; Allain, S.; Scott, C.P.; Cugy, P.; Barbier, D. High manganese austenitic twinning induced plasticity steels: A review of the microstructure properties relationships. *Curr. Opin. Solid State Mater. Sci.* **2011**, *15*, 141–168. [CrossRef]

2. De Cooman, B.C.; Estrin, Y.; Kim, S.K. Twinning-induced plasticity (TWIP) steels. *Acta Mater.* **2018**, *142*, 283–362. [CrossRef]

3. Kusakin, P.S.; Kaibyshev, R.O. High-Mn twinning-induced plasticity steels: Microstructure and mechanical properties. *Rev. Adv. Mater. Sci.* **2016**, *44*, 326–360.

4. Saeed-Akbari, A.; Mosecker, L.; Schwedt, A.; Bleck, W. Characterization and prediction of flow behavior in high-manganese twinning induced plasticity steels: Part I. mechanism maps and work-hardening behavior. *Metall. Mater. Trans. A* **2012**, *43*, 1688–1704. [CrossRef]

5. Kusakin, P.; Belyakov, A.; Haase, C.; Kaibyshev, R.; Molodov, D.A. Microstructure evolution and strengthening mechanisms of Fe–23Mn–0.3C–1.5Al TWIP steel during cold rolling. *Mater. Sci. Eng. A* **2014**, *617*, 52–60. [CrossRef]

6. Yanushkevich, Z.; Belyakov, A.; Haase, C.; Molodov, D.A.; Kaibyshev, R. Structural/textural changes and strengthening of an advanced high-Mn steel subjected to cold rolling. *Mater. Sci. Eng. A* **2016**, *651*, 763–773. [CrossRef]

7. Sakai, T.; Belyakov, A.; Kaibyshev, R.; Miura, H.; Jonas, J.J. Dynamic and post-dynamic recrystallization under hot, cold and severe plastic deformation conditions. *Prog. Mater. Sci.* **2014**, *60*, 130–207. [CrossRef]

8. Tikhonova, M.; Belyakov, A.; Kaibyshev, R. Strain-induced grain evolution in an austenitic stainless steel under warm multiple forging. *Mater. Sci. Eng. A* **2013**, *564*, 413–422. [CrossRef]

9. Tikhonova, M.; Enikeev, N.; Valiev, R.Z.; Belyakov, A.; Kaibyshev, R. Submicrocrystalline austenitic stainless steel processed by cold or warm high pressure torsion. *Mater. Sci. Forum* **2016**, *838*, 398–403. [CrossRef]

10. Torganchuk, V.; Belyakov, A.; Kaibyshev, R. Effect of rolling temperature on microstructure and mechanical properties of 18%Mn TWIP/TRIP steels. *Mater. Sci. Eng. A* **2017**, *708*, 110–117. [CrossRef]

11. Morozova, A.; Kaibyshev, R. Grain refinement and strengthening of a Cu–0.1Cr–0.06Zr alloy subjected to equal channel angular pressing. *Philos. Mag.* **2017**, *97*, 2053–2076. [CrossRef]

12. Yanushkevich, Z.; Dobatkin, S.V.; Belyakov, A.; Kaibyshev, R. Hall-Petch relationship for austenitic stainless steels processed by large strain warm rolling. *Acta Mater.* **2017**, *136*, 39–48. [CrossRef]

13. Choudhary, B.K.; Isaac Samuel, E.; Bhanu Sankara Rao, K.; Mannan, S.L. Tensile stress-strain and work hardening behaviour of 316LN austenitic stainless steel. *Mater. Sci. Technol.* **2001**, *17*, 223–231. [CrossRef]

14. Ludwigson, D.C. Modified stress-strain relation for fcc metals and alloys. *Metal. Trans.* **1972**, *2*, 2825–2828. [CrossRef]

15. Kusakin, P.; Belyakov, A.; Molodov, D.A.; Kaibyshev, R. On the effect of chemical composition on yield strength of TWIP steels. *Mater. Sci. Eng. A* **2017**, *687*, 82–84. [CrossRef]

16. Mannan, S.L.; Samuel, K.G.; Rodriguez, P. Stress-strain relation for 316 stainless steel at 300 K. *Scr. Metall.* **1982**, *16*, 255–257. [CrossRef]

17. Satyanarayana, D.V.V.; Malakondaiah, G.; Sarma, D.S. Analysis of flow behaviour of an aluminium containing austenitic steel. *Mater. Sci. Eng. A* **2007**, *452*, 244–253. [CrossRef]

18. Milititsky, M.; De Wispelaere, N.; Petrov, R.; Ramos, J.E.; Reguly, A.; Hanninen, H. Characterization of the mechanical properties of low-nickel austenitic stainless steels. *Mater. Sci. Eng. A* **2008**, *498*, 289–295. [CrossRef]

19. Liang, Z.Y.; Li, Y.Z.; Huang, M.X. The respective hardening contributions of dislocations and twins to the flow stress of a twinning-induced plasticity steel. *Scr. Mater.* **2016**, *112*, 28–31. [CrossRef]

20. Song, W.; Ingendahl, T.; Bleck, W. Control of strain hardening behavior in high-Mn austenitic steels. *Acta Metall. Sin. (Engl. Lett.)* **2014**, *27*, 546–555. [CrossRef]

21. Lee, Y.-K.; Han, J. Current opinion in medium manganese steel. *Mater. Sci. Technol.* **2015**, *31*, 843–856. [CrossRef]

Article

Effect of Si and Zr on the Microstructure and Properties of Al-Fe-Si-Zr Alloys

Anna Morozova [1], Anna Mogucheva [1,*], Dmitriy Bukin [1], Olga Lukianova [1], Natalya Korotkova [2], Nikolay Belov [2] and Rustam Kaibyshev [1]

[1]   Laboratory of Mechanical Properties of Nanostructured Materials and Superalloys, Belgorod State
     University, Belgorod 308015, Russia; morozova_ai@bsu.edu.ru (A.M.); bukin-15@mail.ru (D.B.);
     lukyanova_oa@bsu.edu.ru (O.L.); rustam_kaibyshev@bsu.edu.ru (R.K.)
[2]   Department of Casting Technologies, Moscow Institute of steel and Alloys, Moscow 119049, Russia;
     darkhopex@mail.ru (N.K.); nikolay-belov@yandex.ru (N.B.)
*   Correspondence: mogucheva@bsu.edu.ru; Tel.: +7-4722-585417

Received: 10 October 2017; Accepted: 8 November 2017; Published: 11 November 2017

Abstract: The effects of Si and Zr on the microstructure, microhardness and electrical conductivity of Al-Fe-Si-Zr alloys were studied. An increase in the Zr content over 0.3 wt. % leads to the formation of primary Al_3Zr inclusions and also decreases mechanical properties. Therefore, the Zr content should not be more than 0.3 wt. %, although the smaller content is insufficient for the strengthening by the secondary Al_3Zr precipitates. The present results indicate that high content of Si significantly affects the hardness and electrical conductivity of the investigated alloys. However, the absence of Si led to the formation of harmful needle-shaped Al_3Fe particles in the microstructure of the investigated alloys after annealing. Therefore, the optimum amount of Si was 0.25–0.50 wt. % due to the formation of the Al_8Fe_2Si phase with the preferable platelet morphology. The maximum microhardness and strengthening effects in Al-1% Fe-0.25% Si-0.3% Zr were observed after annealing at 400–450 °C due to the formation of nanosized coherent Al_3Zr ($L1_2$) dispersoids. The effect of the increasing of the electrical conductivity can be explained by the decomposition of the solid solution. Thus, Al-1% Fe-0.25% Si-0.3% Zr alloy annealed at 450 °C has been studied in detail as the most attractive with respect to the special focus on transmission line applications.

Keywords: aluminum alloys; Al-Fe-Si-Zr system; microstructure; hardness; electrical conductivity

1. Introduction

Al-Fe-Si alloys belonging to the 1XXX series are commonly used in a wide range of different applications including electrical conductors [1]. The fruitful combination of the light weight and reasonable electrical conductivity makes these alloys preferable for the production of wires for overhead power transmission lines compared with Cu [2]. The high electrical conductivity of these alloys approaching the theoretical limit for Al of 62% IACS (The International Annealed Copper Standard defines 17.241 nm as 100%) is attributed to the very low solubility of Si and Fe in aluminum. The tensile properties of these alloys are low. At present, there is a strong requirement for increased strength and thermal stability of Al alloys used for transmission lines. These alloys have to withstand a high operating temperature of ~150°. Such a transition metal as Zr is used to improve the strength at room temperature and to provide the retention of the structure and tensile properties at elevated temperatures due to the formation of Al_3Zr dispersoids during homogenizing treatment [2]. Zr is one of the most promising additives for developing castable, precipitation-hardened alloys with good coarsening and creep resistance and usually used as a recrystallization inhibitor and grain refiner in commercial Al alloys. In addition, the solubility of this alloying element in Al is negligible, and therefore, Al-Zr alloys may exhibit high strength, thermal stability and electrical conductivity, concurrently.

A detailed description of the ternary Al-Fe-Si phase diagram can be found in [3]. It is possible to expect the formation of such equilibrium intermetallics as $Al_{13}Fe_4$ (θ phase), Al_3Fe phase, α-AlFeSi and β-AlFeSi via peritectic (θ and α-AlFeSi) or eutectic (β-AlFeSi) reactions in the aluminum corner of the equilibrium Al-Fe-Si phase. It is know that needle-shaped Al_3Fe particles dramatically decrease the technological ductility of aluminum alloys [3]. The content of Si is added to prevent suppression of Al_3Fe particles formation and to provide the α-AlFeSi (Al_8Fe_2Si) and β-AlFeSi (Al_5FeSi) phases' precipitation into these alloys [3]. However, there is very limited information in the literature on the effectiveness of the influence of Si and Zr on the structure and properties in these alloys. β-AlFeSi is the most harmful phase decreasing the mechanical properties of Al alloys as described in [4]. There are three major types of β-AlFeSi intermetallics with orthorhombic, tetragonal and monoclinic crystal structures. As described in [5], such elements as Mn, Co, Sr or reduction of the solidification rate are widely used to form the α-AlFeSi preferred phase.

Al_3Zr may exist in three main forms: the stable, tetragonal ($D0_{23}$) modification, the metastable cubic ($L1_2$) modification and in the primary form. Increased Zr content or casting defects can lead to the formation of the primary Al_3Zr inclusions. However, a significant precipitation hardening can be provided by a very small addition of Zr. This can be explained by the precipitation of metastable Al_3Zr particles during post-solidification heat treatment. The metastable Al_3Zr phase has the same structure as the Al matrix. The coherence of the metastable Al_3Zr particles with the aluminum matrix leads to better thermal stability and an appreciable precipitation hardening effect. The $L1_2$ structure of the metastable Al_3Zr particles transforms into the complex tetragonal $D0_{23}$ stable form after annealing at temperatures up to 500 °C [6,7]. Stable Al_3Zr precipitates are semi-coherent with the Al matrix. The growth of the metastable Al_3Zr precipitates and the loss of their coherence reduce the hardening effect. Features of the microstructure and mechanical properties of aluminum alloys with the addition of Si, Fe and various contents of Zr are described by Belov et al. [2,8–10].

At present, the implementation of the inert anode technique for the production of primary Al requires the development of a new class of Al-Fe-Si-Zr alloys, which have to combine the low cost of Al-Fe-Si alloys with the high strength and sufficient thermal stability of Al-Zr alloys. The key aim of the present study is to optimize the alloying content and the heat treatment to produce an aluminum alloy belonging to the Al-Fe-Si-Zr system with high electrical conductivity and high performance properties. The concurrent influence of Si and Zr on the microstructure, electrical conductivity and microhardness of several Al-Fe-Si-Zr alloys has been investigated. The effect of casting conditions and multistage annealing on these properties was also examined taking into account the feasibility to use these alloys for high-volume production of wires for transmission lines.

2. Materials and Methods

Experimental alloys of the Al-Fe-Si-Zr system (Table 1) were prepared in an electric furnace LAC PT 90/13 (LAC, s.r.o., Rajhrad, Czexh Republic) in graphite-chamotte crucibles using 99.99% pure Al and master alloys containing 0.15 wt. %, 0.3 wt. %, 0.45 wt. % and 0.6 wt. % Zr and 0.25 wt. %, 0.5 wt. %, 0.75 wt. % and 1 wt. % Si, respectively. The casting temperature for all alloys is marked in green, and the calculated liquidus temperature (Thermo-Calc software AB, TTAL5 database, Version 5.0.4.75, Thermo-Calc Software, Stockholm, Swedeny, 2010) is marked in red in Figure 1, respectively.

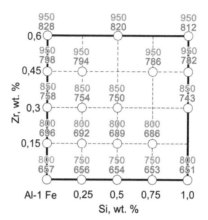

Figure 1. Quadrilateral scheme of Al-1Fe-Si-Zr.

The cast samples were annealed during 3 h in an electric furnace at a temperature ranging from 200 °C to 600 °C using stepped modes as given in Table 1.

Table 1. Annealing regimes of Al-Fe-Si-Zr alloys.

Designation	S200	S250	S300	S350	S400	S450	S500	S550	S600
Annealing regime	200 °C	S200 + 250 °C	S250 + 300 °C	S300 + 350 °C	S350 + 400 °C	S400 + 450 °C	S450 + 500 °C	S500 + 550 °C	S550 + 600 °C

The microstructures of the as-cast and homogenized samples were examined using an Axio Observer MAT and TESCAN VEGA 3 and Quanta 200 scanning electron microscopes (SEM) (FEI, Hillsboro, OR, USA). For microstructure examination, the samples were polished in a perchloric acid-ethanol electrolyte (6 C_2H_5OH/1 $HClO_4$/1 glycerol) and additionally oxidized in a 5% water solution of HBF_4 at constant voltage of 20 V to reveal the grain structure. Foils for transmission electron microscopy were electrolytically thinned in a perchloric acid-ethanol solution and examined using a JEOL JEM-2100 transmission electron microscope (TEM) (JEOL Ltd., Tokyo, Japan) with a double-tilt stage at 200 KV.

The electrical conductivity was measured by the method of eddy currents using a Constanta K6 device (Constanta, Sankt Peterburg, Russia). The measurements of the Vickers microhardness were performed at room temperature on metallographically-polished sections using 50-N loads and a dwell time of 15 s.

3. Results

Typical microstructures that evolved in the Al-Fe-Si-Zr alloys with different contents of Si ranging from 0 wt. % to 1 wt. % are shown in Figure 2. A eutectic included supersaturated aluminum matrix (Al) and the Al_6Fe-phase have been observed along the boundaries of dendritic cells in the considered alloys with the absence of Si (Figure 2a); while (Al) + Al_8Fe_2Si eutectic in the investigated Al alloys with 0.25–1 wt. % Si was associated with an increase in Si content, as shown in Figure 2b–d. In the Al-1% Fe-1% Si-0.3% Zr is observed the precipitating of particles (see Figure 3d), which Belov et al. [11] defined as a Si-rich inclusion. Such a precipitate decreases the electrical conductivity in aluminum alloys [12].

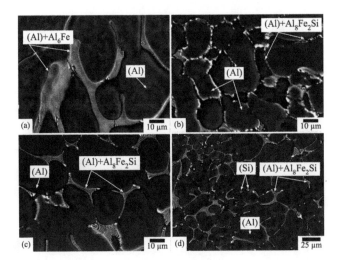

Figure 2. Microstructure of as-cast alloys with different contents of Si: (**a**) Al-1% Fe-0.3% Zr; (**b**) Al-1% Fe-0.25% Si-0.3% Zr; (**c**) Al-1% Fe-0.5% Si-0.3% Zr; (**d**) Al-1% Fe-1% Si-0.3% Zr.

3.1. Scanning Electron Microscope

Figure 3 shows the SEM of the as-cast microstructure of Al-Fe-Si-Zr alloys with different content of Zr and fixed content of Fe 1 wt. % and Si 0.5 wt. %. The observed microstructure consists of α-Al dendrites (matrix) and different Fe-Si-rich intermetallic phases distributed along the aluminum dendrite. The microstructures of the described as-cast alloys Al-1%Fe-0.5%Si with Zr content from 0 wt. % to 0.3 wt. % were nearly the same and consist of the solid solution of aluminum (Al) and a eutectic (Al) + Al$_8$Fe$_2$Si along the boundaries of dendritic cells (Figure 3a–c). Lack of the primary crystals of Al$_3$Zr clearly shows full incorporation of Zr into the solid solution and the optimum rate of crystallization for chemical composition. The primary Al$_3$Zr crystals with a cross-shaped morphology were found in the microstructure of the investigated alloys with 0.45–0.6 wt. % of Zr (Figure 3d).

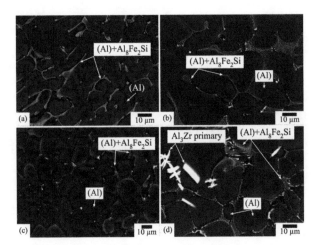

Figure 3. Microstructure of as-cast alloys with different contents of Zr: (**a**) Al-1%Fe-0.5%Si; (**b**) Al-1% Fe-0.5% Si-0.15% Zr; (**c**) Al-1% Fe-0.5% Si-0.3%Z r; (**d**) Al-1% Fe-0.5% Si-0.6% Zr.

The annealing process is a key factor that determines the objectives to produce Al alloys containing up to 0.6 wt. % Zr. Figure 4a shows that the metastable phase Al_6Fe is transformed into a stable needle-like Al_3Fe phase during the heat treatment, which considerably reduces the technological ductility, relative elongation and structural strength [11]. Therefore, the harmful needle-shaped Al_3Fe was observed in the microstructure of the investigated Al-1% Fe-0.3% Zr alloy [1]. An increase in the Si content leads to a decrease in the Al_3Fe and improvement of Al_8Fe_2Si (30%–33% Fe, 6%–12% Si [3]) content, as can be seen in Figure 4b,c.

Figure 4b shows the stable $D0_{23}$ Al_3Zr particles predominantly at the center of the dendritic cells formed as a result of transformation from the metastable Al_3Zr $L1_2$ particles for Al-Fe-Si-Zr alloys with 0.15 wt. % Zr during the heat treatment at the maximum annealing temperature 600 °C. The number of precipitates is lower as compared with alloys with higher Zr content.

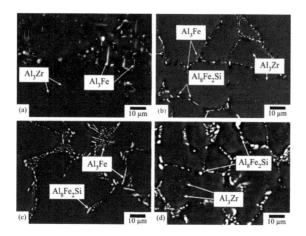

Figure 4. Microstructure of Al-1%Fe-Si-0.3%Zr alloys annealed at S600 with different contents of Si: (**a**) Al-1% Fe-0.3% Zr; (**b**) Al-1% Fe-0.25% Si-0.3% Zr; (**c**) Al-1% Fe-0.5% Si-0.3% Zr; (**d**) Al-1% Fe-1% Si-0.3% Zr.

3.2. Transmission Electron Microscope

Typical fine structures of the Al-1%Fe-0.25%Si-0.3%Zr alloy are shown in Figures 5 and 6 for the S450 peak-aging condition and the S600 overaging condition, respectively. The high magnification of SEM and TEM micrographs of the Al-Fe-Si-Zr alloys with 0.25 wt. % Si and 0.3 wt. % Zr in Figure 5 suggest that small, spheroidal, coherent Al_3Zr ($L1_2$) precipitates are confined to the dendritic cells, which are surrounded by precipitate-free interdendritic channels. All fine Al_3Zr ($L1_2$) particles inside the marked red circle in Figure 5 are characterized by a cubic structure with $a = 4.09$ and an average size of about 10 nm. The orientation relationship between the matrix and Al_3Zr ($L1_2$) precipitates is the cube-cube: (020) Al | | (020) Al_3Zr, (002) Al | | (002) Al_3Zr, [022] Al | | [022] Al_3Zr.

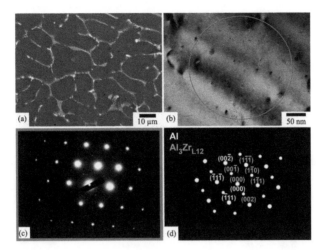

Figure 5. Transmission electron microscope (TEM) micrographs of Al-1% Fe-0.25% Si-0.3% Zr annealed at S450 with coherent L12 Al_3Zr nano-particles: (**a**) scanning electron microscopes (SEM); (**b**) TEM; (**c**) diffraction; (**d**) schematic diffraction pattern with indexes.

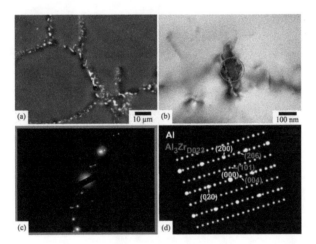

Figure 6. TEM micrographs of Al-1% Fe-0.25% Si-0.3% Zr annealed at S600 with incoherent $D0_{23}$ Al_3Zr particles: (**a**) SEM; (**b**) TEM; (**c**) diffraction; (**d**) schematic diffraction pattern with indexes.

The annealing at S600 (Figure 6) leads to the particle growth in direction [002], and the transformation cubic $L1_2$ precipitates into the stable complex tetragonal $D0_{23}$-phase with the cell parameters $a = b = 0.3999$ nm, $c = 1.7283$ nm. The Al_3Zr ($D0_{23}$) is characterized by the incoherent boundaries and has an orientation relationship with the matrix: (002) Al || (020) Al_3Zr, (020) Al || (002) Al_3Zr, [100] Al || [001] Al_3Zr. A good correlation of the twice interplanar spacing (002) in aluminum (d (002) Al = 0.2025 nm [13]) and interplanar spacing (001) in the Al_3Zr ($D0_{23}$) phase (d(004) Al_3Zr = 0.4321 nm [14]) provides the particles growth along the [002] matrix direction, and the plate-like shape diameter of Al_3ZrD0_{23}-particles is 250 nm, with a thickness of 50 nm.

3.3. Hardness and Electrical Conductivity

Figure 7a–d show the temperature-hardness dependences of the investigated Al alloys after annealing at different temperatures. It is shown that the shapes of the curves were typical and the same in all cases. The Si content led to improved hardness in the alloys. On the other hand, an increase in the annealing temperature decreased hardness.

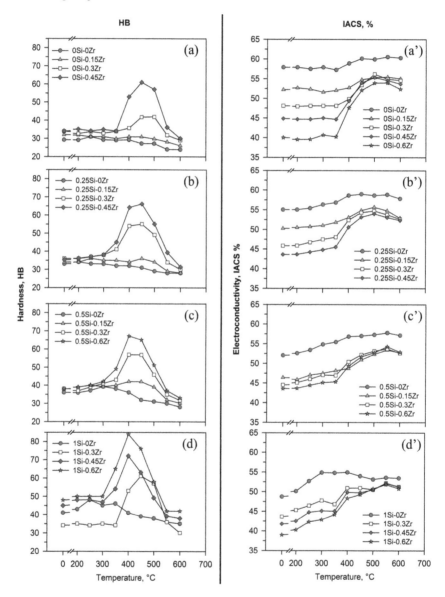

Figure 7. Temperature dependence of the microhardness (**a–d**) and the electrical conductivity (**a–d'**) for Al-1% Fe-Si-Zr alloys with a fixed content of Si and different contents of Zr: (**a, a'**) 0 wt. % Si; (**b, b'**) 0.25 wt. % Si; (**c, c'**) 0.5 wt. % Si; (**d, d'**) 1.0 wt. % Si.

The Zr content leads to an increase in the hardness of the studied alloys. There are no significant changes in the hardness with an increase in the temperature up to 350 °C. The highest hardness is observed in the temperature range from 400 °C to 450 °C. A further increase in the annealing temperature up to 600 °C is accompanied by a decrease in the hardness. Note that the peak-hardness of the Al-Fe-Si-Zr alloys that determines the collective Si and Zr influence improves with its content increase.

The effect of Si and Zr content on the electrical conductivity is presented in Figure 7a′–d′. The increase of Si-content led to the electrical conductivity degradation of as-cast, as well as annealed alloys. The increase of the annealing temperature is accompanied by the small electrical conductivity improvement. The peak electrical conductivity decreases with the increase of Si content, as well. The electrical conductivity decreases significantly with an increase in the Zr content in the temperature range from 20 °C to 350 °C. The highest electrical conductivity was also observed at the range from 450 °C to 500°C. A further increase in the testing temperature up to 500 °C is accompanied by an increase in the conductivity.

4. Discussion

Figures 4 and 5 show that the annealing promoted the phase transformation from the metastable Al_6Fe or α-AlFeSi to the Al_3Fe equilibrium phase and induced a significant change in solute levels in the solid solution. The SEM investigation of the studied Al-Fe-Si-Zr alloys demonstrates that the Fe-containing intermetallics in Al-Fe-Si alloys are both of the so-called α-phases forming the Chinese script-like morphology and sometimes even as polyhedral crystals, as shown in [15,16].

Most of the Fe combines with both aluminum and Si to form secondary intermetallic phases because of the low solid solubility of Fe in aluminum (i.e., max. 0.05% at 650 °C). The equilibrium intermetallic Al_3Fe phase can form at slow solidification rates. However, depending on the alloy composition, cooling rate and presence of trace elements, a wide range of intermediate intermetallic phases, such as Al_mFe, Al_9Fe, Al_6Fe, Al_xFe and α-AlFeSi (Al_8Fe_2Si), can form and are considered metastable Fe-rich phases in the literature [17,18].

An increase in the Si content leads to an acceleration of the solid solution decomposition and its maximum replacement to the lower temperature region (Figure 8). A further increase in Si content did not lead to a change in the speed and temperature of the described peak, so the Si content equal to 0.25 wt. % provides the best combination of hardness and electrical conductivity. The same effect of Si has been described also in [11].

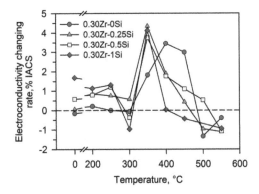

Figure 8. Temperature dependence of the electrical conductivity change of Al-1% Fe-Si-0.3% Zr alloys.

It should be noted that the electrical conductivity of the studied alloys without Zr varies insignificantly in all investigated temperature ranges. An increase in the Si content led to a decrease in the electrical conductivity of the investigated alloys. By contrast, an increase in the Si content led to an increase in the hardness, as shown in Figure 7. The high content of Si should have led to the formation of Si-inclusion in the structure (Figure 3d) and contributed to a decrease in the electrical conductivity of the alloys, as can be seen in Figure 8c,d. The negative effects of Si on the Al alloys' properties are described in the literature [19]. Thus, the addition of the Si in the Al-1% Fe alloy leads to the increase of the hardness, while the electrical conductivity decreases, as can see in Figure 7.

It is known that Zr exists in four forms at different heat processing and heat-treatment stages, i.e., solid solution in matrix, coarse primary Al_3Zr phase, metastable Al_3Zr phase, as well as equilibrium Al_3Zr phase [20]. SEM and TEM micrographs in Figure 5 show the small spheroidal coherent Al_3Zr ($L1_2$) precipitates in the dendritic cells surrounded by precipitate-free interdendritic channels. An increase in the Zr content of more than 0.3 wt. % leads to the formation of primary Al_3Zr particles in the as-cast structure and negatively affects the properties of the investigated alloys. The same results are reported in [21–23]. Furthermore, the appearance of the primary Al_3Zr can also be associated with such disadvantages of the alloy production regime as casting temperature and cooling rate. In the cast structure of the investigated alloys with 0.45–0.6 wt. % Zr, there are primary Al_3Zr crystals, which on the one hand speaks to an insufficient cooling rate ensuring the entry of Zr into the solid solution of aluminum. It is also obvious that the formation of primary Al_3Zr particles leads to a decrease of the volume fraction of secondary precipitations of Al_3Zr (nanoparticles), which negatively affect the thermal stability according to the Zener drag force equitation [24].

It is clearly seen in Figure 4 that Al_3Zr precipitates are forming at the interdendritic channels that result from Zr segregation at the dendrite cores. Thus, dendritically-distributed Al_3Zr precipitates are also a significant problem in commercial wrought alloys, where Zr is added as a recrystallization inhibitor.

The strong age hardening response of the Zr-containing alloy is due to the precipitation of small (10 nm) coherent Al_3Zr ($L1_2$) precipitates, shown by SEM (Figure 4b,c) and TEM (Figure 5). The decrease in the hardness with an increase in temperature can be explained by the precipitation hardening due to the Orowan mechanism. The shear stress required for a dislocation to loop around a precipitate particle is inversely proportional to the edge-to-edge distance between the particles, which was first described by Orowan as shown in Equation (1) [25]:

This is an example of the equation:

$$\Delta\tau = G_B/L, \tag{1}$$

where G_B is the Burgers vector and L is the distance between two particles. According to Equation (1), the strengthening effect from the metastable nanoscale Al_3Zr $L1_2$ dispersoids with an average size of 10 nm was 17-times higher compared to the stable incoherent plate-like $D0_{23}$ Al_3Zr particles with a length equal to 200 nm and a thickness of about 50 nm (Figure 6). Kendig et al. also reported that the maximum hardening is achieved with the size of precipitates Al_3Zr at 5–10 nm [26].

Figure 8 depicts that the electrical conductivity decreases significantly with an increase in the Zr content in the range from 20 °C to 350 °C. By contrast, a further increase in the testing temperature up to 500 °C is accompanied by an increase in the electrical conductivity. This increase in the electrical conductivity with an increasing in temperature can be explained by the decomposition of the solid solution and the diffusion of Zr atoms into particles leading to a decrease of the number of point defects in the matrix, less electron scattering and a decrease in the electrical conductivity. Furthermore, an increase in the Si amount leads to the movement in the maximum of the observed curves to the lower temperature region, which means a decrease in the temperature of the solution decomposition.

The structural and phase transformations during the annealing process were estimated through conductivity and hardness measurements. It is well known that Zr dissolved in α (Al) strongly reduces electrical conductivity. Belov et al. presented the influence of annealing cycles up to 650 °C on the specific conductivity and hardness (HV) of hot-rolled sheets of Al alloys containing up to 0.5 wt. % Zr (mass fraction). It is demonstrated that the conductivity depends on the content of Zr in the Al solid solution, which is the minimum after holding at 450 °C for 3 h. On the other hand, the hardness of the alloy is mainly caused by the amount of nanoparticles of the L1$_2$ (Al$_3$Zr) phase that defines the retention of strain hardening. Chen et al. described an aluminum alloy with 1 wt. % of Zr and 1.7 wt. % of V, which gives an increase to about 5 vol. % of the metastable L1$_2$ phase after aging. It is known that the formation of stable incoherent particles is unexpected and undesirable as this meant that there was less Zr available in the solid solution to form nanoscale secondary Al$_3$Zr precipitates, which are desirable in order to limit grain growth.

It can be seen in Figure 7 that the best combination of electrical conductivity and hardness values can be reached with an acceptable holding time at a temperature of about 450 °C. Electrical conductivity reaches the maximum value at 450 °C and can be explained by the maximum depletion of Zr from the Al solid solution, as was shown in our previous work [17,27]. The maximum electrical conductivity is ensured by all the Zr in Al$_3$Zr particles. It can be realized using the high-temperature S600 annealing. However, use of the lower annealing temperatures in view of providing the optimal hardening due to the formation of the metastable nanosized Al$_3$Zr is also necessary. However, the decomposition of the solid solution at 350 °C required more than 100 h for the slow diffusion of Zr in aluminum according to Knipling [7,18]. It should be noted that conductivity strongly depends not only on the temperature of the annealing, but also on the holding time, as well. Al-Mg-Zr alloys with an ultrafine-grained microstructure combined with a high electrical conductivity (over 57% The International Annealed Copper Standard) after annealing, severe plastic deformation (SPD) via equal channel angular pressing-conforming (ECAP-C), followed by cold drawing have been also described in detail in [25].

Thus, the content of Si and Zr provides the best combination (conjunction) of hardness and electrical conductivity of Al-1% Fe-Si-Zr alloys as can be clearly seen from the summary graphs shown in Figure 9. A compromise of the high conductivity and hardness can be achieved if the temperature of heat treatment is in the range of 450 °C–500 °C and the content of the Si 0.25 wt. %–0.50 wt. %, with the Zr content no more than 0.3 wt. %, respectively, as can be concluded from the obtained results.

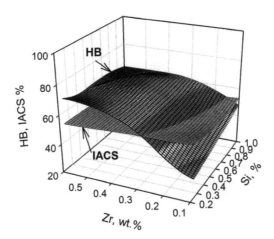

Figure 9. Complex effect of Si and Zr on the electrical conductivity (The International Annealed Copper Standard (IACS)) and hardness (HB) of the Al-1% Fe-Si-Zr alloys.

5. Conclusions

In summary, the relationships between the hardness and electrical conductivity were examined for several Al-Si-Fe-Zr alloys. The influence of the annealing temperature and chemical composition on the microstructure, electrical conductivity and hardness of Al-Fe-Si-Zr alloys with a different content of Si ranging from 0 wt. % to 1 wt. %, various content of Zr from 0 wt. % to 0.6 wt. % and fixed content of Fe of 1 wt. % has been studied in the present work. The following results were obtained:

1. The Al_8Fe_2Si phase with a Chinese script-like morphology was formed for the investigated alloy at 0.25 wt. % of Si, which ensures manufacturability with possible further deformation. An increase in the Si content leads to an increase of Si content in the solid solution and negatively affects the electrical conductivity of the alloys and increases the hardness. Therefore, the favorable content of Si was equal to 0.25 wt. %–0.50 wt. %. It was also described that Si accelerates the decomposition of the solid solution for alloys with a content of Zr equal to 0.3 wt. %.

2. An increase in the content of Zr over 0.3 wt. % in the investigated alloys leads to the formation of the primary Al_3Zr particles, which substantially reduce the hardness; while, 0.15 wt. % of Zr was insufficient for the formation of the secondary Al_3Zr precipitates. The microstructure evaluation showed that with an increase in the amount of Zr from 0.15 wt. % to 0.3 wt. %, an increase in the amount of Al_3Zr dispersoids in the dendrites was observed. The microstructure of the considered alloys with a fixed content of Si equal to 0.5 wt. % and Fe equal to 1 wt. % mostly consists of the solid solution of aluminum (Al) and a eutectic Al_8Fe_2Si + (Al) along the boundaries of dendritic cells. Therefore, the preferable content of Zr was equal to 0.3 wt. %.

3. It was shown that an increase in the Zr content led to an increase in the hardness of the studied alloys. The highest hardness is observed after annealing at 450 °C. A further increase of the temperature leads to a decrease in the hardness.

4. It was clearly shown that small coherent Al_3Zr ($L1_2$) precipitates formed in the Al-1% Fe-0.25% Si-0.3% Zr alloy after S450 annealing. The maximum hardness of the investigated alloy was observed after S450 annealing because of the maximum strengthening effect due to the formation of these particles. It was shown that these particles are effective barriers to grain-boundary migration according to Orowan and transformed into stable incoherent Al_3Zr ($D0_{23}$) precipitates after S600 annealing. The loss of the coherence of Al_3Zr particles led to a decrease in the hardness.

Acknowledgments: The financial support received from the Ministry of Education and Science, Russia, (Belgorod State University Project No. 03.G25.31.0278) is acknowledged. The main results were obtained by using the equipment of the Joint Research Center, «Technology and Materials», Belgorod State National Research University.

Author Contributions: Nikolay Belov and Natalya Korotkova contributed material and realized the electrical conductivity and microhardness testing. Anna Mogucheva, Anna Morozova, Dmitriy Bukin and Olga Lukianova proposed and performed microstructure experiments and realized the data processing. Anna Mogucheva, Anna Morozova and Rustam Kaibyshev discussed and analyzed the obtained results.

Conflicts of Interest: The authors declare no conflict of interest.

References

1. GencalpIrizalp, S.; Saklakoglu, N. Effect of Fe-rich intermetallics on the microstructure and mechanical properties of thixoformed A380 aluminum alloy. *Eng. Sci. Technol. Int. J.* **2014**, *17*, 58–62. [CrossRef]
2. Belov, N.A.; Alabin, A.N.; Eskin, D.G.; Istomin-Kastrovskii, V.V. Optimization of hardening of Al–Zr–Sc cast alloys. *J. Mater. Sci.* **2006**, *41*, 5890–5899. [CrossRef]
3. Mondolfo, L.F. *Aluminum alloys: Structure and Properties*; Butterworths: London, UK, 1976; ISBN 9781483144825.
4. Panahi, D. *Precipitation of Intermetallic Phases from Rapidly Solidifying Aluminum Alloys*; McMaster University: Hamilton, ON, Canada, 2009.
5. Kral, M.V.; Nakashima, P.N.H.; Mitchell, D.R.G. Electron microscope studies of Al-Fe-Si intermetallics in an Al-11 Pct Si alloy. *Metall. Mater. Trans. A* **2006**, *37*, 1987–1997. [CrossRef]
6. Ryum, N. Precipitation and recrystallization in an A1-0.5 wt % Zr-alloy. *Acta Metall.* **1969**, *17*, 269–278. [CrossRef]
7. Knipling, K.E.; Dunand, D.C.; Seidman, D.N. Precipitation evolution in Al–Zr and Al–Zr–Ti alloys during aging at 450–600 °C. *Acta Mater.* **2008**, *56*, 1182–1195. [CrossRef]
8. Belov, N.A.; Alabin, A.N. Energy Efficient Technology for Al–Cu–Mn–Zr Sheet Alloys. *Mater. Sci. Forum* **2013**, *765*, 13–17. [CrossRef]

9. Alabin, A.N.; Belov, A.N.; Korotkova, N.O.; Samoshina, M.E. Effect of annealing on the electrical resistivity and strengthening of low-alloy alloys of the Al–Zr–Si system. *Metal Sci. Heat Treat.* **2016**, *58*, 527–531. [CrossRef]
10. Belov, N.A.; Alabin, A.N.; Prokhorov, A.Y. Effect of zirconium additive on the strength and electrical resistivity of cold-rolled aluminum sheets. *Izv. Vysh. Uchebn. Zaved. Tsvetn. Met.* **2009**, *4*, 42–47.
11. Belov, N.A.; Aksenov, A.A.; Eskin, D.G. *Iron in Aluminium Alloys: Impurity and Alloying Element*; Taylor and Francis: London, UK, 2002; pp. 1107–1112. ISBN 0-415-27352-8.
12. Vorontsova, L.A.; Maslov, V.V.; Peshkov, I.B. *Aluminum and Aluminum Alloys in Electrical Products*; Energiya: Moscow, Russia, 1971. (In Russian)
13. Popovic, S.; Grzeta, B.; Ilakovac, V.; Kroggel, R.; Wendrock, G.; Löffler, H. Lattice constant of the FCC Al-rich α-Phase of Al-Zn alloys in equilibrium with GP zones and the β (Zn)–Phase. *Phys. Status Solidi A* **1992**, *130*, 273–292. [CrossRef]
14. Ma, Y.; Romming, C.; Lebech, B.; Gjonnes, J.; Tafto, J. Structure refinement of Al3Zr using single-crystal X-ray diffraction, powder neutron diffraction and CBED. *Acta Crystallogr. B* **1992**, *48*, 11–16. [CrossRef]
15. Shakiba, M.; Parson, N.; Chen, X.-G. Effect of homogenization treatment and silicon content on the microstructure and hot workability of dilute Al–Fe–Si alloys. *Mater. Sci. Eng. A* **2014**, *619*, 180–189. [CrossRef]
16. Taylor, J.A. Iron-Containing intermetallic phases in Al-Si based casting alloys. *Procedia Mater. Sci.* **2012**, *1*, 19–33. [CrossRef]
17. Liu, P.; Thorvaldsson, T.; Dunlop, G.L. Formation of intermetallic compounds during solidification of dilute Al–Fe–Si alloys. *Mater. Sci. Technol.* **1986**, *2*, 1009–1018. [CrossRef]
18. Skjerpe, P. Intermetallic phases formed during DC-casting of an Al-0.25 Wt Pct Fe-0.13 Wt Pct Si alloy. *Metall. Mater. Trans. A* **1987**, *18*, 189–200. [CrossRef]
19. Belov, N.A.; Alabin, A.N.; Matveeva, I.A.; Eskin, D.G. Effect of Zr additions and annealing temperature on electrical conductivity and hardness of hot rolled Al sheets. *Trans. Nonferrous Met. Soc. China* **2015**, *25*, 2817–2826. [CrossRef]
20. Zhang, J.; Ding, D.; Zhang, W.; Kang, S.; Xu, X.; Gao, Y.; Chen, G.; Chen, W.; You, X. Effect of Zr addition on microstructure and properties of Al–Mn–Si–Zn-based alloy. *Trans. Nonferrous Met. Soc. China* **2014**, *24*, 3872–3878. [CrossRef]
21. He, Y.; Zhang, X.; Cao, Z. Effect of minor Sc and Zr addition on grain refinement of as-cast Al-Zn-MgCu alloys. *Chin. Foundry* **2009**, *6*, 214–218.
22. Garcia, D.A.; Dye, D.; Jackson, M.; Grimes, R.; Dashwood, R.J. Development of microstructure and properties during the multiple extrusion and consolidation of Al–4Mg–1Zr. *Mater. Sci. Eng. A* **2010**, *527*, 3358–3364. [CrossRef]
23. Zhang, L.; Eskin, D.G.; Miroux, A.G.; Katgerman, L. On the mechanism of the formation of primary intermetallics under ultrasonic melt treatment in an Al-Zr-Ti alloy. *IOP Conf. Ser. Mater. Sci. Eng.* **2012**, *27*, 012002. [CrossRef]
24. PA, M.; Ferry, M.; Chandra, T. Five decades of the Zener equation. *ISIJ Int.* **1998**, *38*, 913–924. [CrossRef]
25. Hull, D.; Bacon, D.J. *Introduction to Dislocations*; Butterworth-Heineman: Oxford, UK, 1984; ISBN 0750623618.
26. Kendig, K.; Miracle, D. Strengthening mechanisms of an Al-Mg-Sc-Zr alloy. *Acta Mater.* **2002**, *50*, 4165–4175. [CrossRef]
27. Murashkin, M.Y.; Medvedev, A.E.; Kazykhanov, V.U.; Raab, G.I.; Ovid'ko, I.A.; Valiev, R.Z. Microstructure, strength, electrical conductivity and heat resistance of an Al-Mg-Zr alloy after ECAP-conform and cold drawing. *Rev. Adv. Mater. Sci.* **2016**, *47*, 16–25.

Article

The Effect of Eutectic Structure on the Creep Properties of Sn-3.0Ag-0.5Cu and Sn-8.0Sb-3.0Ag Solders

Yujin Park [1], Jung-Hwan Bang [2], Chul Min Oh [3], Won Sik Hong [3] and Namhyun Kang [1],*

[1]   Department of Materials Science and Engineering, Pusan National University, Busan 46241, Korea;
      yujin3584@naver.com
[2]   Micro-Joining Center, Korea Institute of Industrial Technology, Incheon 21999 , Korea; nova75@kitech.re.kr
[3]   Electronic Convergence Materials & Device Research Center, Korea Electronics Technology Institute,
      Seongnam 13509, Korea; cmoh@keti.re.kr (C.M.O.); wshong@keti.re.kr (W.S.H.)
*   Correspondence: nhkang@pusan.ac.kr; Tel.: +82-51-510-3274

Received: 9 November 2017; Accepted: 28 November 2017; Published: 3 December 2017

Abstract: Solder joints are the main weak points of power modules used in harsh environments. For the power module of electric vehicles, the maximum operating temperature of a chip can reach 175 °C under driving conditions. Therefore, it is necessary to study the high-temperature reliability of solder joints. This study investigated the creep properties of Sn-3.0Ag-0.5Cu (SAC305) and Sn-8.0Sb-3.0Ag (SSA8030) solder joints. The creep test was conducted at 175 and 190 °C with the application of 2.45 MPa. The SAC305 solder had superior creep properties to those of SSA8030 solder at 175 °C and at largely the same homologous temperature (T_H~0.91 for SAC305 and T_H~0.92 for SSA8030). Both solders had primary β-Sn and a eutectic mixture of β-Sn and Ag_3Sn. Compared to SSA8030, the SAC305 solder contained ~10% more eutectic structure and contained Ag_3Sn that was 3 times smaller and more round in shape. Furthermore, the SSA8030 solder precipitated SnSb in an elongated fiber shape (1–50 μm in size) after the creep test. Coarse and elongated Ag_3Sn and SnSb of the SSA8030 solder negatively affected crack propagation in the dislocation creep region and decreased the creep resistance.

Keywords: creep; lead-free solder; Sb solder; Sn-8.0Sb-3.0Ag; solder microstructure

1. Introduction

In order to prevent global warming and other environmental problems, electric and hybrid vehicles are being developed to reduce automobile CO_2 emissions. Development of power modules is essential to obtain high efficiency and current conversion in these vehicles. For power modules made from Si chips, the maximum temperature can increase to 175 °C during operation as they require higher performance and a larger capacity than insulated gate bipolar transistor (IGBT) modules. Therefore, high-temperature reliability needs to be thoroughly evaluated for reliable use of power modules [1–4].

Components for power modules such as semiconductors and terminals are attached using solders. Thermal stress is applied by repeated shrinkage and expansion of solder joints, therefore leading to failure during operation. Because the components for modules in automobiles are used in harsher environments than those of normal electronic components, Pb-rich solders with a high melting point and Pb-containing solders with high reliability have been used for these power modules. However, due to a recent environmental regulation for automobiles, i.e., End-of-Life Vehicle (ELV), the use of Pb-containing solders for electrical components was completely prohibited in the European Union (EU), the European Free Trade Association (EFTA), Japan, Korea, and China. Therefore, studies on high-temperature Pb-free solder to substitute for conventional Pb-containing solder have been conducted [5–10].

The creep test is essential in evaluating solder joint properties at high temperature. Diffusion and dislocation movement in a material are activated as temperature increases. Therefore, the strength of the material is decreased and eventually deformation is accelerated leading to a fracture although the applied stress is lower than its yield strength. The creep mechanism is classified with respect to homologous temperature (T_H) and stress. Homologous temperature is the ratio between test temperature and melting temperature, and it is an important factor for the creep test because creep strength is sensitive to the test temperature compared to the melting temperature.

Creep mechanism is classified into grain-boundary sliding, dislocation creep, and diffusion creep. Creep deformation induced by grain-boundary sliding occurs when the high stress is applied. On the other hand, creep occurring under low stress is diffusion creep and it can be divided into Coble creep and Nabarro-Herring creep. The creep mechanism most frequently studied is dislocation creep occurring at relatively high temperatures and in moderate stress regions [11–16]. Dislocation creep is caused by movement such as glide or climbing of dislocation and corresponds to Low temperature creep, High Temperature creep, and Power-law break down. The strengthening mechanism of the solder is very important in understanding its high-temperature properties. The strengthening mechanisms that can be used for solidification of solder are solid-solution strengthening and precipitation strengthening. Therefore, the study of solid solution and precipitation is important in understanding dislocation creep because lattice deformation due to solid solution and precipitation interfere with the movement of the dislocation.

There are many studies on Pb-free solders, but very few studies on their creep properties have been reported for Pb-free solders used in high-temperature environments [17]. Recently, researchers have studied nano-indentation requiring a short creep test time [18,19]. However, nano-indentation creep is influenced only by microstructures near the dent area, and complex microstructures such as primary phase, eutectic structure, and precipitation need to be considered for a full understanding of creep properties. Pb-containing solders have a sufficient database of creep properties and are primarily composed of Sn. Pb-free solder also has an Sn-rich composition and the creep mechanism map of Sn–Pb solder has been used for Pb-free solder [20]. This study measured the creep properties of Sn-3.0Ag-0.5Cu (SAC305), the most commonly used solder, and Sn-8.0Sb-3.0Ag (SSA8030), a high melting point solder, and analyzed the microstructure of the joint to understand the creep mechanism.

2. Materials and Methods

2.1. Preparation of Creep Specimen

Figure 1 shows a schematic diagram of the power module used in this the study. Solder joints where Si chips and terminals are attached to direct bonded copper (DBC) are the main cause of failure. The terminal used for the power module was fixed to a plastic case, followed by attachment to the DBC using two types of solder. The terminals were coated with Sn and the surface of the DBC coupons were treated with organic solderability preservative (OSP) such as alkylbenzimidazole and diphenylimidazole. SAC305 and SSA8030 used for the solder have a melting temperature of 217 and 232 °C, respectively. The fixed terminals are vulnerable to stress when the solder solidify for reflow soldering and the assembled parts to vehicles are in operation. This becomes a reason for creep fracture at the operating temperature of 175 °C. Figure 1 shows the creep test with the terminal coupons.

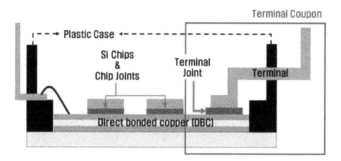

Figure 1. Schematic diagram of power module showing the terminal joint.

2.2. Creep Test

Figure 2 shows a strain measured for the terminal due to expansion and contraction during reflow soldering. A strain gauge was attached to the terminal during terminal soldering to the DBC. The strain of the terminal was measured three times in total, and it showed maximum strains of 1.07×10^{-3}, 7.91×10^{-4}, and 8.14×10^{-4}, respectively. Because the average strain was 8.92×10^{-4} and the length of the terminal (L) was 4 mm, the average-maximum displacement (ΔL) corresponding to the average strain was calculated to be 3.57 μm using Equation (1):

$$\text{strain} = \frac{\Delta L(\text{displacement})}{L(\text{length})}. \tag{1}$$

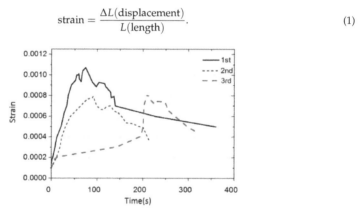

Figure 2. Strain measured for soldering the terminals to the direct bonded copper (DBC).

We applied the calculated displacement (ΔL) to the force–displacement relation obtained through the experiment of the shear stress at the terminal joint, and calculated the stress of the solder joint caused by the displacement. Figure 3 indicates the shear force and displacement behavior measured for the terminal-solder joint. The solid black line of Figure 3 is the shear force and displacement measured for the terminal joint. The red dotted line shows the regression relationship between the displacement and the force in the elastic region. The slope of red dotted line (40.86) shows a different value as compared with the shear modulus of bulk solder (16.60 GPa). The calculated regression equation is as follows:

$$y(\text{force}) = 40.86 \times \Delta L(\text{displacement}) + 1636.8. \tag{2}$$

Substituting the displacement of 3.57 μm into Equation (2), the stress was calculated to be 2.18 MPa and the load of 1.7 kg is obtained by multiplying the force with the joining area (8 mm²). When the stress of 2.18 MPa is applied at 175 °C, dislocation creep will occur at the SAC305 and SSA8030 terminal joints. In order to reliably proceed with the creep test in the dislocation creep area, we applied a stress of 2.45 MPa to the terminal using a weight of 2 kg.

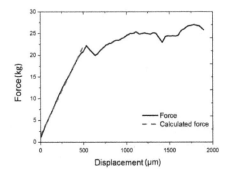

Figure 3. Shear force–displacement behavior measured for terminal-solder joints and the regression line calculated in the elastic region.

The conditions of the creep test such as stress, temperature, and T_H are shown in Table 1. We determined creep failure from the time when the terminal was electrically disconnected completely from the DBC. To observe the failure time, we used a multi-channel multimeter to check electrical resistance every second, and the test was repeated five times in each conditions. We conducted the creep test at maximum operating temperatures of 175 and 190 °C to consider the influence of T_H on creep behavior.

Table 1. Conditions of the creep test.

Solders	SAC305		SSA8030	
Melting temperature	217 °C		234 °C	
Creep test temperature	175 °C	190 °C	175 °C	190 °C
T_H	0.91	0.94	0.88	0.92
Stress		2.45 MPa		

2.3. Microstructure

For microstructural observation, specimens were cold-mounted with conductive copper resin. The mounted specimens were subjected to a final polishing using 0.05 μm colloidal silica. The microstructure was observed using the back-scattered electron (BSE) mode of a field-emission scanning electron microscopy (FE-SEM, TESCAN, Brno, Czech Republic) with no etching. We measured the composition of the phase using energy dispersive spectroscopy (EDS) and field-emission electron probe micro analysis (FE-EPMA, JEOL, Tokyo, Japan). In order to observe the crystal orientation and stress distribution, we applied argon ion milling treatment on the specimens after final polishing, followed by electron back-scattered diffraction (EBSD). The amount and shape of the microstructures were measured from 5 to 10 images, and the average value was calculated using the i-solution DT program (IMT i-solution Inc., TX, USA).

3. Results

3.1. Effect of Solder Composition and Test Temperature on Creep Failure

Figure 4 shows the creep failure time for the SAC305 and SSA8030 solders at 175 and 190 °C. At 175 °C, the failure time of the SAC305 and SSA8030 solder was 207 min and 84 min, respectively. At 190 °C, the failure time of the SAC305 solder was 66 min and that of the SSA8030 solder was 47 min.

The SAC305 solder, compared to the SSA8030 solder, exhibited a better creep property at both 175 and 190 °C. Considering the relative ratio of test temperature with respect to melting temperature, for the same T_H values of 0.91 (175 °C for SAC305) and 0.92 (190 °C for SSA8030), the failure time was recorded to be 207 and 47 min, respectively. Therefore, the creep property of the SAC305 was significantly superior to that of the SSA8030 for the approximately same T_H level.

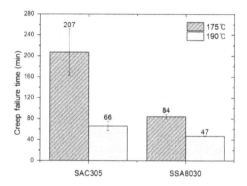

Figure 4. Effects of solders and test temperature on creep failure time. The bars correspond to standard deviations.

Figure 5 shows the fracture surface after the creep test at 175 °C. The SAC305 solder exhibited an elongated dimple fracture parallel to the direction of applied load (Figure 5a). Figure 5b shows the fracture surface at high magnification ($\times$10,000) for the SAC305 solder joint. We measured a dimple size of approximately 10 μm and noted that Ag_3Sn had precipitated on the dimple surface. For the SSA8030 solder, it was difficult to distinguish the dimple shape clearly at low magnification (Figure 5c) because it is more vulnerable to oxidation than the SAC305 [21]. However, in Figure 5d, dimples approximately 10 μm in size were observed and SnSb and Ag_3Sn were precipitated on the dimple surface. Based on EDS analysis, Ag_3Sn grew in a round shape, as shown in Figure 5b,d. However, SnSb was precipitated in an irregular and sharp shape (Figure 5d) on the SSA8030. Coarse SnSb precipitates of large size (~50 μm) and rough surface were also observed as shown in Figure 5c.

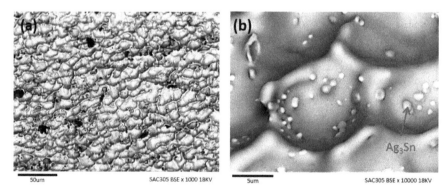

Figure 5. *Cont.*

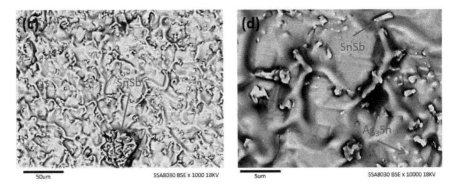

Figure 5. Fracture surface after creep testing at 175 °C: (**a,b**) SAC305 solder joints; (**c,d**) SSA8030 solder joints.

3.2. Effect of Solder Composition on Microstructure

Figure 6a,c shows the microstructures in BSE mode before and after the creep experiment of the SAC305 solder joint, respectively, and Figure 6b,d shows those of the SSA8030 solder joint, respectively. The light gray microstructure is the primary β-Sn, the dark gray is the eutectic mixture of β-Sn and Ag_3Sn, and the black is the Cu_6Sn_5. The SSA8030 and SAC305 solders had the same type of microstructure. The SSA8030 had no Cu content but produced Cu_6Sn_5 by diffusion of Cu from the DBC substrate during reflow soldering. The amount of Cu_6Sn_5 increased after the creep test for the SAC305 and SSA8030 solders (Figure 6c,d). The SSA8030 solder indicated a coarser Cu_6Sn_5 than the SAC305. After the creep test, we measured the area fraction of Cu_6Sn_5 in the SAC305 solder and SSA8030 solder to be 2.0% and 3.9%, respectively. Although the SAC305 solder had less Cu_6Sn_5 than the SSA8030 solder, the SAC305 solder had smaller and uniformly distributed Cu_6Sn_5. Both of the solders had largely the same width of β-Sn dendrites, and their widths were approximately 12 μm, which was similar to the dimple size (~10 μm) of the fracture surface shown in Figure 5b,d. In other words, decohesion occurred at the boundary due to the strength difference between the eutectic region reinforced by Ag_3Sn and the relatively low strength of the β-Sn dendrite region. Finally, a dimple fracture formed by the cavity growth [22]. We also confirmed it by the Ag_3Sn detected on the dimple surface (Figure 5b,d).

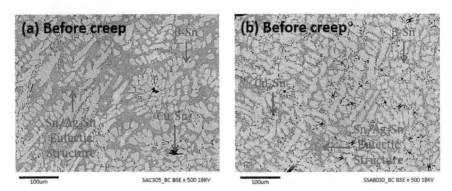

Figure 6. *Cont.*

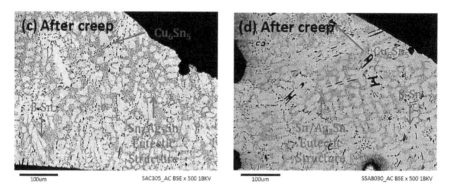

Figure 6. Microstructure before and after the creep test, respectively, for (**a,c**) SAC305 and (**b,d**) SSA8030 solders.

Figure 7a shows the EBSD phase map with Sn in red and Cu_6Sn_5 in green. Black dots in the red region were confirmed as Ag_3Sn precipitate using EDS. Figure 7b is the kernel average misorientation (KAM) map that shows the stress field through a variation of the local crystal orientation [23]. We observed a stress field of green and red around the interface where Cu_6Sn_5 and Ag_3Sn formed. Because the stress field prevents the movement of dislocation and plastic deformation, the precipitation of Cu_6Sn_5 and Ag_3Sn is related to the creep strength.

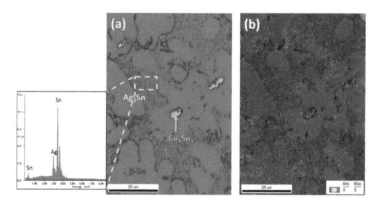

Figure 7. EBSD images of SAC305 before the creep test: (**a**) inverse pole figure (IPF); (**b**) kernel average misorientation (KAM).

Unlike Sn-37Pb solders producing a lamella-eutectic structure, Pb-free solders produce dispersed Ag_3Sn with a fibrous form around the primary β-Sn [24,25]. Ag_3Sn is a thermally stable compound because of the limited solubility of Ag in β-Sn [26,27]; therefore, it strengthens the eutectic structure. Ag_3Sn is one of the components in the eutectic mixture for Sn-rich Pb-free solders containing Ag content. Figure 8 shows the amount of eutectic structure for the SAC305 and SSA8030 solders. Before the creep test, the SAC305 had the largest amount of eutectic structure (58.5%). After the creep test, the SAC305 solder contained a larger amount of eutectic structure than the SSA8030 solder. The creep test at 175 °C decreased the amount of eutectic structure for both solders. This is because of the Ostwald Ripening effect in which intermetallic compounds (IMCs) smaller than the critical size are resolved, and they combine with the large IMCs, increasing the size of the IMCs [28].

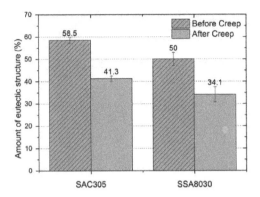

Figure 8. The amount of eutectic microstructure before and after creep testing at 175 °C. The bars correspond to standard deviations.

4. Discussion

Figure 9 shows the eutectic structure observed in detail. The white precipitates are Ag_3Sn and the black are Cu_6Sn_5. The SAC305 shows Ag_3Sn of smaller size and of a rounder shape than that of the SSA8030. Furthermore, the SSA8030 produced very small SnSb IMCs (dark gray precipitates in Figure 9d) with some large SnSb IMCs after the creep test. For the precipitates of small size and round shape, their strength increases because they reduce the inter back stress formation and the dislocation movement [29]. Therefore, we measured the size and shape of Ag_3Sn quantitatively and it is shown in Table 2. The precipitate size was measured from the long axis of the precipitates, and the roundness (R) was calculated following ISO16112 using Equation (3) as follows [30]:

$$R = \frac{A/l_m^2}{\pi/4} = \frac{A}{A_m}.$$ (3)

R is roundness, l_m is the maximum diameter of the precipitate, A_m is an area of the circle whose diameter is l_m, and A is the actual size of the precipitate.

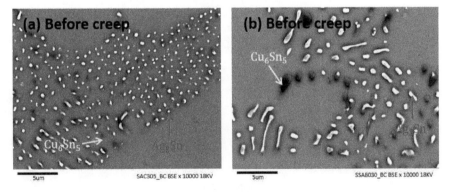

Figure 9. *Cont.*

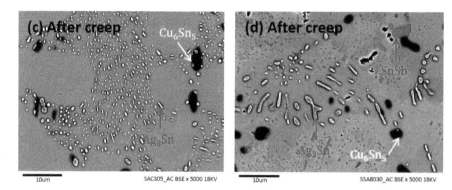

Figure 9. Intermetallic compounds before and after creep testing at 175 °C, respectively, for (a,c) SAC305 and (b,d) SSA8030 solders.

Table 2. Size and shape of Ag₃Sn in eutectic microstructure before and after creep test.

Creep	Solders	Fraction of Ag₃Sn (%)	Longer axis of Ag₃Sn (μm)	Roundness (R)	Δσₚ (MPa)
Before Creep	SAC305	4.7	0.37	0.88	35.58
	SSA8030	5.7	1.16	0.70	11.08
After Creep	SAC305	4.4	0.48	0.94	24.56
	SSA8030	3.4	1.19	0.79	8.56

The Ag₃Sn in the SAC305 solder had an average size of approximately 0.37 μm and was in a round shape regardless of the creep test. After the creep test at 175 °C, the average size of Ag₃Sn increased to 0.48 μm with no variation in Ag₃Sn fraction. Meanwhile, Ag₃Sn in the SSA8030 solder had an elongated shape and an average size of 1.16 and 1.19 μm, respectively, before and after the creep test. The SSA8030 showed no variation in the average size of Ag₃Sn regardless of the creep test. The SAC305 showed Ag₃Sn that was 3 times smaller and rounder than that of the SSA8030 regardless of the creep test. Moreover, the maximum size of Ag₃Sn in the SSA8030 solder was approximately 7 μm. The coarse Ag₃Sn negatively affects mechanical strength [31]. Fine and abundant Ag₃Sn hinders the movement of dislocation and increases creep properties [32]. In order to analyze quantitatively the mechanical strength variation by precipitate, Equations (4) and (5) were used as follows [33]:

$$\Delta\sigma_P = 5.9\left(f^{\frac{1}{2}}/x\right) \times [\ln(x/2.5E - 4] \tag{4}$$

$$x = D \times (2/3)^{1/2}. \tag{5}$$

where f is the volume fraction of the precipitate, x is the mean planar intercept diameter (μm), and D is the mean diameter (μm) of the precipitate. We calculated the strength ($\Delta\sigma_P$) due to the precipitates by substituting the size and amount of Ag₃Sn measured in Figure 9 and Table 2. Because the SAC305 solder had small and round Ag₃Sn, the strength ($\Delta\sigma_P$) due to precipitates was 3 times that of the SSA8030 solder (Table 2). In Equation (4), the volume fraction (f) of the precipitate has a power of 0.5 and it has less influence on the strength than the size factor (x) of the precipitate. Specifically, large and elongated SnSb IMCs were precipitated in the SSA8030 solder after the creep test (Figures 5d and 9d). Therefore, we conducted additional analysis to investigate the effect of the large and elongated SnSb on creep failure.

Figure 10a,b shows the FE-EPMA images of the SSA8030 solder after the creep test at 175 °C. Sn was the most abundant element for β-Sn and the eutectic mixture (Figure 10a). SnSb was distributed irregularly in β-Sn and the eutectic mixture as a red phase, and rich Sb content in β-Sn dendrites was also observed as a light blue phase (Figure 10b). Sb elements were solved in the Sn-rich phase and produced SnSb during the creep test at 175 °C. Moreover, the creep crack propagated along the coarse SnSb having a size of approximately 50 μm. Figure 10c is the enlarged view of crack propagation through the coarse SnSb IMCs. In the image quality (IQ) map of EBSD, the coarse SnSb was composed of grain boundaries. The crack propagated along the grain boundary of SnSb. Thus, the coarse SnSb precipitates acted as a crack propagation path and decreased the creep properties of the SSA8030 solder.

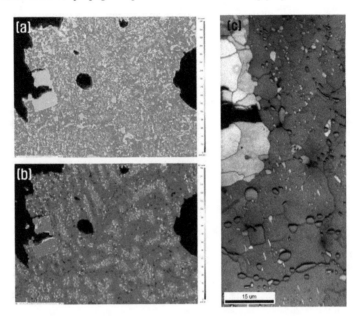

Figure 10. Microstructure of SSA8030 solders after the creep test at 175 °C: (**a,b**) Sn and Sb mapping using FE-EPMA, respectively; (**c**) IQ map of EBSD.

The power modules of electric vehicles require a high operating temperature, so the creep properties of terminal-solder joints need to be investigated. However, the size of the terminal-solder joint is so small that the relation of grain boundary and dislocation to the creep properties cannot be observed. For future study, therefore, high-temperature reliability in the scale of bulk solder must be explored so that the creep mechanism of grin-boundary sliding, dislocation creep, and diffusion creep can be understood. Creep properties obtained from bulk solder should be helpful in understanding the real properties of terminal-solder joints.

5. Conclusions

This study investigated the creep properties of the SAC305 and SSA8030 solder joints for the power modules of electric vehicles. The detailed effects of solder composition and microstructure on creep failure time are summarized as follows:

1. During reflow soldering, the terminal joint of the power module corresponded to the dislocation creep region because a stress of 2.18 MPa was applied at the maximum operating temperature of 175 °C.
2. Both solders had primary β-Sn and a eutectic mixture of β-Sn and Ag_3Sn. SAC305 solder contained 58.5% eutectic structure before the creep test after which it decreased to 41.3%. For the SSA8030 solder, the amount of eutectic structure also decreased after the creep test, from 50% to 34.1%. The amount of eutectic structure in the SAC305 solder was larger than in the SSA8030 solder.

3. The SAC305 solder had small and round Ag$_3$Sn; therefore, the strength ($\Delta\sigma_P$) due to the precipitates was 3 times greater than that of the SSA8030 solder. The SAC305 solder, compared with the SSA8030 solder, lasted a longer time at 175 °C and at largely the same homologous temperature (T_H~0.91 for SAC305 and T_H~0.92 for SSA8030).

4. During the creep test, Sb atoms solved in β-Sn precipitated into SnSb in the SSA8030 solder. The coarse SnSb (~50 μm) precipitated and became a crack propagation path, therefore decreasing the creep property.

Acknowledgments: This work was supported by the Technology Innovation Program (grant No. 10051318, Development of SiC Automotive OBC Power Module with Environment-Friendly High-Temperature Bonding Materials) funded by the Ministry of Trade, Industry & Energy (MOTIE, Korea) and the National Research Foundation of Korea (NRF) grant funded by the Korea government (MSIT) through GCRC-SOP (grant no. 2011-0030013).

Author Contributions: Y. Park, W.S. Hong and C.M. Oh conceived and designed the experiments; Y. Park performed the experiments; Y. Park and N. Kang analyzed the data; J. Bang contributed reagents/materials/analysis tools; Y. Park and N. Kang wrote the paper.

Conflicts of Interest: The authors declare no conflict of interest.

References

1. Feller, L.; Hartmann, S.; Schneider, D. Lifetime analysis of solder joints in high power IGBT module for increasing the reliability for operation at 150 °C. *Microelectron. Reliab.* **2008**, *48*, 1161–1166. [CrossRef]
2. Liang, Z. Status and trend of automotive power packaging. *ISPSD* **2012**, *24*, 325–331. [CrossRef]
3. McCluskey, F.P.; Dash, M.; Wang, Z.; Huff, D. Reliability of high temperature solder alternatives. *Microelectron. Reliab.* **2006**, *46*, 1910–1914. [CrossRef]
4. Yoon, J.W.; Bang, J.H.; Ko, Y.H.; Yoo, S.H.; Kim, J.K.; Lee, C.W. Power module packaging technology with extended reliability for electric vehicle applications. *J. Microelectron. Packag. Soc.* **2014**, *21*, 1–13. [CrossRef]
5. Santos, W.L.R.; Brito, C.; Quaresma, J.M.V.; Spinelli, J.E.; Garcia, A. Plate-like cell growth during directional solidification of a Zn-20 wt %Sn high-temperature lead-free solder alloy. *Mater. Sci. Eng. B* **2014**, *182*, 29–36. [CrossRef]
6. Ji, H.; Qiao, Y.; Li, M. Rapid formation of intermetallic joints through ultrasonic-assisted die bonding with Sn-0.7Cu solder for high temperature packaging application. *Scr. Mater.* **2016**, *110*, 19–23. [CrossRef]
7. Nahavandi, M.; Hanim, M.A.A.; Ismarrubie, A.N.; Hajalilou, A.; Rohaizuan, R.; Fadzli, M.Z.S. Effects of silver and antimony content in lead-free high-temperature solders of Bi-Ag and Bi-Sb on copper substrate. *J. Electron. Mater.* **2014**, *43*, 579–585. [CrossRef]
8. Heo, M.; Kang, N.; Park, S.; Kim, J.; Hong, W. Kinetics of intermetallic compounds growth induced by electromigration of Sn-0.7Cu solder. *Korean J. Met. Mater.* **2016**, *54*, 908–915. [CrossRef]
9. Kim, B.; Lee, C.; Lee, D.; Kang, N. Effect of Sb addition on Bi-2.6Ag-0.1Cu solders for high-temperature applications. *J. Alloy Compd.* **2014**, *592*, 207–212. [CrossRef]
10. Sakai, S.; Yoshida, H.; Hiratsuka, J.; Vandecasteele, C.; Kohlmeyer, R.; Rotter, V.S.; Passarini, F.; Santini, A.; Peeler, M.; Li, J.; et al. An international comparative study of end-of-life vehicle (ELV) recycling systems. *J Mater. Cycles Waste Manag.* **2014**, *16*, 1–20. [CrossRef]
11. Dasgupta, A.; Pecht, M. Material failure mechanisms and damage models. *IEEE Trans. Reliab.* **1991**, *40*, 531–536. [CrossRef]
12. Dasgupta, A.; Hu, J.M. Failure mechanism models for plastic deformations. *IEEE Trans. Reliab.* **1992**, *41*, 168–174. [CrossRef]
13. Dieter, G.E. *Mechanical Metallurgy (Graded)*; McGraw-Hill: New York, NY, USA, 1986.
14. Coble, R.L. A model for boundary diffusion controlled creep in polycrystalline materials. *J. Appl. Phys.* **1963**, *34*, 1679–1682. [CrossRef]
15. Garofalo, F. *Fundamentals of Creep and Creep-Rupture in Metals*; McMillan Series in Material Science; McMillan: Basingstoke, UK, 1965.
16. Raj, R.; Ashby, M.F. On grain boundary sliding and diffusional creep. *Metall. Mater. Trans. B* **1971**, *2*, 1113–1127. [CrossRef]

17. Fahim, A.; Ahmed, S.; Chowdhury, M.R.; Suhling, J.C.; Lall, P. High temperature creep response of lead free solders. In Proceedings of the 15th IEEE Intersociety Conference on ITherm, Las Vegas, NV, USA, 31 May–3 June 2016; pp. 1218–1224. [CrossRef]

18. Cordova, M.E.; Shen, Y.L. Indentation versus uniaxial power-law creep: A numerical assessment. *J. Mater. Sci.* **2015**, *50*, 1394–1400. [CrossRef]

19. Hasnine, M.; Suhling, J.C.; Prorok, B.C.; Bozack, M.J.; Lall, P. Nano mechanical characterization of lead free solder joints. *MEMS Nanotechnol.* **2014**, *5*, 11–22. [CrossRef]

20. Xiag, Q.; Armstrong, W.D. Tensile creep and microstructural characterization of bulk Sn3.9Ag0.6Cu lead-free solder. *J. Electron. Mater.* **2005**, *34*, 196–211. [CrossRef]

21. Chen, F.; Du, Y.; Zeng, R.; Gan, G.; Du, C. Thermodynamics of oxidation on Pb-free solders at elevated temperature. *Mater. Sci. Forum* **2009**, *610–613*, 526–530. [CrossRef]

22. Ochoa, F.; Deng, X.; Chawla, N. Effects of cooling rate on creep behavior of a Sn3.5Ag alloy. *J. Electron. Mater.* **2004**, *33*, 1596–1607. [CrossRef]

23. Han, I.; Eom, J.; Yun, J.; Lee, B.; Kang, C. Microstructure and hardness of 1st layer with crystallographic orientation of solidification structure in multipass weld using high Mn-Ni flux cored wire. *J. Weld. Join.* **2016**, *34*, 77–82. [CrossRef]

24. Lee, J.; Kim, H.; Lee, Y.; Choi, Y. Interfacial Properties with Kind of Surface Finish and Sn-Ag Based Lead-free Solder. *J. Weld. Join.* **2009**, *27*, 20–24. [CrossRef]

25. Park, J.; Kim, M.; Oh, C.; Do, S.; Seo, J.; Kim, D.; Hong, W. Solder join fatigue life of flexible impact sensor module for automotive electronics. *Korean J. Met. Mater.* **2017**, *55*, 232–240. [CrossRef]

26. Seo, S.; Kang, S.; Shih, D.; Lee, H. The evolution of microstructure and microhardness of Sn-Ag and Sn-Cu solders during high temperature aging. *Microelectron. Reliab.* **2009**, *49*, 288–295. [CrossRef]

27. Hong, W.; Kim, W.; Park, N.; Kim, K. Activation energy for intermetallic compound formation of Sn-40Pb/Cu and Sn-3.0Ag-0.5Cu/Cu Solder joints. *J. Weld. Join.* **2007**, *25*, 184–190. [CrossRef]

28. Morando, C.; Fornaro, O.; Garbellini, O.; Palacio, H. Microstructure evolution during the aging at elevated temperature of Sn-Ag-Cu solder alloys. *Procedia Mater. Sci.* **2012**, *1*, 80–86. [CrossRef]

29. Kerr, M.; Chawla, N. Creep deformation behavior of Sn-3.5Ag solder/Cu couple at small length scales. *Acta Mater.* **2004**, *52*, 4527–4535. [CrossRef]

30. ISO. *ISO/TC25 Cast Irons and Pig Irons*; ISO: Geneva, Switzerland; Available online: https://www.iso.org/committee/47206/x/catalogue/ (accessed on 3 December 2017).

31. Suganuma, K.; Huh, S.; Kim, K.; Nakase, H.; Nakamura, Y. Effect of Ag content on Properties of Sn-Ag Binary Alloy Solder. *Mater. Trans. JIM* **2001**, *42*, 286–291. [CrossRef]

32. Masayoshi, S.; Noboru, H.; Hirohiko, W.; Masayuki, Y. High Temperature Creep Properties of Sn-3.5Ag and Sn-5Sb Lead-free Solder Alloys. *Trans. JWRI* **2011**, *40*, 49–54.

33. Wang, R.; Garcia, C.I.; Hua, M.; Deardo, A. Microstructure and precipitation behavior of Nb, Ti complex Microalloyed steel produced by compact strip processing. *ISIJ Int.* **2006**, *46*, 1345–1353. [CrossRef]

 *metals*

Article

Effect of Tungsten on Creep Behavior of 9%Cr–3%Co Martensitic Steels

Alexandra Fedoseeva, Nadezhda Dudova, Rustam Kaibyshev and Andrey Belyakov *

Laboratory of Mechanical Properties of Nanostructured Materials and Superalloys, Belgorod National Research University, Pobeda 85, Belgorod 308015, Russia; fedoseeva@bsu.edu.ru (A.F.); dudova@bsu.edu.ru (N.D.); rustam_kaibyshev@bsu.edu.ru (R.K.)
* Correspondence: belyakov@bsu.edu.ru; Tel.: +7-4722-58-54-57

Received: 20 November 2017; Accepted: 11 December 2017; Published: 18 December 2017

Abstract: The effect of increasing tungsten content from 2 to 3 wt % on the creep rupture strength of a 3 wt % Co-modified P92-type steel was studied. Creep tests were carried out at a temperature of 650 °C under applied stresses ranging from 100 to 220 MPa with a step of 20 MPa. It was found that an increase in W content from 2 to 3 wt % resulted in a +15% and +14% increase in the creep rupture strength in the short-term region (up to 10^3 h) and long-term one (up to 10^4 h), respectively. On the other hand, in the long-term creep region, the effect of W on creep strength diminished with increasing rupture time, up to complete disappearance at 10^5 h, because of depletion of excess W from the solid solution in the form of precipitation of the Laves phase particles. An increase in W content led to the increased amount of Laves phase and rapid coarsening of these particles under long-term creep. The contribution of W to the enhancement of creep resistance has short-term character.

Keywords: martensitic steels; creep; precipitation; electron microscopy

1. Introduction

The heat resistant steels with 9–12%Cr are widely used as structural materials for boilers, main steam pipes, and turbines of fossil power plants with increased thermal efficiency [1,2]. The excellent creep resistance of these steels is attributed to the tempered martensite lath structure (TMLS) consisting of prior austenite grains (PAG), packets, blocks, and laths, and containing a high density of separate dislocations and a dispersion of secondary phase particles [1–5]. Stability of TMLS is provided by $M_{23}C_6$-type carbides and MX (where M is V and/or Nb, and X is C and/or N) carbonitrides precipitated during tempering at boundaries and within ferritic matrix, respectively, and boundary Laves phase particles precipitated during creep [1,2,4–12]. MX carbonitrides, which are highly effective in pinning of lattice dislocations, play a vital role in superior long-term creep resistance of the high chromium martensitic steels, whereas boundary $M_{23}C_6$ carbides and Laves phase particles exerting a high Zener drag force stabilize the TMLS [1,2,5,7–14]. This dispersion of secondary phase particles withstands short-term creep [4,9,15]. The Laves phase provides effective stabilization of TMLS and therefore promotes creep resistance, although their effect on the rearrangement of lattice dislocation is negligibly small [9,11,15]. However, the particles of Laves phase grow with a high rate under creep condition and, at present, these boundary precipitations are considered to be responsible for the creep ductility of the P92-type steels during long-term aging [16]. The enhanced long-term creep strength could be achieved by hindering this microstructural evolution. An effective way to achieve this goal is to slow down the diffusion-controlled processes, such as the climb of dislocations, knitting reaction between dislocation and lath boundaries, particle coarsening, etc., by such substitutional additives as Co, W, and Mo [1,9,17–19].

It was recently shown that cobalt addition significantly hinders the coarsening of $M_{23}C_6$ carbides and MX carbonitrides under creep conditions, which results in the superior creep resistance of martensitic steels [9,13,18,20]. This positive effect of Co is attributed to hindering diffusion within ferrite [9,18]. Efficiency of W as an alloying element in hindering diffusion is much higher than that of Co. As a result, W and Mo are known as effective alloying additives to enhance creep resistance of high chromium martensitic steels [1,19]. These elements provide an effective solid solution strengthening [1]. It was shown [21] that addition of 1% W gives +35 MPa increase in the creep rupture strength at 600 °C for 1000 h. However, in contrast with cobalt, the tungsten and molybdenum have limited solubility within ferrite, and their excessive content leads to precipitation of such W- and Mo-rich particles as Laves phase $Fe_2(W,Mo)$ or M_6C carbides [1,7,9–12,22,23]. This depletion does not occur in the 9%Cr steel containing no or low amount of W [7]. Depletion of solid solution by these elements highly deteriorates the creep resistance [7,10,11]. It is worth noting that, at present, the most of experimental data on the effect of W on creep behavior were obtained for cobalt-free high chromium martensitic steels. The aim of the present work is to report the effect of W addition on the creep strength and microstructure evolution during creep of two 9%Cr martensitic steels containing 3%Co and distinguished by W content.

2. Materials and Methods

Two Co-modified P92-type steels with 2 and 3 wt % W denoted here as the 9Cr2W and the 9Cr3Wsteels, respectively, were produced by air melting as 40 kg ingots. Chemical compositions of these steels, measured by a FOUNDRY-MASTER UVR optical emission spectrometer (Oxford Instruments, Ambingdon, UK) are presented in Table 1.

Table 1. Chemical composition of the steels studied (wt %).

Steel	Fe	C	Cr	Co.	Mo	W	V	Nb	B	N	Si	Mn
9Cr2W	bal.	0.12	9.3	3.1	0.44	2.0	0.2	0.06	0.005	0.05	0.08	0.2
9Cr3W	bal.	0.12	9.5	3.2	0.45	3.1	0.2	0.06	0.005	0.05	0.06	0.2

Square bars with cross-section of 13×13 mm^2 were cast and hot-forged in the temperature interval 1150–1050 °C after homogenization annealing at 1100 °C for 1 h by the Central Research Institute for Machine-Building Technology, Moscow, Russia. Both steels were solution-treated at 1050 °C for 30 min, cooled in air, and subsequently tempered at 750 °C for 3 h. Tensile tests were carried out on specimens having a cross section of 1.4×3 mm^2 and a 16 mm gauge length using an Instron 5882 Universal Testing Machine (Instron, Norwood, MA, USA) at room temperature and at 650 °C with a strain rate of 2×10^{-3} s^{-1}. Flat specimens with a gauge length of 25 mm and a cross section of 7×3 mm^2 (for 220–140 MPa) and cylindrical specimens with a gauge length of 100 mm and a diameter of 10 mm (for 120–80 MPa) were subjected to creep tests until rupture. The creep tests were carried out in the air at 650 °C under different initial stresses ranging from 80 to 220 MPa with a step of 20 MPa. The 100,000 h creep rupture strength was estimated by extrapolation of the experimental data using the Larson–Miller equation [24]:

$$P = T(\lg\tau + 36) \times 10^{-3} \tag{1}$$

where P is the parameter of Larson–Miller; T is temperature, (K); τ is time to rupture, (h).

The structural characterization was carried out using a transmission electron microscope, JEOL-2100, (TEM) (JEOL Ltd., Tokyo, Japan) with an INCA energy dispersive X-ray spectroscope (EDS) (Oxford Instruments, Abingdon, UK) and scanning electron microscope, Quanta 600FEG, (SEM) (FEI, Hillsboro, OR, USA) on ruptured creep specimens in the gauge sections corresponding to uniform deformation in the middle between grip portion and fracture surface. The size distribution and mean radius of the secondary phase particles were estimated by counting of 150 to 250 particles per specimen on at least 15 arbitrarily selected typical TEM images for each data point. The error bars are given according to the standard deviation. Identification of the precipitates was done on the basis of the combination of EDS composition measurements of the metallic elements and indexing of electron diffraction patterns by TEM. The subgrain sizes were evaluated on TEM micrographs by the linear intercept method including all clearly visible (sub)boundaries. The dislocation densities in the grain and subgrain interiors were estimated as a number of intersections of individual dislocations with upper or down foil surfaces per unit area on at least six arbitrarily selected typical TEM images for each data point [25]. The dislocation observation was carried out under multiple-beam conditions with large excitation vectors for several diffracted planes for each TEM image. The W-rich M_6C carbides and Laves phase particles could be clearly distinguished from other precipitates by their bright contrast in the back scattered electron (BSE) image (Z-contrast) [26]. M_6C carbides and Laves phase particles were separated from each other by EDS composition measurements by TEM and particle size distribution [27]. The volume fractions of the precipitated phases were calculated by the Thermo-Calc software (Version 5.0.4 75, Thermo-Calc software AB, Stockholm, Sweden, 2010) using the TCFE7 database for the following compositions of steels (in wt %): 0.1%C-9.4%Cr-0.5%Mo-2.0 (or 3.0)%W-3.0%Co-0.2%V-0.05%Nb-0.05%N-0.005%B and Fe-balance. The following phases were chosen independently for calculation: austenite (FCC_A1), ferrite (BCC_A2), cementite, $M_{23}C_6$ carbide, M_7C_3 carbide, M_6C carbide, and Laves phase ($Fe_2(W, Mo)$) (C14).

3. Results

3.1. Tempered Martensite Lath Structure

After tempering at 750 °C, TMLS forms in both steels. However, in the 9Cr2W steel, the additional formation of subgrains was observed (Figure 1a), whereas TMLS is dominant in the 9Cr3W steel (Figure 1b). The average sizes of PAGs were 11 and 20 μm for the 9Cr2W and 9Cr3W steels, respectively. More details about the microstructure of the steels studied after normalization at 1050 °C and tempering at 750 °C can be found elsewhere [27].

Figure 1. Mixed lath structure and subgrain one (**a**) in the 9Cr2W steel and homogeneous tempered martensite lath structure in the 9Cr3W steel (**b**) after normalization at 1050 °C and tempering at 750 °C.

The lath thickness was approximately 0.4 μm for both steels. The high dislocation density of approximately 2×10^{14} m^{-2} was observed within the lath and subgrain interiors. In the structure of both steels, $M_{23}C_6$ carbides located on the boundaries of PAGs, packets, blocks, and laths, and MX-type carbonitrides uniformly distributed within the martensitic laths were observed. The mean size of $M_{23}C_6$ carbides was about 90 nm. V-rich MX carbonitrides with a "wing" shape [8,15] have a mean longitudinal size of 20 nm. Nb-rich MX carbonitrides with a round shape have an average size of 40 nm. Dimensions of these particles in both steels were the same. The W-rich precipitates of M_6C carbide (Fe_3W_3C) and Laves phase (Fe_2W) were found in the 9Cr3W steel alongside the $M_{23}C_6$ and MX particles [27]. Therefore, the solubility of (3%W + 0.5%Mo) excesses the thermodynamically equilibrium solubility limit even at the tempering temperature of 750 °C. The 1 wt %W additive provides the precipitation of the W-rich M_6C carbides and Laves phase even under tempering. No formation of thermodynamically stable W-rich Laves phases was reported in the conventional 9%Cr martensitic steels after tempering [1,2,7–12], and the appearance of the less stable W-rich M_6C carbides was found only in a 10%Cr–2%W steel [23]. Under tempering, the partial transformation of M_6C carbides into Laves phase particles may occur if M_6C carbides are occupied by other M_6C_6 carbides and do not have access to W segregation in the vicinity of PAG–lath boundaries [27].

3.2. Tensile Test

The W effect on engineering stress-strain curves is shown in Figure 2, and yield stress (YS), ultimate tensile stress (UTS), and ductility δ are summarized in Table 2.

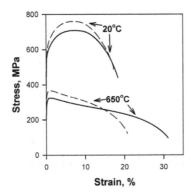

Figure 2. Tensile behavior of the 9Cr2W (solid lines) and 9Cr3W (dash lines) steels after heat treatment consisting of normalizing at 1050 °C and tempering at 750 °C. Tensile tests were carried out at room temperature and at 650 °C (creep test temperature).

Table 2. Values of yield stress (YS), ultimate tensile stress (UTS), and ductility (δ), obtained under tension, at room temperature and 650 °C for the 9Cr2W and the 9Cr3W steels.

Steel	Temperature Test	YS, MPa	UTS, MPa	δ, %
9Cr2W	20 °C	560	708	19
	650 °C	295	325	32
9Cr3W	20 °C	570	760	18
	650 °C	340	370	21

The shapes of the engineering σ–ε curves at room and elevated test temperatures for both steels were nearly the same, whereas YS and UTS are higher and δ is smaller for the 9Cr3W steel. The σ–ε curves at elevated test temperature showed the continuous yielding for both steels. After a short stage of extensive strain hardening, the apparent steady-state flow appeared and occurred up to necking. Next, post-uniform necking elongation took place up to fracture. Ductility at room test temperature was nearly the same for both steels, while at elevated temperature ductility of the 9Cr2W steel was +52% more than for 9Cr3W steel. Increments of +2% and +7% in YS and UTS at room temperature and of +15% and +14% in YS and UTS at elevated test temperature were observed for the steel with increased W content. It is obvious that these increments are provided by solid solution strengthening due to increasing W content up to 3%.

3.3. Creep Behavior

Figure 3a shows the creep rupture data of the steels at a temperature of 650 °C. In general, the creep rupture time of both Co-modified steels is significantly longer in comparison with the P92 steel, that is indicative of the positive effect of Co on the creep strength [1,5,15,28].

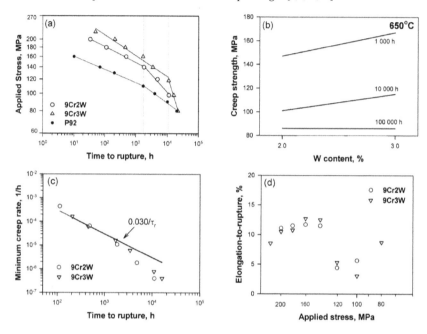

Figure 3. (**a**) Time to rupture vs. stress curves for the 9Cr2W and 9Cr3W steels in comparison with data for a P92 steel [2,7]; (**b**) the 1000 h, 10,000 h, and 100,000 h creep rupture strengths of steels at 650 °C as a function of W content; (**c**) minimum creep rate as a function of time to rupture for the 9Cr2W and 9Cr3W steels; and (**d**) applied stress vs. elongation-to-rupture for the 9Cr2W and 9Cr3W steels. The dotted lines in (**a**) indicate the time to rupture corresponding to the creep strength breakdown.

In the short-term region up to 10^3 and 10^4 h, the creep strength increase is +15% and +14% from 145 to 167 MPa and from 101 to 115 MPa, respectively, due to increased W content from 2 to 3 wt %. In the long-term creep region, the effect of W additives on the creep strength tends to diminish. For both steels the 100,000 h creep rupture strength of ~85 MPa predicted through the Larson–Miller parameter (Figure 3b) is nearly the same, and, therefore, the positive effect of W disappears. However, this value is 15% higher than the creep rupture strength of ~72 MPa for the P92 steel predicted through the Larson–Miller parameter from data published in previous works [28,29].

It was recently shown that creep strength breakdown is a tertiary creep phenomenon [7,13]. The Monkman–Grant relationship relating the rupture time, τ_r, to the minimum or steady-state creep rate is described as the Equation (2):

$$\tau_r = (c' / \dot{\varepsilon}_{min})^{m'}, \tag{2}$$

where c' and m' are constants. This relationship is used for the prediction of creep life of heat-resistant steels [1,13]. Analysis of Equation (2) for the studied steels (Figure 3c) shows that this approach is suitable for describing the relation of rupture time with the offset strain rate, $\dot{\varepsilon}_{min}$. For short-term conditions, which corresponds to τ_r less than approximately 2000–3500 h, τ_r is inversely proportional to $\dot{\varepsilon}_{min}$ (Figure 3c) at $m' = 1$ [13]. The constant c' is 3.0×10^{-2}. For long-term conditions, which correspond to a τ_r greater than approximately 2000–3500 h, the relationship between τ_r and $\dot{\varepsilon}_{min}$

deviates downward. The transition from short-term creep to a long-term one appears as the deviation from the linear dependence described by Equation (2), which indicates $m' < 1$ [13]. Loss of ductility occurs at low stresses of 120–100 MPa in both steels (Figure 3d). For the 9Cr2W steel, a decrease in the elongation-to-rupture correlates with the creep strength breakdown appearance in Figure 3a, while for the 9Cr3W steel, the changes in elongation-to-rupture have irregular character. For high stresses from 220 to 140 MPa, elongation-to-rupture increases from 8% to 13%, then remarkably reduces to 3–5% at low stresses of 120 and 100 MPa, and then increases to 8% at 80 MPa. The relation of loss of ductility and creep strength breakdown is not revealed for the 9Cr3W steel. In contrast with the dependencies of the applied stress vs. the rupture time (Figure 3a), there was no distinct inflection point for the transition from the short-term region to the long-term one by the shapes of aforementioned curves. The minimum creep rate decreased from approximately 10^{-7} to 10^{-10} s^{-1} with a decrease in the applied stress from 220 to 100 MPa. There was a linear dependence between the minimum creep rate and the applied stress (Figure 4). The experimental data obey a power law relationship throughout the whole range of the applied stress of the usual form [1,7,13]:

$$\dot{\varepsilon}_{min} = A \times \sigma^n \exp\left(\frac{-Q}{RT}\right), \tag{3}$$

where $\dot{\varepsilon}_{min}$ is the minimum creep rate, σ is the applied stress, Q is the activation energy for a plastic deformation, R is the gas constant, T is the absolute temperature, A is a constant, and n is the "apparent" stress exponent.

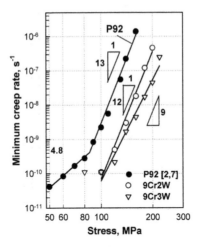

Figure 4. Minimum creep rate as a function of time to rupture for the 9Cr2W and 9Cr3W steels in comparison with data for the P92 steel [2,7].

For the applied stresses from 220 to 100 MPa, these plots provide the best linear fit with a regression coefficient of 0.98 for $n = 12$ and 9 for the 9Cr2W and 9Cr3W steels, respectively. This n value at all tested stress regimes remains constant. The steady-state creep of the 9Cr2W and 9Cr3W steels was controlled using the same process for the short- and long-term regions at a creep rate ranging from 10^{-6} to 10^{-10} s^{-1}. This clearly indicates that there is no creep strength breakdown during the steady-state creep of both steels [7,13]. However, for the 9Cr3W steel, minimum creep rate at 80 MPa is similar with that at 100 MPa, which indicates the change in the process that controls the steady-state creep.

3.4. Crept Microstructures

SEM-BSE and TEM images of both steels after creep rupture tests are shown in Figure 5.

Laves phase, enriched by W and Mo, could be distinguished as the white particles in Z-contrast, and Cr-enriched $M_{23}C_6$ carbides as the grey particles on the grey background of matrix [10,30]. The short-term creep tests ($\leq$2000 h) did not change the lath structure in both steels (Figure 5a,b). W-rich particles exhibited nearly round shape. After the long-term creep rupture tests ($\geq$2000 h) the lath thickness increased (Figure 5c,d), and transformation of the lath structure into subgrain structure took place in both steels. The well-defined subgrain structure evolved in the 9Cr2W steel, only. TEM studies support this conclusion (Figure 5e,f). In the 9Cr2W steel, the subgrains rapidly grew from 0.6 to 1.5 μm after 2000 h (Figure 6a), whereas in the 9Cr3W steel, this size insignificantly changed from 0.6 to 0.7 μm (Figure 6b). It should be noted that upon further creep tests from 2000 to 10,000 h, the subgrain sizes remained almost unchanged in both steels (Figure 6a,b).

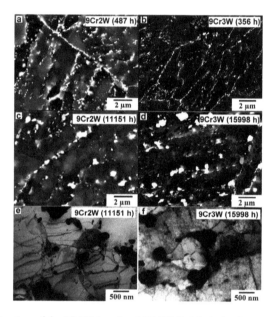

Figure 5. Microstructure of the 9Cr2W (**a,c,e**) and 9Cr3W (**b,d,f**) steels after creep rupture tests at 650 °C under the stress of: (**a**) 160 MPa, 487 h; (**b**) 180 MPa, 356 h; (**c,e**) 100 MPa, 11,151 h; (**d,f**) 100 MPa, 15,998 h; obtained by SEM (**a–d**) and TEM (**e,f**).

There is a difference in the effect of creep on the distributions of second phase particles in the two steels. In the 9Cr3W steel, the fine particles densely distributed along the boundaries can be seen after short-term creep (Figure 5b), whereas in the 9Cr2W steel, these precipitates were slightly coarser, and their density was less (Figure 5a).

$M_{23}C_6$ carbides. In the 9Cr2W steel, coarsening of $M_{23}C_6$ carbides occurred faster than in the 9Cr3W steel during creep tests (Figure 6c,d). Mean size of these carbides increased to 200–250 nm after 1000 h in the 9Cr2W steel (Figure 6c), whereas in the 9Cr3W steel, the size of these carbides remained less than 100 nm up to rupture times of ~5000 h (Figure 6d). Therefore, the W addition enhanced the coarsening resistance of $M_{23}C_6$ carbides in short-term creep conditions. At the same time, under long-term creep conditions, the average dimension of $M_{23}C_6$ carbides and their morphology became essentially the same in both steels. The average Cr and W contents in the $M_{23}C_6$ carbides were the same in both steels and tended to slightly increase with increasing rupture time (Figure 7a). In contrast, the portion of Fe decreased with increasing rupture time. Under long-term exposure, the chemical composition of $M_{23}C_6$ carbides tended to approach the thermodynamically equilibrium composition (Table 3).

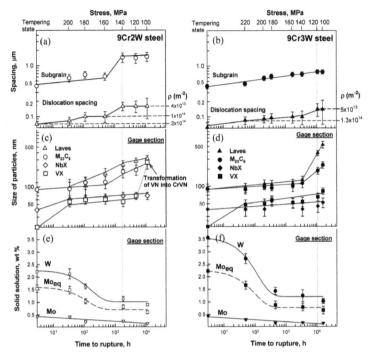

Figure 6. Mean size of spacing (**a,b**), mean size of particles of different phases (**c,d**), and change in W and Mo content in the solid solution (**e,f**) as function of time in the 9Cr2W (**a,c,e**) and 9Cr3W (**b,d,f**) steels during creep tests at 650 °C under the stresses of 100–220 MPa. The vertical dotted lines indicate the time to rupture corresponding to the creep strength breakdown. Mo$_{eq}$ = Mo + 0.5W.

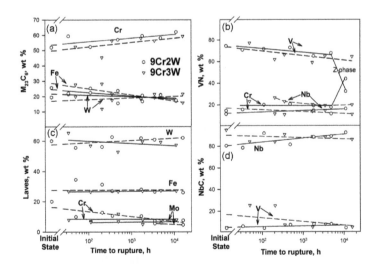

Figure 7. Change in the chemical compositions of (**a**) M$_{23}$C$_6$ carbides; (**b**) V-rich MX; (**c**) Laves phase particles; and (**d**) Nb-rich MX during creep tests at 650 °C under the stresses of 100–220 MPa.

Table 3. Weight fraction of elements in the different phases in the steels studied at 650 and 750 °C as calculated by Thermo-Calc.

Element		Cr	Fe	W	V	Nb	N	C	Mo	Co.
					650 °C (Creep)					
MX	9Cr2W	-	-	-	61.98	17.67	18.25	0.51	-	-
	9Cr3W	-	-	-	61.96	18.59	17.78	0.44	-	-
$M_{23}C_6$	9Cr2W	60.54	14.08	7.50	0.44	-	-	5.00	12.25	-
	9Cr3W	60.32	13.95	8.59	-	-	-	5.00	11.24	-
Laves	9Cr2W	6.26	32.57	56.99	-	-	-	-	3.99	-
	9Cr3W	6.42	32.20	57.77	-	-	-	-	3.42	-
Solid solution	9Cr2W	8.39	87.59	0.68	0.02	-	-	-	0.18	3.13
	9Cr3W	8.47	87.16	0.68	0.04	-	-	-	0.16	3.13
					750 °C (Tempering)					
MX	9Cr2W	-	-	-	61.29	17.99	17.73	0.77	-	-
	9Cr3W	-	-	-	59.50	19.38	17.78	0.70	-	-
$M_{23}C_6$	9Cr2W	51.78	20.61	13.92	-	-	-	4.83	8.03	-
	9Cr3W	51.32	20.30	16.83				4.78	5.80	-
Laves	9Cr2W	5.35	33.62	56.49	-	-	-	-	4.31	-
	9Cr3W	5.50	33.27	57.22	-	-	-	-	3.76	-
Solid solution	9Cr2W	8.51	86.42	1.60	0.03	-	-	-	0.37	3.00
	9Cr3W	8.60	85.98	1.63	0.04	-	-	-	0.32	3.00

Laves phase. The precipitation behavior of Laves phase in the two steels was quite different (Figure 6c,d). In the 9Cr2W steel, the precipitation of Laves phase particles started to occur with a high rate under creep condition. Next, these particles gradually grew with a high rate up to 250 nm at creep rupture time of 11,151 h (Figure 6c). In contrast, in the 9Cr3W steel, the particles of Laves phase precipitated under tempering remained their size of ~100 nm up to 5000 h of creep tests. Then, the rapid coarsening of Laves phase started to occur, and their average size attained ~550 nm at creep rupture time of ~16,000 h (Figure 6d). Thus, high W content slowed down the coarsening of Laves phase and provided the improved coarsening resistance of Laves phase up to about 10,000 h. However, upon further creep, the extensive precipitation of Laves phase induced their growth with an increased rate. The final average size of Laves phase particles in the 9Cr3W steel became even higher than that in the 9Cr2W steel (530 nm and 312 nm, respectively). In the 9Cr3W steel, the volume fraction of Laves phase was significantly higher (2.394%) in comparison with the 9Cr2W steel (1.315%) as calculated by Thermo-Calc software (Table 4). It is indicative of the full depletion of excess content of W at 650 °C from the solid solution under long-term creep.

Table 4. The volume fraction of second phases in the steels at 650 °C as calculated by Thermo-Calc.

Phase	Volume Fraction (%)	
	9Cr2W Steel	9Cr3W Steel
$M_{23}C_6$	1.958	1.966
MX	0.246	0.238
Laves phase	1.315	2.394

In both steels, Fe and Cr contents insignificantly decreased in the Laves phase, and Mo content slightly increased during creep (Figure 7c). W content decreased in the 9Cr2W steel, whereas in the 3 wt % W steel it increased. However, these chemical composition changes of the Laves phase particles were insignificant. Therefore, the Laves phase particles initially precipitated in accordance with the thermodynamically equilibrium content (Table 3), and their coarsening was not associated with the changes in their chemical composition.

No evidence for M_6C carbides presence in the 9Cr3W steel was found after creep tests. These carbides were replaced by more stable Laves phase under short-term creep as in other high-chromium martensitic steels [22].

Nb-rich MX carbonitrides. Thermo-Calc calculation predicted the existence of a unified (V,Nb)N nitride with a very low C content at 650 °C only in both steels (Table 4), whereas Nb-rich and V-rich separation of MX phase remained under creep condition at this temperature. In the 9Cr3W steel, Nb-rich MX carbonitrides grew from 40 to 50 nm under long-term creep condition (Figure 6d). In contrast, in the 9Cr2W steel, they rapidly grew from 40 to 60 nm under short-term creep conditions and then to 70 nm under long-term creep conditions (Figure 6c). Thus, W additives hindered the coarsening of Nb-rich MX carbonitrides. Effect of W additions is more pronounced under short-term creep. Nb and V content in the Nb-rich MX particles slightly increased and decreased, respectively, with increasing rupture time (Figure 7d) in opposition to thermodynamically equilibrium values (Table 3).

V-rich MX carbonitrides. The average size of V-rich MX particles attained 60 nm after long-term creep tests in both steels. In the 9Cr2W steel, the well-known replacement of V-rich MX carbonitrides by Z-phase (CrVN nitride), which is in fact the most thermodynamically stable nitride [31,32], started to occur after ~5000 h. The full transformation of all V-rich MX carbonitrides was found after 10,000 h of creep tests. The mean size of Z-phase was about 300 nm. Therefore, the size of V-rich carbonitrides increased by a factor of ~4.5 and they could not more contribute to the creep resistance of the 9Cr2W steel (Figure 6c). +1 wt %W additive shifted the onset of this transformation from 5000 h to 16,000 h; the separate Z-phase particles in the 9Cr3W steel were revealed after 16,000 h [33]. Nb content in the V-rich MX particles slightly increased with rupture time in both steel (Figure 7b). V content in the V-rich MX particles tended to decrease in the 9Cr3W steel with rupture time. At rupture time $\leq$5000 h, in the 9Cr2W steel, the V and Cr contents were essentially independent on the rupture time. Therefore, no gradual increase in Cr content within the V-rich MX particles resulting in transformation of their cubic lattice into tetragonal lattice of Z-phase [20,32,34] was detected.

Solid solution. Solid solution of the 9Cr3W steel (Figure 6f) was more enriched by W than that of the 9Cr2W steel (Figure 6e) during short-term creep up to time t of ~500–700 h. After this exposure, there is no significant difference in W content in the solid solution of two steels.

Depletion of W and Mo from the solid solution took place with increasing rupture time. However, the rate of the W depletion was higher in the 9Cr3W steel under short-term creep conditions. Depletion of W, $f(W)$, from the solid solution could be described as:

$$f(W) \sim 1.3exp(-0.0059t) \quad \text{for the 9Cr2W steel,} \tag{4}$$

$$\text{and } f(W) \sim 2.4exp(-0.0083t) \quad \text{for the 9Cr3W steel.} \tag{5}$$

Depletion of Mo, $f(Mo)$, from the solid solution could be described as:

$$f(Mo) \sim -0.08t \quad \text{for both steels.} \tag{6}$$

This finding indicates the strong dependence of W content on the rupture time up to an achievement of the thermodynamically equilibrium value of W in the ferritic matrix (Table 3). At rupture time $\geq$700 h, the $Mo_{eq} = \Sigma(Mo + 0.5W)$ [35] was the same for both steels since no significant difference in W and Mo content was found in the ferritic matrix of both steels by TEM.

4. Discussion

4.1. Contribution of W to the Solid Solution Strengthening

The experimental data showed that increasing W content from 2 to 3 wt % improves the creep resistance of the 0.1C-9Cr-3Co-0.5Mo-VNbBN steel at 650 °C under short-term creep conditions due to solid solution strengthening as main hardening mechanism. As a result, the TMLS essentially remains in the 9Cr3W steel, whereas a well-defined subgrain structure evolves in the 9Cr2W steel. Despite this fact, the creep strength of the 9Cr3W steel approaches to the level of the 9Cr2W steel with increasing rupture time up to 10^5 h according to estimation by Larson–Miller parameter. It is worth noting that effect of Co on the microstructural evolution under creep is nearly the same [18]. However, both Co and W additions could not prevent the breakdown of creep strength taking place at a rupture time of ~2000 h for the P92 steel and the 9Cr2W steel, and at a rupture time of ~10,000 for the 9Cr3W steel (Figure 3a,c).

Mo_{eq} content in the solid solution of both steels is significantly higher than the solubility limit for these elements at 650 °C and the precipitation of Laves phase provides approaching of W and Mo contents to their thermodynamically equilibrium levels under creep condition. It is known [1,7,10–12,15,22] that the precipitation of Laves phase during creep is hindered by low rate of these substitutional elements diffusion. The continuous decrease in W content in the solid solution indicates the continuous decomposition of the supersaturated solid solution accompanied by the precipitation of the fine Laves phase particles during creep tests. Under long-term condition, the volume fraction of Laves phase attains the thermodynamically equilibrium value (Table 4). The same Mo_{eq} value in the solid solution of both steels is attained after 700 h of short-term creep exposure. This indicates no advantage in solid solution strengthening for the 9Cr3W steel after 700 h of creep test.

However, the 9Cr3W steel demonstrates +14% increments in creep strength even after 700 h of creep testing. Increasing W content up to 3% contributes to creep strength not only by solid solution strengthening but by creation the preconditions for improved creep strength, such as homogeneous TMLS, increased dislocation density, and narrow size distributions of boundary particles after tempering, which provide advanced creep strength upon creep time more than 700 h. Therefore, under short-term creep, the W addition slows down the rearrangement of lattice dislocations by climb that may prevent the aforementioned transformation of lath boundaries to sub-boundaries. However, under long-term conditions, a saturation of the solid solution by W and Mo is the same in both steels and the retardation of dislocation climb by W addition could not provide the stabilization of TMLS. It is obvious that the influence of W content on the stability of TMLS under long-term creep conditions is attributed to its effect on a dispersion of secondary phase.

4.2. Contribution of W to the Particle Strengthening

The experimental data showed that increasing W content from 2 to 3 wt % improves the creep resistance of the steel in the short-term conditions due to sluggish kinetics of the coarsening of $M_{23}C_6$ carbides, precipitation of fine Laves phase particles, and preventing transformation of TMLS into subgrain structure. The effect of W on the coarsening behavior of $M_{23}C_6$ carbides and Laves phase was considered in the companion work [36] in sufficient details. In that study, we briefly summarize that 1 wt % W addition affects the size distribution of these two types of particles after tempering that leads to the difference in the coarsening behavior between two steels [36] during creep. In the work [37], it was found that the coarsening of $M_{23}C_6$ carbides correlates with the changes in their chemical composition. It is worth noting that the coarsening of $M_{23}C_6$ carbides in both steels is accompanied by an increase in Cr and W contents and a decrease in Fe content (Figure 6a). In work [37], it has been observed that $M_{23}C_6$ carbides grow during 50,000 h up to the chemical equilibrium. In both steels, the character of evolution of chemical composition of $M_{23}C_6$ carbides is essentially the same. The Cr and (W + Mo) contents in $M_{23}C_6$ carbides increase at the expense of Fe and reach the thermodynamically equilibrium values (Table 3). It is worth noting that Thermo-Calc calculation at 650 °C predicts the Cr and (W + Mo) content in $M_{23}C_6$ carbides higher and lower, respectively, than that at 750 °C. Under tempering condition, only Cr content attained thermodynamically equilibrium value in $M_{23}C_6$ carbides. Cr atoms diffuse slightly slower than Fe atoms in ferrite, and the diffusion rate of W and Mo atoms is the lowest in comparison with Cr [38]. This is the reason for approaching the thermodynamically equilibrium chemical composition of $M_{23}C_6$ carbides during a long-term exposure.

In both steels, the various second phase particles pin the lath boundaries. MX particles homogeneously distributed within the lath exert Zener drag pressure, which can be evaluated as [9,15,23,39,40]:

$$P_Z = \frac{3\gamma F_V}{d}, \tag{7}$$

where γ is the boundary surface energy per unit area (0.153 J m^{-2}) [15,40], F_v is the volume fraction of particles calculated by Thermo-Calc, and d is a mean size of particles, m.

The boundary particles of $M_{23}C_6$ and Laves phases exert Zener drag pressure estimated as [9,15,23,39,40]:

$$P_B = \frac{\gamma F_{vB} D}{d^2}, \tag{8}$$

where D is subgrain size or lath width, μm, and F_{vB} is the fraction of particles located at the boundaries. The pinning pressures were calculated separately for $M_{23}C_6$ carbides and Laves phase according with:

$$P_B = \frac{\gamma F_{vB} D_0}{d_0^2} \cdot \frac{\beta_i}{\beta_0}, \tag{9}$$

where β_i is density of $M_{23}C_6$ carbides or Laves phase, located along boundaries, for each applied stress. Calculation of Zener drag pressure for the different kinds of particles was considered in the previous work [9] in details.

+1 wt % W addition as well as an applied stress affect the particles sizes and their volume fractions (Figure 6 and Table 3). As a result, there is a difference in the pinning pressure between the two steels, and Zener drag pressure depends on the creep test time (Figure 8). Both steels contain essentially the same volume fraction of MX carbonitrides, and the pinning pressure from these particles (P_Z) is the same for both steels. At high applied stress, MX carbonitrides give a minor contribution to the total Zener pressure due to the low volume fraction of these particles and random distribution within subgrain and laths.

In the 9Cr2W steel, the $M_{23}C_6$ carbides gives the main contribution to overall Zener drag force at any applied stress [23], whereas in the 9Cr3W steel the highest Zener drag is exerted by Laves phase particles at the applied stresses more than 100 MPa [39]. Only at an applied stress of 100 MPa the Zener drag exerted by $M_{23}C_6$ carbides in this steel (9Cr3W) becomes higher than that exerted by Laves phase particles, and the P_B and P_Z values originated from Laves phase and MX carbonitrides, respectively, are the same at this condition (Figure 8b). In contrast, in the 9Cr2W steel, the P_Z value drops at an applied stress of 100 MPa, owing to transformation of V-rich MX carbonitrides to Z-phase, and the pinning pressures from the boundary $M_{23}C_6$ carbides (P_B) and MX carbonitrides with a random particle distribution (P_Z) are similar at an applied stress of 120 MPa (Figure 8a).

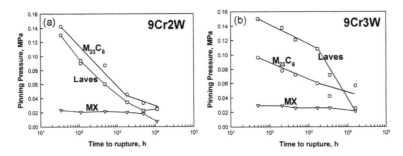

Figure 8. Change in the pinning pressures from different kinds of particles on the grain and lath boundaries of the (**a**) 9Cr2W steel and (**b**) 9Cr3W steel during creep tests at 650 °C under the stresses of 100–220 MPa.

Thus, the Zener pressure exerted by boundary particles remains for a long time until the particle coarsening occurs. Size stabilization of the Laves phase particles in the 9Cr3W steel provides a significant contribution to high Zener pressure, whereas their rapid coarsening removes this effect. Thus, Laves phase particles give short-term contribution to the precipitation strengthening. The decrease in Zener pressure exerted by Laves phase does not lead to the transformation of TMLS into subgrain structure as in the 9Cr2W steel because $M_{23}C_6$ carbides are able to provide the high level of Zener pressure [23] (more than 0.05 MPa) at the low stresses in the steel with 3 wt % W.

Therefore, there is no positive effect of increased W content on the creep resistance of the Co-modified P92 steel under long-term creep conditions. An alloying of 0.1C-9Cr-3Co-0.5Mo-VNbBN martensitic steel by 3%W does not seem justified for the applying at long-term creep condition (for 10^5 h) at 650 °C. Therefore, the W content of ~2 wt % is optimal for the Co-containing high chromium martensitic steels.

5. Conclusions

The microstructures of two 0.1C-9Cr-3Co-0.5Mo-VNbBN martensitic steels with different content of W (2 wt % and 3 wt %) in the tempered and crept at 650 °C under stresses of 100–220 MPa conditions were studied. The main results can be summarized as follows:

1. The structure of both steels tempered at 750 °C for 3 h is the tempered martensite lath structure with the lath thickness of 0.4 μm. An increased W content leads to the formation of W-rich Laves phase particles and M_6C carbides on the boundaries in addition to the $M_{23}C_6$ carbides located also on the boundaries and MX carbonitrides distributed uniformly within the ferritic matrix.
2. The steel with 3 wt % W content demonstrates a +15% increase in the 10,000 h creep rupture strength at 650 °C due to hindering the coarsening of the $M_{23}C_6$ carbides, MX carbonitrides and Laves phase particles. Tungsten also slows down the transformation of V-rich MX carbonitrides into Z-phase particles.
3. An increase in W content in the steels provides an increase in the amount of Laves phase. Under long-term conditions, the depletion of excess W from the solid solution leads to the rapid coarsening of Laves phase particles; the contribution of this phase to Zener drag has a short-term character.
4. The predicted long-term creep rupture strength for 100,000 h is about 85 MPa for both steels. This value is independent on W content due to depletion of its excess from the solid solution up to thermodynamically equilibrium value due to Laves phase precipitation. Therefore, there is no positive effect of increased W content on the creep resistance of the Co-modified P92 steel under long-term creep conditions. An alloying of 0.1C-9Cr-3Co-0.5Mo-VNbBN martensitic steel by 3%W does not seem justified for long-term creep condition (for 10^5 h) at 650 °C. Therefore, the W content of ~2 wt % is optimal for the Co-containing high chromium martensitic steels.

Acknowledgments: This study was financially supported by the Ministry of Education and Science of Russian Federation, under project of Government Task No. 11.2868.2017/PCh. The authors are grateful to V. Skorobogatykh and I. Shchenkova, Central Research Institute for Machine-Building Technology, for supplying the test material and to the staff of the Joint Research Center, "Technology and Materials", Belgorod National Research University, for their assistance with instrumental analysis.

Author Contributions: A.F., N.D., A.B. and R.K. formulated the original problem, designed the study, developed the methodology, and wrote the manuscript. A.F. and N.D. performed the experiment, collected data, and assisted with data analysis. A.B. and R.K. provided direction, guidance, and interpretation of data.

Conflicts of Interest: The authors declare no conflict of interest.

References

1. Abe, F.; Kern, T.U.; Viswanathan, R. *Creep Resistant Steels*; Part I; Woodhead Publishing in Materials: Cambridge, UK, 2008; p. 800.

2. Kaybyshev, R.O.; Skorobogatykh, V.N.; Shchenkova, I.A. New martensitic steels for fossil power plant: Creep resistance. *Phys. Met. Metallogr.* **2010**, *109*, 186–200. [CrossRef]

3. Kitahara, H.; Ueji, R.; Tsuji, N.; Minamino, Y. Crystallographic features of lath martensite in low-carbon steel. *Acta Mater.* **2006**, *54*, 1279–1288. [CrossRef]

4. Ghassemi-Armaki, H.; Chen, R.; Maruyama, K.; Igarashi, M. Premature creep failure in strength enhanced high Cr ferritic steels caused by static recovery of tempered martensite lath structures. *Mater. Sci. Eng. A* **2010**, *527*, 6581–6588. [CrossRef]

5. Abe, F. Analysis of creep rates of tempered martensitic 9% Cr steel based on microstructure evolution. *Mater. Sci. Eng. A* **2009**, *510*, 64–69. [CrossRef]

6. Kostka, A.; Tak, K.-G.; Hellmig, R.J.; Estrin, Y.; Eggeler, G. On the contribution of carbides and micrograin boundaries to the creep strength of tempered martensite ferritic steels. *Acta Mater.* **2007**, *55*, 539–550. [CrossRef]

7. Abe, F. Effect of fine precipitation and subsequent coarsening of Fe2W laves phase on the creep deformation behavior of tempered martensitic 9Cr-W steels. *Metall. Mater. Trans. A* **2005**, *36*, 321–331. [CrossRef]

8. Taneike, M.; Sawada, K.; Abe, F. Effect of carbon concentration on precipitation behavior of $M_{23}C_6$ carbides and MX carbonitrides in martensitic 9Cr steel during heat treatment. *Metall. Mater. Trans. A* **2004**, *35*, 1255–1261. [CrossRef]

9. Dudova, N.; Plotnikova, A.; Molodov, D.; Belyakov, A.; Kaibyshev, R. Structural changes of tempered martensitic 9%Cr-2%W-3%Co steel during creep at 650 °C. *Mater. Sci. Eng. A* **2012**, *534*, 632–639. [CrossRef]

10. Kipelova, A.; Belyakov, A.; Kaibyshev, R. Laves phase evolution in a modified P911 heat resistant steel during creep at 923K. *Mater. Sci. Eng. A* **2012**, *532*, 71–77. [CrossRef]

11. Fedorova, I.; Belyakov, A.; Kozlov, P.; Skorobogatykh, V.; Shenkova, I.; Kaibyshev, R. Laves-phase precipitates in a low-carbon 9% Cr martensitic steel during aging and creep at 923K. *Mater. Sci. Eng. A* **2014**, *615*, 153–163. [CrossRef]

12. Isik, M.I.; Kostka, A.; Yardley, V.A.; Pradeep, K.G.; Duarte, M.J.; Choi, P.P.; Raabe, D.; Eggeler, G. The nucleation of Mo-rich Laves phase particles adjacent to $M_{23}C_6$ micrograin boundary carbides in 12% Cr tempered martensite ferritic steels. *Acta Mater.* **2015**, *90*, 94–104. [CrossRef]

13. Ghassemi-Armaki, H.; Chen, R.; Maruyama, K.; Igarashi, M. Creep behavior and degradation of subgrain structures pinned by nanoscale precipitates in strength-enhanced 5 to 12 Pct Cr ferritic steels. *Metall. Mater. Trans. A* **2011**, *42*, 3084–3094. [CrossRef]

14. Yin, F.-S.; Tian, L.-Q.; Xue, B.; Jiang, X.-B.; Zhou, L. Effect of Carbon Content on Microstructure and Mechanical Properties of 9 to 12 pct Cr Ferritic/Martensitic Heat-Resistant Steels. *Metall. Mater. Trans. A* **2012**, *43*, 2203–2209. [CrossRef]

15. Dudko, V.; Belyakov, A.; Molodov, D.; Kaibyshev, R. Microstructure evolution and pinning of boundaries by precipitates in a 9 pct. Cr heat resistant steel during creep. *Metall. Mater. Trans. A* **2013**, *44*, S162–S172. [CrossRef]

16. Zhong, W.; Wang, W.; Yang, X.; Li, W.; Yan, W.; Sha, W.; Wang, W.; Shan, Y.; Yang, K. Relationship between Laves phase and the impact brittleness of P92 steel reevaluated. *Mater. Sci. Eng. A* **2015**, *639*, 252–258. [CrossRef]

17. Helis, L.; Toda, Y.; Hara, T.; Miyazaki, H.; Abe, F. Effect of cobalt on the microstructure of tempered martensitic 9Cr steel for ultra-supercritical power plants. *Mater. Sci. Eng. A* **2009**, *510*, 88–94. [CrossRef]

18. Kipelova, A.; Odnobokova, M.; Belyakov, A.; Kaibyshev, R. Effect of Co on creep behavior of a P911 steel. *Metall. Mater. Trans. A* **2013**, *44*, 577–583. [CrossRef]

19. Sawada, K.; Takeda, M.; Maruyama, K.; Ishii, R.; Yamada, M.; Nagae, Y.; Komine, R. Effect of W on recovery of lath structure during creep of high chromium martensitic steels. *Mater. Sci. Eng. A* **1999**, *267*, 19–25. [CrossRef]

20. Fedoseeva, A.; Dudova, N.; Kaibyshev, R. Creep strength breakdown and microstructure evolution in a 3%Co modified P92 steel. *Mater. Sci. Eng. A* **2016**, *654*, 1–12. [CrossRef]

21. Tsuchida, Y.; Okamoto, K. Improvement of creep rupture strength of high Cr ferritic steel by addition of W. *ISIJ Int.* **1995**, *35*, 317–323. [CrossRef]

22. Li, Q. Precipitation of Fe2W laves phase and modeling of its direct influence on the strength of a 12Cr-2W steel. *Metall. Mater. Trans. A* **2006**, *37*, 89–97. [CrossRef]

23. Dudova, N.; Kaibyshev, R. On the precipitation sequence in a 10% Cr steel under tempering. *ISIJ Int.* **2011**, *51*, 826–831. [CrossRef]

24. Wilshire, B.; Scharning, P. Prediction of long term creep data for forged 1Cr-1Mo-0.25V steel. *Mater. Sci. Technol.* **2008**, *24*, 1–9. [CrossRef]

25. Hirsch, P.B.; Howie, A.; Nicholson, R.B.; Pashley, D.W.; Whelan, M.J. *Electron Microscopy of Thin Crystals*, 2nd ed.; Krieger: New York, NY, USA, 1977; p. 563.

26. Dimmler, G.; Weinert, P.; Kozeschnik, E.; Cerjak, H. Quantification of the Laves phase in advanced 9–12% Cr steels using a standard SEM. *Mater. Charact.* **2003**, *51*, 341–352. [CrossRef]

27. Fedoseeva, A.; Dudova, N.; Glatzel, U.; Kaibyshev, R. Effect of W on tempering behaviour of a 3% Co modified P92 steel. *J. Mater. Sci.* **2016**, *51*, 9424–9439. [CrossRef]

28. Kimura, K.; Toda, Y.; Kushima, H.; Sawada, K. Creep strength of high chromium steel with ferrite matrix. *Int. J. Press. Vessels Pip.* **2010**, *87*, 282–288. [CrossRef]

29. Yoshizawa, M.; Igarashi, M.; Moriguchi, K.; Iseda, A.; GhassemiArmaki, H.; Maruyama, K. Effect of precipitates on long-term creep deformation properties of P92 and P122 type advanced ferritic steels for USC power plants. *Mater. Sci. Eng. A* **2009**, *510*, 162–168. [CrossRef]

30. Hattestrand, A.; Andren, H.O. Evaluation of particle size distributions of precipitates in a 9% chromium steel using energy filtered transmission electron microscopy. *Micron* **2001**, *32*, 489–498. [CrossRef]

31. Cipolla, L.; Danielsen, H.K.; Venditti, D.; Di Nunzio, P.E.; Hald, J.; Somers, M.A.J. Conversion of MX nitrides to Z-phase in a martensitic 12% Cr steel. *Acta Mater.* **2010**, *58*, 669–679. [CrossRef]

32. Danielsen, H.K.; Di Nunzio, P.E.; Hald, J. Kinetics of Z-Phase Precipitation in 9 to 12 pct Cr Steels. *Metall. Mater. Trans. A* **2013**, *44*, 2445–2452. [CrossRef]

33. Fedoseeva, A.; Dudova, N.; Kaibyshev, R. Effect of Tungsten on a Dispersion of M(C,N) Carbonitrides in 9% Cr Steels Under Creep Conditions. *Trans. Indian Inst. Met.* **2016**, *69*, 211–215. [CrossRef]

34. Kaibyshev, R.O.; Skorobogatykh, V.N.; Shchenkova, I.A. Formation of the Z-phase and prospects of martensitic steels with 11% Cr for operation above 590 °C. *Met. Sci. Heat Treat.* **2010**, *52*, 90–99. [CrossRef]

35. Klueh, R.L. Elevated temperature ferritic and martensitic steels and their application to future nuclear reactors. *Int. Mater. Rev.* **2005**, *50*, 287–310. [CrossRef]

36. Fedoseeva, A.; Dudova, N.; Kaibyshev, R. Effect of stresses on the structural changes in high-chromium steel upon creep. *Phys. Met. Metall.* **2017**, *118*, 591–600. [CrossRef]

37. Ghassemi-Armaki, H.; Chen, R.; Kano, S.; Maruyama, K.; Hasegawa, Y.; Igarashi, M. Strain-induced coarsening of nanoscale precipitates in strength enhanced high Cr ferritic steels. *Mater. Sci. Eng. A* **2012**, *532*, 373–380. [CrossRef]

38. Mehrer, H.; Stolica, N.; Stolwijk, N.A. Landolt Bornstein- Numerical Data and Functional Relationships in Science and Technology, New Series, Group III: Crystals and Solid State Physics. In *Diffusion in Solid Metals and Alloys*; Springer: Berlin/Heidelberg, Germany, 1990; Volume 26, pp. 47–48. ISBN 978-3-540-50886-1.

39. Fedoseeva, A.; Dudova, N.; Kaibyshev, R. Creep behavior and microstructure of a 9Cr-3Co-3W martensitic steel. *J. Mater. Sci.* **2017**, *52*, 2974–2988. [CrossRef]

40. Humphreys, F.J.; Hatherly, M. *Recrystallization and Related Annealing Phenomena*, 2nd ed.; Elsevier: Atlanta, GA, USA, 2004; pp. 91–112.

Article

Study on the Control of Rare Earth Metals and Their Behaviors in the Industrial Practical Production of Q420q Structural Bridge Steel Plate

Rensheng Chu [1,2], Yong Fan [2,3,*], Zhanjun Li [1], Jingang Liu [1], Na Yin [1] and Ning Hao [1]

[1]   Shougang Research Institute of Technology, Beijing 100043, China; churensheng@163.com (R.C.);
      lizhanjun_2000@163.com (Z.L.); liujg9916@163.com (J.L.); annayin1986@126.com (N.Y.);
      ustbwht@163.com (N.H.)
[2]   Department of Materials Engineering, KU Leuven, Kasteelpark Arenberg 44, B-3001 Heverlee, Belgium
[3]   Institut für Eisen- und Stahltechnologie, TU Bergakademie Freiberg, 09599 Freiberg, Germany
*   Correspondence: Yong.Fan@extern.tu-freiberg.de

Received: 24 February 2018; Accepted: 31 March 2018; Published: 5 April 2018

Abstract: Rare earth (RE) addition can refine and change the shape/distribution of inclusions in steel to improve its strength and toughness. In this paper, the control of RE, specifically Ce and La, and their behaviors in the practical industrial production of high-strength structural steel with 420 MPa yield strength were studied. In particular, the interactions between RE and Al, Nb, S, O were investigated, with the aim of improving the steel toughness and welding performance. The impact energy of the plate with RE is approximately 50 J higher than the regular plate without RE. The toughness of the plate from ladle furnace (LF) refining with RE addition is better than the one from Ruhrstahl and Hereaeus (RH) refining. The RE inclusions could induce the intragranular ferrite and refine the grain size to the preferred size. After welding at the heat input of 200 kJ/cm, the grain size at the heat affected zone was found to be the finest in the plate from the LF process with RE addition. Notably, the microstructure of ferrite was quasi-polygonal.

Keywords: structural steel plate; nonmetallic inclusions; rare earth control

---

1. Introduction

Structural steels are applied in various infrastructure, such as bridges, shipbuilding, and construction. Nowadays, the performance requirements of structural steel are getting higher and higher under various loading conditions such as earthquake, fire, wind, etc. [1–3]. Among the various performance properties, the most important requirements are low temperature impact toughness and welding. There are many laboratory studies on the effects of rare earth (RE) metals addition to the toughness of steels [4–8]. RE can refine and change the shape/distribution of inclusions in steel to improve strength and toughness. The addition of lanthanide metals to continuously cast steel is particularly advantageous because of their ability to refine as-cast structures, reduce segregation, and increase hot ductility at temperatures just below solidification [9]. People studied the optimum mass ratio of La and Ce in La + Ce combined treatment, and the best incubation time of Le + Ce (3:1) treatment [10]. There were some works on the efficiency of Ti-bearing inclusions in refining the grain size, and dealing with Al and Mn contents in weld metal inclusions [11,12]. Moreover, some research has focused on the effect of RE in high-heat-input welding, and the induction of acicular ferrite behavior [13–15]. During World War II, it was found that the addition of RE metals to steel could significantly improve its properties. Since then, RE metals have been widely applied in steel manufacturing. The RE nitride and the austenite grain size could be controlled to around 10 mm, when the amount of RE metals is more than 50 ppm [12,16–19]. RE can spheroidize inclusions that mainly rely on the combination of sulfur, reducing MnS formation, which results in RE sulfides and oxysulfide [13]. This leads to more applications in the special steel field [20]. It is shown that Ce is effective in the range 0.04–0.2 mass pct, but there is a rare report on the interactions between RE and other elements in the inclusions when S is less than 0.003 mass pct and RE is less than 0.005 mass pct, aiming to improve the toughness of steel. Due to the low amount of RE that is added, the production cost remains low. In this paper, we studied the control of RE, specifically Ce and La, and their behaviors in the practical industrial production of high-strength structural steel with 420 MPa yield strength, especially the interactions between RE and Al, Nb, S, O in the steel plate, with the aim of improving the steel toughness. Tests and analysis in the steelmaking process were tracked by some manufacturers in the production of structural steel of 420 MPa.

2. Experimental

The steelmaking process of high-strength structural steel with 420 MPa yield strength is shown in Figure 1. Through the RE treatment combined with heat treatment and controlled rolling, it was possible to control the cooling stability, which could limit the yield ratio to less than 0.85. Moreover, the toughness and welding performance of the final product could be enhanced.

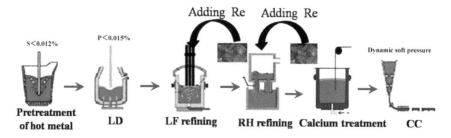

Figure 1. Steelmaking process of structural steel of 420 MPa plate.

The RE content in the alloy used in the present research was >95 mass pct, with a ratio of Ce and La in 2 for structural steel of 420 MPa plate production. The 2:1 ratio of Ce and La is nearly the same as their original ratio in nature. Their use avoids the complicated process of RE purification, which saves the overall cost of the whole process. The RE alloy was purified by refining treatment, and it was added by vacuum packaging. The chemical composition of structural steel of 420 MPa plate was shown in Table 1. The steelmaking process for Q420q steel involved pretreatment of hot metal, top blown oxygen converter (LD), ladle furnace (LF) refining, Ruhrstahl and Hereaeus (RH) refining, calcium treatment, and continuous casting (CC). In this steelmaking process, deoxidation was realized by aluminum after the LD process. The deep desulfurization and adjustment of alloying composition were performed in LF refining. The main function of RH refining was degassing and removal of inclusions. Hence, the protection of the casting was enhanced to prevent the secondary oxidation. Moreover, the process of calcium treatment was also used to control the morphology of inclusions. In addition, the sampling of liquid steel used a special "sampler", which is a high-temperature-prove alloy steel sampler with a cork binding on it. This sampler could prevent the contamination of liquid steel and slag.

Table 1. The chemical composition of the high-strength structural steel with 420 MPa yield strength, mass pct.

The Position for RE Alloy Added	C	Si	Mn	P	S	(Nb + Ti + etc.)	La	Ce
LF refining	0.144	0.28	1.46	0.014	0.0019	<1	<0.0010	<0.0020
RH refining	0.148	0.27	1.5	0.014	0.0016	<1	<0.0030	<0.0060

The inclusion behaviors of the steel, such as particle size, composition, and species were investigated via an automated scanning electron microscope (ASPEX explorer, Hillsboro, OR, USA). We could find the inclusion by EDS scanning automatically, and calculate the inclusion's center. Then, the equipment would scan towards eight dimensions from the center to obtain the spinning line. The inclusion structure and size would be determined via this information. The samples were cut into 25 mm × 25 mm × 20 mm shapes by line cutting, followed by coarse and fine grinding, and polishing.

The composition and type of inclusions were shown with the use of CaO-MgO-Al_2O_3 and CaO-Al_2O_3-CaS ternary phase diagrams. The chemical composition of inclusions was determined by EDS (S3400N) analysis, from which they were converted into the content of oxide and sulfide. The impact toughness testing was performed with a pendulum impact hammer (SANS, JB-300B, Shenzhen, China), which could generate a standard strike energy of 300 J, using the Charpy V-notched impact test method for metals. The high heat input welding experiments were realized with an electric gas welding tester (255 A, 34 V). The steel plate sample is 20 mm in thickness. Flux cored wire was used in the test, and the welding angle was 17°.

3. Results and Discussion

The chemical composition of the high-strength structural steel with 420 MPa yield strength was shown in Table 1. Nb and Ti are present. However, it was known that they have small effect on the oxide metallurgy, especially with high aluminum. The main inclusion of our present study is calcium aluminate, which contains rare Ti. Moreover, the Nb addition improved the strength, and the addition of RE would also cause some transformation to Nb. This will also be discussed in detail in this paper.

Component distribution of inclusions in structural steel of 420 MPa plate with RE in some practical industrial production was shown in Figure 2. It could be found that the component distribution of inclusions, in most cases, was calcium aluminate. It is known that most of the calcium aluminate inclusions have a high melting point >1873 K (1600 °C) [21]. The inclusions with high melting point exist in the form of spinel or calcium aluminate wrapped with CaS. The size of most of the inclusions was between 1 and 20 μm, and the most significant inclusions in class B are calcium aluminate with a low melting point <1773 K (1500 °C), [22] which could reach maximum 180 μm in size. As to the inclusions from LF refining process with RE addition, they are Al_2O_3-CaS inclusions in Al_2O_3-CaS-CaO system. We could not find calcium aluminate with low melting point. The size of most of the inclusions size is between 1 and 10 μm. The same phenomenon was also observed from the RH refining process with RE addition. However, some calcium aluminate inclusions were found with a low melting point. The size of most of the inclusions is between 1 and 15 μm. The average sizes of inclusions in a plate are 2.75 μm (LF with RE) and 3.11 μm (RH with RE), which are less than 4.42 μm of the regular structural steel without RE. The addition of RE would decrease the size of the inclusions.

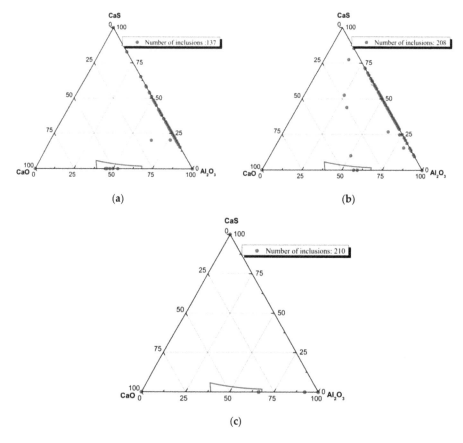

Figure 2. Component distribution of inclusions in structural steel. ((**a**) Inclusions of regular structural steel without rare earth (RE); (**b**) inclusions in structural steel from ladle furnace (LF) refining with RE addition; (**c**) inclusions in structural steel from Ruhrstahl and Hereaeus (RH) refining with RE addition).

3.1. Comparison of Inclusions Resulted from Ladle Furnace (LF) and Ruhrstahl and Hereaeus (RH) Processes

Figure 3 shows the size distribution of inclusions in structural steel of 420 MPa plate from different processes. It was resulted from the scan of the plate cross-section, from center to margin, with ASPEX explorer. It could be found that the inclusions that are >10 μm are concentrated in the center. They were calcium aluminate with high melting point ($CaO \cdot 2Al_2O_3$ and $CaO \cdot 6Al_2O_3$). These inclusions could become B type stringer inclusions (mainly calcium aluminate) after the hot rolling process, especially in bulky plate steels. The calcium aluminate with low melting point ($12CaO \cdot 7Al_2O_3$) was easy to transform during hot rolling, and this would also become a B type stringer inclusion [23,24].

The large-scale B type non-metallic inclusions distributed along the rolling direction in the steel plate would severely damage the serving performance of steel. In the regular plate without RE, the size of most of the inclusions is between 1 and 5 μm and they are distributed evenly. The number of inclusions of size in 5–10 μm is 45, slightly less than that of 87 in 1–5 μm. The size distribution in 5–10 μm is more uniform. The number of inclusions >10 μm is 5, with a bad distribution, especially in the thickness direction. The number of inclusions in plate produced from LF refining with RE addition is 208. The inclusions are evenly distributed through thickness direction. There are only

48 RE inclusions, and the rest are calcium aluminate inclusions. The RE inclusions in a plate produced from RH refining with RE addition are clearly increased. All the inclusions are the mixture of RE and calcium aluminate.

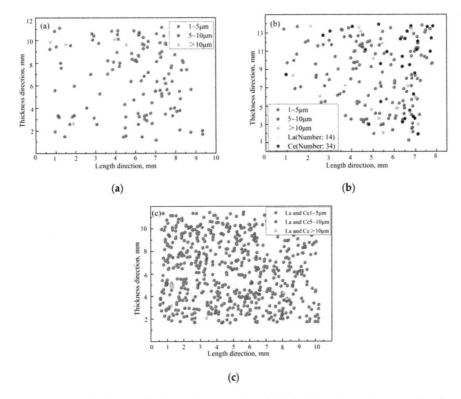

(a)

(b)

(c)

Figure 3. Size distribution of inclusions in structural steel. ((**a**) inclusions in regular structural steel without RE; (**b**) inclusions in structural steel from LF refining with RE addition; (c) inclusions in structural steel from RH refining with RE addition).

The quantity of RE inclusions from the LF refining process was less than that from RH refining. The main reason for this was that the RE content of steel is low in the LF refining process due to the removal of RE inclusions from the slag. We also found inclusions such as La and Ce, which occasionally appear alone or coexist together. The composition of the inclusions in structural steel from LF refining process with RE addition was shown in Table 2. It was also found that there are many Nb precipitates in the inclusions. When the inclusions contained Nb, there would be no existence of RE in the inclusions (Figure 4).

Table 2. The composition of the inclusions from LF refining process with RE addition, mass pct.

Element	Mg	Al	S	Ca	Nb	La	Ce
Average content	0.4	3.14	1.33	0.75	17.84	5.54	7.9

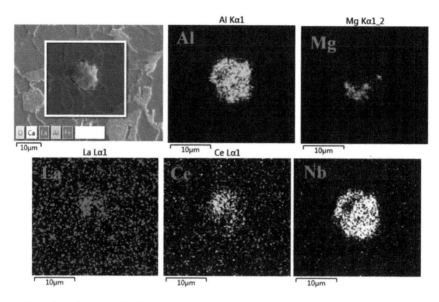

Figure 4. Surface scan of inclusion in structural steel from LF refining process with RE addition.

It could be seen that the quantity of RE inclusions from RH refining process is more than that from LF refining, due to the higher content of RE. The inclusions of RE all coexist together in the inclusions in the whole bulk of sample. The composition of the inclusions from RH process with RE addition is shown in Table 3. It was also found that there is no Nb precipitation in the inclusions, and the content of Al and Mg in the inclusions is also drastically reduced (Figure 5).

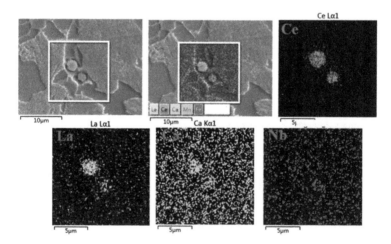

Figure 5. Surface scan of inclusion in structural steel from RH refining process with RE addition.

Table 3. The composition of the inclusions from RH refining process with RE, mass pct.

Element	Mg	Al	S	Ca	Nb	La	Ce
Average content	0.01	0.005	5.19	0.43	0.00	12.42	18.73

There is a mechanism of inclusions formation between strong deoxidation elements, and alloying elements by adding RE in the different refining process. The inclusions from LF refining process contained aluminum, which shows the reverse phenomenon of the RH refining process. As a robust deoxidizing element, calcium could attract aluminum oxide to become calcium aluminate inclusions when RE is not included. This showed that RE also has this ability, like calcium, which could reduce aluminum oxide inclusions or calcium aluminate inclusions.

From the composition analysis, it was found that the Nb precipitation was remarkably reduced. The reason for this might be that RE firstly forms inclusions with other nonmetallic elements such as O and S, which left little chance for Nb to form inclusions during the cooling process. In general, the Nb is precipitated during the cooling, and forms NbCN. However, there is only a small fraction of inclusions which can be neglected. The Nb inclusions are all small in size, and would become the nucleation particles. With the addition of RE, the precipitation of Nb takes place along with the formation of inclusions. Nb usually coexists with O/S/Ca/Al in LF process with RE addition. However, this coexistence disappeared in RH process with RE addition. RE, instead of Nb, is presented in the calcium aluminate inclusions. It could also be observed that the S content was increased in the inclusions. The reason for this is that the binding capacity was robust between RE and S. Meanwhile, CaS could also be formed. The temperature of RH is lower than that in the LF process which is more conducive to the precipitation of sulfide.

There might be another mechanism affecting the formation of RE inclusions. The RE inclusions formation begin with a nucleation from the oxide and sulfide inclusion with high melting point. The RE with high activity would be absorbed into that hard core. Moreover, due to the chemical site, there would be a density difference between the nucleus and the RE atoms, which enhanced the diffusion of RE to the hard core. This phenomenon contributed to the formation of the oxide and sulfide inclusion with RE. In addition, the size of inclusion has a relationship with RE aggregation. The more RE present, the higher the content of the other absorbing element.

3.2. Toughness and Welding Performance

It can be seen from Figure 6 that the impact energy of the plate with RE is approximately 50 J more than the one without RE. The toughness of the plate with RE from LF process is slightly more than that from RH process. RE has a positive effect on the toughness of the plate. The product performance can meet the requirements of high-heat-input welding.

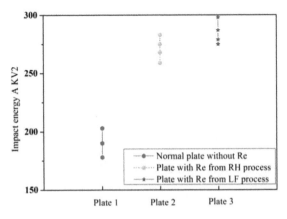

Figure 6. Impact energy of different plate.

The grain sizes of the plate are 8.63 μm (LF with RE), 9.93 μm (RH with RE), and 10.32 μm (regular plate without RE). The refinement of grain size would increase the toughness of the plate. A conclusion can be drawn from Figure 7 that the plate from the LF process with RE at the coarse-grain area is the finest with a high heat input of 200 kJ/cm. It also can be seen that the grain size of a regular plate without RE is the largest during the high heat input welding. As for the toughness performance, the plate with RE from LF process is not very different to that from the RH process. However, the difference in the grain size is more substantial after the high heat input welding. All these reasons show that the RE inclusions could induce the intragranular ferrite to refine the grains to the preferred size (Figure 8). The number of inclusions in the selected format is more than that in RH process.

Figure 7. Microstructure of different plate in high heat input welding.

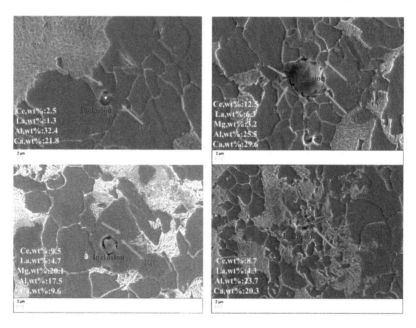

Figure 8. Inclusions induce intragranular ferrite to refine the grains.

The type of ferrite is not acicular but quasi-polygonal, more like a lath bainite structure. The grain is refined by inducing acicular ferrite in the general oxide metallurgy. However, the RE inclusions changed their structure to quasi-polygonal. In the current welding process, the steel plate would generate grain boundary ferrite, and coarsening during the high heat input welding. Therefore, we tried to lessen the effect from grain boundary ferrite to improve the roughness. It was clear that the RE inclusion induced the quasi-polygonal ferrite to refine the grain, and increased the toughness.

From Figure 9, it is possible to see that the inclusions of RE can induce the intragranular ferrite. The inclusions produced a depleted manganese zone, a depleted manganese zone would result in the formation of the intragranular ferrite. These inclusions of preferable size could induce the intragranular ferrite and refine the grain size. The intragranular ferrite that they influenced is not acicular, but quasi-polygonal. The ability to refine the grains of quasi-polygonal ferrite is weaker than that of acicular ferrite.

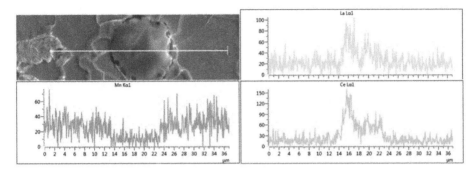

Figure 9. Inclusion particle produces depleted manganese area.

4. Conclusions

(1) Regarding the structural steel of 420 MPa plate without RE, most of the inclusions were calcium aluminate with a high melting point >1873 K (1600 °C), and other inclusions were calcium aluminate with a low melting point <1773 K (1500 °C). The most significant inclusions in class B could reach a size of 180 μm. The addition of RE in the LF refining process found that no low melting point calcium aluminate inclusions were precipitated. The addition RE in the RH refining process showed the presence of low melting point inclusion.

(2) The quantity of RE inclusions from the LF process was low. La and Ce inclusions occasionally appeared alone or coexist together. There is no Nb content in the RE inclusions, and only Nb mono-precipitates exist. The La and Ce inclusions from the RH process would always coexist together. Nb precipitation did not appear, and Al was also drastically reduced, while S increased in the inclusions.

(3) RE has a positive effect on the toughness of the steel plate. The impact energy of the plate with RE is approximately 50 J higher than for a regular plate without RE. The toughness of the plate with RE from LF refining is better than that from RH refining.

(4) RE inclusions could induce the intragranular ferrite, and refine the grains to the preferred size. After the welding at a heat input of 200 kJ/cm, the finest grain size at the heat affected zone was found in the plate from LF process with RE addition. The microstructure of ferrite was quasi-polygonal.

Acknowledgments: The author Yong Fan thanks the Japan Society for the Promotion of Science (JSPS) and Alexander von Humboldt (AvH) Foundation for their financial support.

Author Contributions: Rensheng Chu, and Yong Fan proposed and performed the whole researches, and wrote the main manuscript. Zhanjun Li, Jingang Liu, Na Yin, and Ning Hao helped with the experimental and the modification of the whole manuscript.

Conflicts of Interest: The authors declare no competing interests.

References

1. Naser, M.Z.; Kodur, V.K.R. Comparative fire behavior of composite girders under flexural and shear loading. *Thin-Walled Struct.* **2017**, *116*, 82–90. [CrossRef]
2. Kodur, V.K.; Aziz, E.M.; Naser, M.Z. Strategies for enhancing fire performance of steel bridges. *Eng. Struct.* **2017**, *131*, 446–458. [CrossRef]

3. Gowda, S.; Hotz, C.; Manigandan, K.; Srivatsan, T.S.; Patnaik, A.; Payer, J. Quasi-Static, Cyclic Fatigue and Fracture Behavior of Alloy Steel for Structural Applications: Influence of Orientation. *Mater. Perform. Charact.* **2016**, *5*, 148–163. [CrossRef]

4. Jiang, Q.C.; Liang, H.Q.; Sui, H.L. Effect of Y-Ce Complex Modification on Thermal Fatigue Behavior of High Cr Cast Hot Working Die Steels. *ISIJ Int.* **2004**, *44*, 1762–1766. [CrossRef]

5. Lan, J.; He, J.; Ding, W.; Wang, Q.; Zhu, Y. Effect of Rare Earth Metals on the Microstructure and Impact Toughness of a Cast 0.4C–5Cr–1.2Mo–1.0V Steel. *ISIJ Int.* **2000**, *40*, 1275–1282. [CrossRef]

6. Guo, M.; Suito, H. Influence of Dissolved Cerium and Primary Inclusion Particles of Ce_2O_3 and CeS on Solidification Behavior of Fe-0.20 mass%C-0.02 mass%P Alloy. *ISIJ Int.* **1999**, *39*, 722–729. [CrossRef]

7. Waudby, P.E. Rare earth additions to steel. *Int. Met. Rev.* **1978**, *23*, 74–98. [CrossRef]

8. Luyckx, L. Current trends in the use of rare earths in steelmaking. *Electr. Furn. Conf.* **1973**, *31*, 5–7.

9. Wilson, W.G.; Heaslip, L.J.; Sommerville, I.D. Rare Earth Additions in Continuously Cast Steel. *JOM* **1985**, *37*, 36–41. [CrossRef]

10. Song, M.-M.; Song, B.; Xin, W.-B.; Sun, G.-L.; Song, G.-Y.; Hu, C.-L. Effects of rare earth addition on microstructure of C–Mn steel. *Ironmak. Steelmak. Process.* **2015**, *8*, 594–599. [CrossRef]

11. Mills, A.R.; Thewlis, G.; Whiteman, J.A. Nature of inclusions in steel weld metals and their influence on formation of acicular ferrite. *Mater. Sci. Tech.* **1987**, *12*, 1051–1061. [CrossRef]

12. Zhang, Z.; Farrar, R.A. Role of non-metallic inclusions in formation of acicular ferrite in low alloy weld metals. *Mater. Sci. Tech.* **1996**, *12*, 237–260. [CrossRef]

13. Song, M.; Song, B.; Zhang, S.; Xue, Z.; Yang, Z.; Xu, R. Role of Lanthanum Addition on Acicular Ferrite Transformation in C–Mn Steel. *ISIJ Int.* **2017**, *57*, 1261–1267. [CrossRef]

14. Thewlis, G. Effect of cerium sulphide particle dispersions on acicular ferrite microstructure development in steels. *Mater. Sci. Technol.* **2006**, *22*, 153–166. [CrossRef]

15. Wen, B.; Song, B. In Situ Observation of the Evolution of Intragranular Acicular Ferrite at Ce–Containing Inclusions in 16Mn Steel. *Steel Res. Int.* **2012**, *83*, 487–495.

16. Pan, F.; Zhang, J.; Chen, H.-L.; Su, Y.-H.; Kuo, C.-L.; Su, Y.-H.; Chen, S.-H.; Lin, K.-J.; Hsieh, P.-H.; Hwang, W.-S. Effects of Rare Earth Metals on Steel Microstructures. *Materials* **2016**, *9*, 417. [CrossRef] [PubMed]

17. Lee, J. Evaluation of the nucleation potential of intragranular acicular ferrite in steel weldments. *Acta Metall. Mater.* **1994**, *42*, 3291–3298. [CrossRef]

18. Hsu, T.Y. Effects of Rare Earth Element on Isothermal and Martensitic Transformations in Low Carbon Steels. *ISIJ Int.* **1998**, *38*, 1153–1164. [CrossRef]

19. Nako, H.; Okazaki, Y.; Speer, J.G. Acicular Ferrite Formation on Ti-Rare Earth Metal-Zr Complex Oxides. *ISIJ Int.* **2015**, *55*, 250–256. [CrossRef]

20. Gao, J.; Fu, P.; Liu, H.; Li, D. Effects of Rare Earth on the Microstructure and Impact Toughness of H13. *Steel. Met.* **2015**, *5*, 383–394. [CrossRef]

21. Ye, G.; Jonsson, P.; Lund, T. Thermodynamics and Kinetics of the Modification of Al_2O_3 Inclusions. *ISIJ Int.* **1996**, *36*, S105–S108. [CrossRef]

22. Dawson, S.; Mountford, N.D.G.; Sommerville, I.D.; McLean, A. The Evaluation of Metal Cleanliness in the Steel Industry. IV. Preferential Dissolution Techniques. *Ironmak. Steelmak.* **1988**, *15*, 54–55.

23. Zhao, D.W.; Li, H.B.; Bao, C.; Yang, J. Inclusion Evolution during Modification of Alumina Inclusions by Calcium in Liquid Steel and Deformation during Hot Rolling Process. *ISIJ Int.* **2015**, *55*, 2125–2134. [CrossRef]

24. Zhao, D.W.; Li, H.B.; Cui, Y.; Yang, J. Control of Inclusion Composition in Calcium Treated Aluminum Killed Steels. *ISIJ Int.* **2016**, *56*, 1181–1187. [CrossRef]

Article

Effect of Al$_2$O$_3$ Nanoparticles as Reinforcement on the Tensile Behavior of Al-12Si Composites

Pan Ma [1], Yandong Jia [2,*], Prashanth Konda Gokuldoss [3,4,*], Zhishui Yu [1], Shanglei Yang [1], Jian Zhao [1] and Chonggui Li [1]

[1] School of Materials Engineering, Shanghai University of Engineering Science, 201620 Shanghai, China; mapanhit@hotmail.com (P.M.); yuzhishui@sues.edu.cn (Z.Y.); yslei@126.com (S.Y.); zhaojianhit@163.com (J.Z.); chongguili@sues.edu.cn (C.L.)

[2] School of Materials Science and Engineering, Shanghai University, 200444 Shanghai, China

[3] Department of Manufacturing and Civil Engineering, Norwegian University of Science and Technology, 222815 Gjøvik, Norway

[4] Erich Schmid Institute of Materials Science, Austrian Academy of Sciences, A-8700 Leoben, Austria

* Correspondence: yandongjia@shu.edu.cn (Y.J.); kgprashanth@gmail.com (P.K.G.); Tel.: +86-137-6107-8756 (Y.J.); +47-973-646-67 (P.K.G.)

Received: 16 August 2017; Accepted: 7 September 2017; Published: 10 September 2017

Abstract: Al$_2$O$_3$ nanoparticle-reinforced Al-12Si matrix composites were successfully fabricated by hot pressing and subsequent hot extrusion. The influence of weight fraction of Al$_2$O$_3$ particles on the microstructure, mechanical properties, and the corresponding strengthening mechanisms were investigated in detail. The Al$_2$O$_3$ particles are uniformly distributed in the matrix, when 2 and 5 wt. % of Al$_2$O$_3$ particles were added to the Al-12Si matrix. Significant agglomeration can be found in composites with 10 wt. % addition of Al$_2$O$_3$ nanoparticles. The maximum hardness, the yield strength, and tensile strength were obtained for the composite with 5 wt. % Al$_2$O$_3$ addition, which showed an increase of about ~11%, 23%, and 26%, respectively, compared with the Al-12Si matrix. Meanwhile, the elongation increased to about ~30%. The contribution of different mechanisms including Orowan strengthening, thermal mismatch strengthening, and load transfer strengthening were analyzed. It was shown that the thermal mismatch strengthening has a more significant contribution to strengthening these composites than the Orowan and load transfer strengthening mechanisms.

Keywords: Al metal matrix composites; microstructure; mechanical properties; strengthening mechanism

---

1. Introduction

Al metal matrix composites (MMCs) have gained considerable attention as the ideal candidates for potential lightweight materials used in automobiles and other structural applications because of their attractive properties, such as low density, high specific stiffness, and superior wear resistance [1–6]. Recently, nano-sized reinforcements have been examined for the fabrication of Al-based composites. It has been shown that the addition of a small amount of fine ceramic particles improves the strength significantly [7,8]. Al$_2$O$_3$ particles with high specific stiffness and superior high temperature properties are used as inert ceramic reinforcement phases in MMCs [9,10]. A large number of studies have been performed on the fabrication and characterization of nano-sized Al$_2$O$_3$-reinforced Al matrix composites. Karbalaei Akbari et al. [11] studied nano-sized Al$_2$O$_3$-reinforced A356 alloy with stirring casting, and agglomerated nanoparticles were observed on dendrites in the fracture surface of the Al-Al$_2$O$_3$ reinforcement samples. Su et al. [12] fabricated Al$_2$O$_3$ nanocomposites by solid-liquid mixed casting combined with ultrasonic treatment and demonstrated that the ultrasonic vibration during the solidification was beneficial to refine the grain structure, and to improve the resulting distribution of Al$_2$O$_3$ nanoparticle in the matrix. Sajjadi et al. [13,14] showed that decreasing the alumina particle size combined with compo-casting process can yield the best mechanical properties for Al-Al$_2$O$_3$ composites. The addition of alumina led to the improvement in yield strength (YS), ultimate tensile strength (UTS), compression strength, and hardness. However, the poor wettability of Al$_2$O$_3$ with molten aluminum makes it difficult to achieve a uniform distribution of Al$_2$O$_3$ nanoparticles in the composite [15].

The results demonstrate that the solidification processes have some advantages such as a wide selection of materials and range of materials that can be processed. Good matrix-particle bonding, ease in controlling the microstructure of the matrix, and flexibility in processing are some attractive features. However, the solidification processes are not suited for the dispersion of fine-sized particles due to the possibility of the formation of agglomeration. Agglomerate formation is due to the van der Waals force of attraction, non-uniform particle distribution, and poor wettability at the interface between metallic and ceramic reinforcements [16–22]. Powder metallurgy (PM) is an alternative fabrication technique the can yield a uniform distribution of reinforcement and flexibility in reinforcement composition and design [18–22]. Al matrix composites reinforced with nano- and submicron-sized Al$_2$O$_3$ particles were prepared by wet attrition milling and subsequent hot extrusion processing by Tabandeh Khorshid et al. [23]. It has been observed that both the hardness and strength of the composites increase and reach a saturation before showing a decline upon increasing the volume fraction of the nanoparticles. Razavi et al. [24] addressed the effect of mechanical milling and reinforcement nanoparticles on the densification response and the compressibility of aluminum powder. However, the effect of reinforcement on the strengthening mechanisms are not well illustrated. To facilitate the development of such composites, it is necessary to fully understand the strengthening mechanism of nano-sized particle reinforcements in the Al matrix composite. In this paper, the nano Al$_2$O$_3$ particle-reinforced Al-12Si matrix composites were prepared by powder metallurgy. Al-12Si alloy was chosen for this study because it is a near eutectic composition with good fluidity that has been widely used for a variety of industrial applications [25–28]. The influence of the weight fraction of Al$_2$O$_3$ particles on the microstructure and mechanical properties of Al alloy matrix was studied in detail. Additionally, the strengthening mechanism operating in these materials was investigated.

2. Materials and Methods

The Al-12Si gas atomized alloy powder (nominal composition of Al-88 wt. % and Si-12 wt. %) was selected as the matrix material. Al$_2$O$_3$ nanoparticles (d50 ~50 nm) of 2%, 5%, and 10% wt. % were used as reinforcement. The powders were milled to have a uniform mixture of reinforcement in the Al-12Si matrix. Milling was performed in a planetary ball mill (Retsch, Haan, Germany) under an argon atmosphere at room temperature for 2 h. The milling speed and ball to powder ratio were 150 rpm and 10:1, respectively. Consolidation (hot pressing) of the homogeneously mixed powders were carried out using hardened steel dies. Samples with a diameter of 10 mm and a height of 10 mm were hot pressed using the following parameters: load—600 MPa, temperature—673 K, and time—15 min. The samples were subsequently hot extruded with a pressure of ~800 MPa at 673 K, and the extrusion ratio was 6:1. As a result, 4-mm diameter rods were obtained after hot extrusion. For comparison, a monolithic Al-12Si matrix alloy (without reinforcement) was also fabricated by the same method.

For metallographic observation, the samples were prepared using conventional grinding and polishing techniques (from 200 grit to 4000 grit papers and then cloth polishing using diamond paste). The samples were then etched in 0.5% Hydroflouricacid (HF) solution. The microstructure of the composites were characterized by optical microscope (OM) and scanning electron microscope (SEM, Gemini1530 Zeiss) - (Carl Zeiss AG, Oberkochen, Germany). The structural characterization was performed by X-ray diffraction using D3290 PANalytical X'pert PRO (Almelo, The Netherlands) equipped with Cu-Kα radiation. Hardness measurements were carried out using a Vickers micro-hardness (Schimadzu, Dusiburg, Germany) tester at a constant load of 200 gf for a dwell time of 15 s. Tensile tests were carried out using an Instron-5869 device (INSTRON GmbH, Darmstadt, Germany) at a rate of 0.5 mm/min, and the strain was calculated using a laser extensometer. The tensile test samples (according to American Society of Testing and Materials (ASTM) standard: ASTM: E8/E8M–13a) were cylindrical tensile bars with a total length of 52 mm that were machined from the extruded samples. The dimensions along the gauge length of the tensile bars were: length—17.5 mm, diameter—3.5 mm, and 4 mm around the grips. Three nominally identical specimens were tested to obtain the average values.

3. Results

3.1. Microstructure Analysis and Phase Identification

SEM micrographs of monolithic Al-12Si and Al-12Si/Al_2O_3 nanocomposites are shown in Figure 1. The extruded sample in Figure 1a reveals that the eutectic Si (~20 μm) is uniformly distributed in the Al matrix. Figure 1b,c show the microstructure of the composites as a function of the increasing content of Al_2O_3 particles. It can be observed from Figure 1b,c that the Al_2O_3 particles have a relative homogeneous distribution without significant agglomeration of the particles. When the reinforcement content reaches 10 wt. %, the agglomeration of the Al_2O_3 particles is quite evident (Figure 1d). The degree of Al_2O_3 agglomeration varies randomly throughout the samples, as marked in the Figure 1d, which may lead to the formation of defects such as pores and may act as stress concentrators, thereby hampering the overall properties of the composites. The high magnification image in Figure 2a shows the presence of agglomeration in the Al-12Si/Al_2O_3 (10 wt. %) nanocomposite. The energy dispersive spectrum analysis (Figure 2b) shows that the small particles within the agglomeration consist mainly of Al and O and the ratio is close to that of Al_2O_3. Since the nanoparticles have high surface energy due to the high surface to volume ratio, the nanoparticles tend to stick with each other and form agglomerates. Hence, the agglomerates increase with greater additions of the nanoparticles (in excess), resulting in an uneven distribution of the reinforcement in the matrix. Nevertheless, no visible flaws or pores were observed at the Al_2O_3/Al interface.

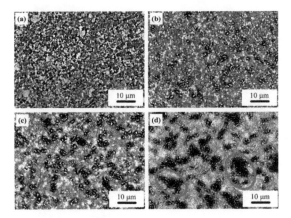

Figure 1. Scanning Electron Microscopy (SEM) images of hot extruded Al-12Si/Al_2O_3 nanocomposites with (**a**) 0, (**b**) 2, (**c**) 5, and (**d**) 10 wt. % reinforcement.

Figure 3 shows the XRD pattern of the composites reinforced with different contents of Al2O3. As observed, the samples mainly consist of three phases: (1) the Al matrix phase; (2) the Si phase from the matrix; and (3) the Al_2O_3 phase from the reinforcement. The diffraction peaks of the Al phase show the highest intensity because of their weight fraction in the composite. At 2 wt. % Al_2O_3 content, the diffraction peaks of v are hard to observe in the XRD pattern due to their low intensity. When the Al_2O_3 content increases to 5 wt. %, the small diffraction peaks of the Al_2O_3 phase are visible, and their intensity increases when the Al_2O_3 content is increased to 10 wt. %. No other visible peaks were observed, suggesting that there is no additional phase formed between the (Al-12Si) matrix and the (Al$_2$O$_3$) reinforcement.

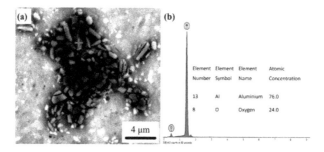

Figure 2. Scanning Electron Microscopy (SEM) image and Energy Dispersive Spectroscopy (EDS) analysis data for Al-12Si/10 wt. % Al$_2$O$_3$ nanocomposite in the as-extruded condition.

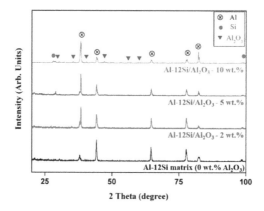

Figure 3. XRD analysis of Al-12Si/Al$_2$O$_3$ nanocomposites with (**a**) 0, (**b**) 2, (**c**) 5, and (**d**) 10 wt. % addition of Al$_2$O$_3$ reinforcement.

3.2. Physical and Mechanical Properties

Both the theoretical and experimental densities for all composite samples are listed in Table 1. The theoretical density of the composites are calculated using the rule of mixtures. High relative densities are achieved for all composites, indicating a good interface between the Al-Si matrix and the Al$_2$O$_3$ reinforcement. The relative density for the composite with 10 wt. % Al$_2$O$_3$ decreases slightly due to the agglomeration of Al$_2$O$_3$ particles, which leads to the presence of some pores.

Table 1. Density of the as-extruded Al-12Si/Al$_2$O$_3$ nanocomposites.

Weight Percent of Al$_2$O$_3$ (%)	Theoretical Density (g/cc)	Measured Density (g/cc)	Relative Density (%)
0	2.61	2.60	99.6
2	2.63	2.61	99.2
5	2.66	2.64	99.5
10	2.70	2.65	98.1

Figure 4 shows the Vickers hardness data for the Al-12Si/Al$_2$O$_3$ nanocomposites. The matrix shows a hardness value of 72 ± 1 HV$_{0.2}$, which increases to 77 ± 1 HV$_{0.2}$ (a change of ~5 HV$_{0.2}$) when 2 wt. % of Al$_2$O$_3$ reinforcement is added to the matrix. The composite reinforced with 5 wt. %

Al$_2$O$_3$ nanoparticles shows a higher hardness (80 $\pm$ 1 HV$_{0.2}$) compared to the other systems. With a further increase in the weight percentage of Al$_2$O$_3$ nanoparticles (to 10 wt. %), the composites show a decrease in the hardness value compared to the former ones due to the porosity and irregularities generated by the agglomeration of nanoparticles. Al$_2$O$_3$ nanoparticles dispersed uniformly in the Al-12Si alloy may significantly improve their hardness properties compared to the base alloy. The hard Al$_2$O$_3$ nanoparticles in the Al matrix act as barriers to the motion of dislocations generated in the matrix and the higher particle density causes Orowan mechanism. However, an increased addition of nanoparticles leads to agglomeration of the reinforcement, resulting in an increased level of porosity in the material and a subsequent degradation of the hardness values.

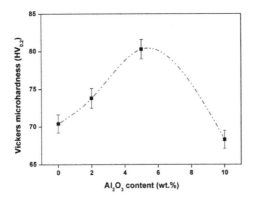

Figure 4. Vickers hardness measurements of Al-12Si/Al$_2$O$_3$ nanocomposites produced by the hot extrusion method.

Figure 5 shows the room temperature tensile stress-strain curves of the nanocomposite, and their properties are summarized in Table 2. It can be observed that the addition of Al$_2$O$_3$ significantly improves the yield strength (YS), ultimate tensile strength (UTS), and elongation of the Al-12Si matrix up to 5 wt. % Al$_2$O$_3$ addition. This is due to the good dispersion and strong interfacial bonding between the Al-12Si matrix and Al$_2$O$_3$ reinforcement. The YS, UTS, and elongation of the Al-12Si matrix in the as-extruded condition is found to be 198 MPa, 228 MPa, and 23%, respectively. Adding 2 wt. % of Al$_2$O$_3$ changes the YS, UTS, and elongation to 228 MPa, 256 MPa, and 24%. With further increasing the Al$_2$O$_3$ addition to 5 wt. % the YS increases to 245 MPa, UTS to 286 MPa, and ductility to 30%. The YS and UTS of the composite drops (to 195 MPa and 221 MPa, respectively) when the reinforcement content reaches 10 wt. %. In summary, the composite reinforced with 5 wt. % Al$_2$O$_3$ nanoparticles shows higher YS and UTS compared to the base alloy and composites reinforced with varying percentages of Al$_2$O$_3$ nanoparticles considered in this study.

Table 2. Tensile properties of the as-extruded Al-12Si/Al$_2$O$_3$ nanoparticle-reinforced composites.

Weight Percent of Al$_2$O$_3$ (%)	Yield Strength (MPa)	Ultimate Tensile Strength (MPa)	Elongation (%)
0	198 $\pm$ 2	228 $\pm$ 2	23 $\pm$ 2
2	228 $\pm$ 2	256 $\pm$ 3	24 $\pm$ 2
5	245 $\pm$ 2	286 $\pm$ 3	30 $\pm$ 3
10	195 $\pm$ 2	221 $\pm$ 2	19 $\pm$ 1

Figure 6 shows the tensile fracture surfaces of Al-12Si/Al$_2$O$_3$ nanocomposites. It can be observed (Figure 6a) that the fracture mode of the matrix Al-12Si alloy is predominantly ductile with several dimples all over the surface. As shown in Figure 6b, 2 wt. % nano Al$_2$O$_3$-reinforced composite also displays dimples like the matrix Al-12Si alloy, except that the length scale of the dimples are different.

Similarly, composites with 5 wt. % Al$_2$O$_3$ also show the presence of dimples again at different length scales than the 2 wt. % Al$_2$O$_3$ composite and Al-12Si-based alloy. When the Al$_2$O$_3$ content is increased to 10 wt. % (Figure 6d), both the number and size of the dimples decrease significantly, leading to a relatively brittle type of failure. Some defects such as cracks and segregation are observed on the fracture surface (along the neighborhood of the Al$_2$O$_3$ reinforcement particles), which also limits both the strength and ductility of the composites. The segregation/agglomeration will prevent an effective bonding between the reinforcement and the matrix and may lead to further defects such as cracks in the composites, and these cracks will inevitably leads to a premature failure of the composites. In addition, the agglomeration/clustering of the reinforcement particles can cause the presence of increased porosity in the composites [29].

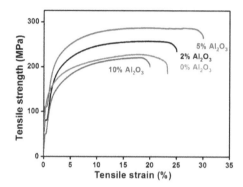

Figure 5. The tensile true stress-true strain curves of Al-12Si/Al$_2$O$_3$ nanocomposites produced by the hot extrusion method.

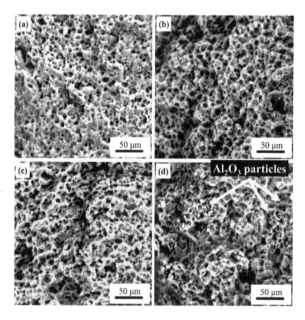

Figure 6. Fracture morphologies of the as-extruded Al-12Si/Al$_2$O$_3$ nanocomposites after room temperature tensile tests.

3.3. Strengthening Mechanism

Unlike tensile strength, yield strength usually shows less dependency on the presence of the stress concentrations in the composites. Therefore, yield strength is more representative of the discussion of the strengthening mechanisms [30]. The strengthening mechanisms operating in the Al-12Si/Al$_2$O$_3$ nanocomposites mainly include the load transfer mechanism, Orowan strengthening mechanism, and thermal mismatch enhancement mechanism. These three mechanisms operate simultaneously [31–33]. The most accepted strengthening mechanism is the direct strengthening mechanism, also known as load transfer mechanism. The strengthening of the composites takes place due to the transfer of load from the weaker matrix to the harder reinforcement through their interface, as proposed by Nardon and Prewo [34].

$$\Delta\sigma_{load} = 0.5f\sigma_m \tag{1}$$

where σ_m is the yield strength of the matrix (MPa), and f is the volume fraction of Al$_2$O$_3$ particles. According to Equation (1), the effect of load transfer is calculated and is given in Table 3. The Orowan strengthening mechanism is very important in metal matrix composites, if the reinforcement particles are fine and the inter-particle spacing is not large. Incremental yield strength of the composite caused by Orowan stress of the Al$_2$O$_3$ particles may be expressed as [35,36]:

$$\Delta\sigma_{orowan} = \frac{0.538G_m b\sqrt{f}}{d} \ln\frac{d}{2b} \tag{2}$$

where Gm is the shear modulus of the matrix.

$$G_m = G_{Al}V_{Al} + G_{Si}V_{Si} \tag{3}$$

where G_{Al}=27 GPa, G_{Si}= 66.8 GPa, B is the Burgers vector of the matrix, b = ((1.414/2) × 0.286 nm) = 0.202 nm, and d and f are the diameter and volume fraction of Al$_2$O$_3$ particle, respectively. This effect is calculated and the results are given in Table 3. As shown in Table 3, the higher the Al$_2$O$_3$ content, the greater the effect of the Orowan mechanism. When a composite is subjected to thermal changes, then the difference in their coefficients of thermal expansion (CTE) between the matrix and reinforced particles will produce a thermal strain and internal stress state changes. In order to accommodate this thermal mismatch effect, dislocations are generated around reinforced particles within the matrix to reduce the stored energy. The magnitude of thermal strain determines the density of dislocations generated within the composites. The increased dislocation density within the matrix leads to the strengthening of the composite. In this study, the large discrepancy of thermal expansion coefficients between the Al$_2$O$_3$ (7.85 × 10^{-6}/K) and the Al-12Si matrix (20.96 × 10^{-6}/K) generate dislocations in the Al matrix. The improvement in yield strength from thermal mismatch can be calculated using Equation (4) [37–39]:

$$\Delta\sigma_{CTE} = \sqrt{3}aGb\sqrt{\rho_{CTE}} \tag{4}$$

where a is a constant, a = 1. ρ_{CTE} is the dislocation density caused by CTE differences, and can be described as follows:

$$\rho_{CTE} = \frac{12\Delta\alpha\Delta Tf}{bd(1-f)} \tag{5}$$

where $\Delta\alpha$ is the CTE difference between the matrix and the reinforcement, ΔT is the temperature difference between the testing temperature and room temperature. This strengthening effect is also shown in Table 3. These strengthening mechanisms are additive, and the total yield strength of the composite could be considered as the sum of different mechanisms rather than the contribution of a single one. In its simplest form, the superposition of the individual strengthening mechanisms may be carried out linearly.

$$\Delta\sigma_C = \Delta\sigma_{Load} + \Delta\sigma_{Orowan} + \Delta\sigma_{CTE} \tag{6}$$

Table 3. Contribution of different strengthening mechanisms on the yield stress of Al-12Si/Al$_2$O$_3$ extruded nanocomposites.

Weight Percent of Al$_2$O$_3$ (%)	$\triangle\sigma_{Load}$ (MPa)	$\triangle\sigma_{Orowan}$ (MPa)	$\triangle\sigma_{CTE}$ (MPa)	$\triangle\sigma_C$ (MPa)
2	1.50	3.85	27.94	33.19
5	4.14	6.13	42.47	52.74
10	6.65	8.75	53.49	68.89

The individual contributions of the described three mechanisms to the yield strength of the composite calculated from Equations (1)–(4) are listed in Table 3 as a function of the weight percentage of Al$_2$O$_3$ nanoparticles. The presented data show that the effect induced by thermal mismatch strengthening ($\triangle\sigma_{CTE}$) makes the largest contribution to the yield strength improvement. The strength increments resulting from the load transfer ($\triangle\sigma_{Load}$) and Orowan strengthening ($\triangle\sigma_{Orowan}$) are very similar and are much lower. Figure 7 shows the variation of experimental YS and the theoretical values calculated using Equation (6). The total improvement from the multiple strengthening mechanisms for yield strength is ~30 MPa and 45 MPa, and the predicted values of the yield strength for the composites are ~33 MPa and 53 MPa, respectively, for the composites with 2 and 5 wt. % Al$_2$O$_3$ addition. This shows that the experimental and theoretical values are very close and hence the strengthening mechanisms based on load sharing, Orowan mechanism, and thermal mismatch strengthening are expected to play a role in these composites. The small discrepancy between the experimental and theoretical values may arise from the defects (such as porosity, agglomeration, and impurities) present in the samples. A large discrepancy was observed for the composite with 10 wt. % Al$_2$O$_3$ as reinforcement, which was mainly caused by the profound agglomeration of the reinforcement in the composite.

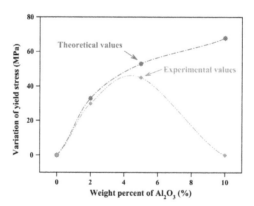

Figure 7. Comparison of the strengthening effect in the as-extruded Al-12Si/Al$_2$O$_3$ nanocomposites between experimental observation and calculated theoretical values.

4. Summary

In this study, Al$_2$O$_3$ nanoparticle-reinforced Al-12Si composites were prepared by hot pressing and subsequent hot extrusion. The microstructure, mechanical properties, and the strengthening mechanisms were investigated in detail. The conclusions are as follows:

(1) The composites reinforced with 2 and 5 wt. % Al$_2$O$_3$ nanoparticles exhibited relatively homogeneous distribution, while the microstructure of the composites with 10 wt. % reinforcement exhibited agglomeration of the Al$_2$O$_3$ particles in the matrix.

(2) The hardness of the composites increased first (for 2 and 5 wt. % reinforcement) and then decreased (for 10 wt. % reinforcement). The strength of the composites also displayed similar trends

to that of the hardness. The maximum values were obtained for the composites with 5 wt. % Al_2O_3. The YS, UTS, and the elongation were 244 MPa, 286 MPa, and ~30%, respectively, with 5 wt. % Al_2O_3 addition, which showed an increase to about ~23%, 26%, and 30%, respectively, compared with the Al-12Si matrix alloy.

(3) The strengthening mechanism analysis shows that the yield strength of the composites increased mostly because of the thermal mismatch enhancement mechanism compared to the other two strengthening mechanisms (load transfer and Orowan strengthening).

Acknowledgments: This work was supported by the National Natural Science Foundation of China (Nos. 51601110, 51601109, 51375294, 51402189), the China Postdoctoral Science Foundation (No. 2016M601563), the Natural Science Foundation of Shanghai (No. 17ZR1440800, 14ZR1418800), the Youth Teacher Development Program of Shanghai Universities (Nos. ZZGCD15100, ZZGCD15036), and the Program of Shanghai University of Engineering Science (2016-45).

Author Contributions: Yandong Jia and Prashanth Konda Gokuldoss formulated the idea. ZhiShui Yu and Jian Zhao helped with the literature survey. Pan Ma, Shanglei Yang and Chonggui Li carried out the experiments, Pan Ma wrote the paper and Prashanth Konda Gokuldoss supervised the research.

Conflicts of Interest: The authors declare no conflict of interest.

References

1. Scudino, S.; Liu, G.; Prashanth, K.G.; Bartusch, B.; Surreddi, K.B.; Murty, B.S.; Eckert, J. Mechanical properties of Al-based metal matrix composites reinforced with Zr-based glassy particles produced by powder metallurgy. *Acta Mater.* **2009**, *57*, 2029–2039. [CrossRef]
2. Bodunrin, M.O.; Alaneme, K.K.; Chown, L.H. Aluminium matrix hybrid composites: A review of reinforcement philosophies; mechanical, corrosion and tribological characteristics. *J. Mater. Res. Technol.* **2015**, *4*, 434–445. [CrossRef]
3. Hunt, W.; Herling, D.R. Aluminum Metal Matrix Composites. *Adv. Mater. Process.* **2004**, *162*, 39–44.
4. Wang, Z.; Prashanth, K.G.; Scudino, S.; Chaubey, A.K.; Sordelet, D.J.; Zhang, W.W.; Li, Y.Y.; Eckert, J. Tensile properties of Al matrix composites reinforced with in-situ devitrified $Al_{84}Gd_6Ni_7Co_3$ glassy particles. *J. Alloy. Compd.* **2014**, *586*, S419–S422. [CrossRef]
5. Prashanth, K.G.; Scudino, S.; Chaubey, A.K.; Lober, L.; Wang, P.; Attar, H.; Schimansky, F.; Pyczak, F.; Eckert, J. Processing of Al-12Si–TNM composites by selective laser melting and evaluation of compressive and wear properties. *J. Mater. Res.* **2016**, *31*, 55–65. [CrossRef]
6. Wang, Z.; Tan, J.; Sun, B.A.; Scudino, S.; Prashanth, K.G.; Zhang, W.W.; Li, Y.Y.; Eckert, J. Fabrication and mechanical properties of Al-based metal matrix composites reinforced with $Mg_{65}Cu_{20}Zn_5Y_{10}$ metallic glass particles. *Mater. Sci. Eng. A* **2014**, *600*, 53–58. [CrossRef]
7. Mazahery, A.; Abdizadeh, H.; Baharvandi, H.R. Development of high- performance A356/nano-Al_2O_3 composites. *Mater. Sci. Eng. A* **2009**, *518*, 61–64. [CrossRef]
8. Tjong, S.C. Novel nanoparticle-reinforced metal matrix composites with enhanced mechanical properties. *Adv. Eng. Mater.* **2007**, *9*, 639–652. [CrossRef]
9. Habibnejad-Korayem, M.; Mahmudi, R.; Poole, W.J. Enhanced properties of Mg-based nano-composites reinforced with Al_2O_3 nano-particles. *Mater. Sci. Eng. A* **2009**, *519*, 198–203. [CrossRef]
10. Shehata, F.; Fathy, A.; Abdelhameed, M.; Moustaf, S.F. Preparation and properties of Al_2O_3 nanoparticle reinforced copper matrix composites by in situ processing. *Mater. Des.* **2009**, *30*, 2756–2762. [CrossRef]
11. Karbalaei Akbari, M.; Mirzaee, O.; Baharvandi, H.R. Fabrication and study on mechanical properties and fracture behavior of nanometric Al_2O_3 particle- reinforced A356 composites focusing on the parameters of vortex method. *Mater. Des.* **2013**, *46*, 199–205. [CrossRef]
12. Su, H.; Gao, W.L.; Feng, Z.H.; Lu, Z. Processing, microstructure and tensile properties of nano-sized Al_2O_3 particle reinforced aluminum matrix composites. *Mater. Des.* **2012**, *36*, 590–596. [CrossRef]
13. Sajjadi, S.A.; Ezatpour, H.R.; Beygi, H. Microstructure and mechanical properties of Al-Al_2O_3 micro and nano composites fabricated by stir casting. *Mater. Sci. Eng. A* **2011**, *528*, 8765–8771. [CrossRef]
14. Sajjadi, S.A.; Ezatpour, H.R.; Torabi Parizi, M. Comparison of microstructure and mechanical properties of A356 aluminum alloy/Al_2O_3 composites fabricated by stir and compo-casting processes. *Mater. Des.* **2012**, *34*, 106–111. [CrossRef]

15. Chou, S.N.; Huang, J.L.; Lii, D.F.; Lu, H.H. The mechanical properties of Al_2O_3/aluminum alloy A356 composite manufactured by squeeze casting. *J. Alloys Compd.* **2006**, *419*, 98–102. [CrossRef]

16. Amirkhanlou, S.; Niroumand, B. Effects of reinforcement distribution on low and high temperature tensile properties of $Al356/SiC_p$ cast composites produced by a novel reinforcement dispersion technique. *Mater. Sci. Eng. A* **2011**, *528*, 7186–7195. [CrossRef]

17. Karbalaei Akbari, M.; Baharvandi, H.R.; Mirzaee, O. Fabrication of nano-sized Al_2O_3 reinforced casting aluminum composite focusing on preparation process of reinforcement powders and evaluation of its properties. *Compos. Part B.* **2013**, *55*, 426–432. [CrossRef]

18. Li, S.F.; Sunb, B.; Imaia, H.; Mimotob, T.; Kondoha, K. Powder metallurgy titanium metal matrix composites reinforced with carbon nanotubes and graphite. *Compos. Part A* **2013**, *48*, 57–66. [CrossRef]

19. Kallip, K.; Kishore Babu, N.; AlOgab, K.A.; Kollo, L.; Maeder, X.; Arroyo, Y.; Leparoux, M. Microstructure and mechanical properties of near net shaped aluminium/alumina nanocomposites fabricated by powder metallurgy. *J. Alloys Compd.* **2017**, *714*, 133–143. [CrossRef]

20. Prashanth, K.G.; Murty, B.S. Production, kinetic study and properties of Fe-based glass and its composites. *Mater. Manuf. Processes.* **2010**, *25*, 592–597. [CrossRef]

21. Wang, Z.; Tan, J.; Scudino, S.; Sun, B.A.; Qu, R.T.; He, J.; Prashanth, K.G.; Zhang, W.W.; Li, Y.Y.; Eckert, J. Mechanical behavior of Al-based matrix composites reinforced with $Mg_{58}Cu_{28.5}Gd_{11}Ag_{2.5}$ metallic glass. *Adv. Powder technol.* **2014**, *25*, 635–639. [CrossRef]

22. Marko, D.; Prashanth, K.G.; Scudino, S.; Wang, Z.; Ellendt, N.; Uhlenwinkel, V.; Eckert, J. Al-based metal matrix composites reinforced with $Fe_{49.9}Co_{35.1}Nb_{7.7}B_{4.5}Si_{2.8}$ glassy powder: Mechanical behavior under tensile loading. *J. Alloys Compd.* **2014**, *615*, S382–S385. [CrossRef]

23. Tabandeh Khorshid, M.; Jenabali Jahromi, S.A.; Moshksar, M.M. Mechanical properties of tri-modal Al matrix composites reinforced by nano- and submicron sized Al_2O_3 particulates developed by wet attrition milling and hot extrusion. *Mater. Des.* **2010**, *31*, 3880–3884. [CrossRef]

24. Razavi, H.Z.; Hafizpour, H.R.; Simchi, A. An investigation on the compressibility of aluminum/nano-alumina composite powder prepared by blending and mechanical milling. *Mater. Sci. Eng. A* **2007**, *454–455*, 89–98. [CrossRef]

25. Prashanth, K.G.; Scudino, S.; Klauss, H.J.; Surreddi, K.B.; Löber, L.; Wang, Z.; Chaubey, A.K.; Kühn, U.; Eckert, J. Microstructure and mechanical properties of Al-12Si produced by selective laser melting: Effect of heat treatment. *Mater. Sci. Eng. A* **2014**, *590*, 153–160. [CrossRef]

26. Suryawanshi, J.; Prashanth, K.G.; Scudino, S.; Eckert, J.; Prakash, O.; Ramamurty, U. Simultaneous enhancements of strength and toughness in an Al-12Si alloy synthesized using selective laser melting. *Acta Mater.* **2016**, *115*, 285–294. [CrossRef]

27. Prashanth, K.G.; Scudino, S.; Eckert, J. Defining the tensile properties of Al-12Si parts produced by selective laser melting. *Acta Mater.* **2017**, *126*, 25–35. [CrossRef]

28. Prashanth, K.G.; Debalina, B.; Wang, Z.; Gostin, P.; Gebert, A.; Calin, M.; Kühn, U.; Kamaraj, M.; Scudino, S.; Eckert, J. Triobological and corrosion properties of Al-12Si produced by selective laser melting. *J. Mater. Res.* **2014**, *29*, 2044–2054. [CrossRef]

29. Li, C.D.; Wang, X.J.; Liu, W.Q.; Wu, K.; Shi, H.L.; Ding, C.; Hu, X.S.; Zheng, M.Y. Microstructure and strengthening mechanism of carbon nanotubes reinforced magnesium matrix composite. *Mater. Sci. Eng. A* **2014**, *597*, 264–269. [CrossRef]

30. Zhang, D.; Zhan, Z. Preparation of graphene nanoplatelets-copper composites by a modified semi-powder method and their mechanical properties. *J. Alloys Compd.* **2016**, *654*, 226–233. [CrossRef]

31. Li, M.Q.; Zhai, H.X.; Huang, Z.Y.; Liu, X.H.; Zhou, Y.; Li, S.B.; Li, C.W. Tensile behavior and strengthening mechanism in ultrafine $TiC_{0.5}$ particle reinforced Cu-Al matrix composites. *J. Alloys Compd.* **2015**, *628*, 186–194. [CrossRef]

32. Park, J.G.; Keum, D.H.; Lee, Y.H. Strengthening mechanisms in carbon nanotube-reinforced aluminum composites. *Carbon.* **2015**, *95*, 690–698. [CrossRef]

33. Bisht, A.; Srivastava, M.; Manoj Kumar, R.; Lahir, I.; Lahiri, D. Strengthening mechanism in graphene nanoplatelets reinforced aluminum composite fabricated through spark plasma sintering. *Mater. Sci. Eng. A* **2017**, *695*, 20–28. [CrossRef]

34. Nardone, V.C.; Prewo, K.M. On the strength of discontinuous silicon carbide reinforced aluminum composites. *Scripta Metall.* **1986**, *20*, 43–48. [CrossRef]

35. Bakshi, S.R.; Agarwal, A. An analysis of the factors affecting strengthening in carbon nanotube reinforced aluminum composites. *Carbon* **2011**, *49*, 533–544. [CrossRef]

36. Ma, F.C.; Zhou, J.J.; Liu, P.; Li, W.; Liu, X.K.; Pan, D.; Lu, W.J.; Zhang, D.; Wu, L.Z.; Wei, X.Q. Strengthening effects of TiC particles and microstructure refinement in insitu TiC-reinforced Ti matrix composites. *Mater. Charact.* **2017**, *127*, 27–34. [CrossRef]

37. Sree Manu, K.M.; Arun Kumar, S.; Rajan, T.P.D.; Riyas Mohammed, M.; Pai, B.C. Effect of alumina nanoparticle on strengthening of Al-Si alloy through dendrite refinement, interfacial bonding and dislocation bowing. *J. Alloys Compd.* **2017**, *712*, 394–405. [CrossRef]

38. Li, J.; Liu, B.; Fang, Q.H.; Huang, Z.W.; Liu, Y.W. Atomic-scale strengthening mechanism of dislocation-obstacle interaction in silicon carbide particle- reinforced copper matrix nanocomposites. *Ceram. Int.* **2017**, *43*, 3839–3846. [CrossRef]

39. Zhao, X.; Lu, C.; Tieu, A.K.; Pei, L.Q.; Zhang, L.; Cheng, K.Y.; Huang, M.H. Strengthening mechanisms and dislocation processes in <111> textured nanotwinned copper. *Mater. Sci. Eng. A* **2016**, *676*, 474–486. [CrossRef]

Article

Precipitation Stages and Reaction Kinetics of AlMgSi Alloys during the Artificial Aging Process Monitored by In-Situ Electrical Resistivity Measurement Method

Hong He [1], Long Zhang [1], Shikang Li [1], Xiaodong Wu [1], Hui Zhang [2] and Luoxing Li [1,*]

[1] State Key Laboratory of Advanced Design and Manufacturing for Vehicle Body,
College of Mechanical and Vehicle Engineering, Hunan University, Changsha 410082, China;
hehong_hnu@hnu.edu.cn (H.H.); long060810304@163.com (L.Z.);
kangkangli2009@126.com (S.L.); dongdong830206@163.com (X.W.)
[2] College of Materials Science and Engineering, Hunan University, Changsha 410082, China;
zhanghui63hunu@163.com
* Correspondence: luoxing_li@hnu.edu.cn; Tel.: +86-731-888-21571

Received: 4 December 2017; Accepted: 5 January 2018; Published: 11 January 2018

Abstract: The precipitation process and reaction kinetics during artificial aging, precipitate microstructure, and mechanical properties after aging of AlMgSi alloys were investigated employing in-situ electrical resistivity measurement, Transmission Electron Microscopy (TEM) observation, and tensile test methods. Three aging stages in sequence, namely formation of GP zones, transition from GP zones to β'' phase, transition from β'' to β' phase, and coarsening of both phases, were clearly distinguished by the variation of the resistivity. It was discussed together with the mechanical properties and precipitate morphology evolution. Fast formation of GP zones and β'' phase leads to an obvious decrease of the resistivity and increase of the mechanical strength. The formation of β'' phase in the second stage, which contributes to the peak aging strength, has much higher reaction kinetics than reactions in the other two stages. All of these stages finished faster with higher reaction kinetics under higher temperatures, due to higher atom diffusion capacity. The results proved that the in-situ electrical resistivity method, as proposed in the current study, is a simple, effective, and convenient technique for real-time monitoring of the precipitation process of AlMgSi alloys. Its further application for industrial production and scientific research is also evaluated.

Keywords: aluminum alloys; aging; precipitation; electrical resistivity; mechanical properties

---

1. Introduction

The light weighting features of the precipitation hardened AlMgSi aluminum alloys trigged tremendous investigation interest in recent years [1–5]. The precipitation behavior during the heat treatment process plays a decisive role in final performance of the AlMgSi alloys. Properties of these alloys, e.g., workability, mechanical strength, electrical conductivity, corrosion resistance etc., are greatly determined by the phase type, morphology, size distribution, and number density of their precipitates [1,3,6–9]. Therefore, a deep understanding of the precipitation behavior during the artificial aging process is crucial.

Significant achievements in the characterization of the precipitation behavior of AlMgSi alloys have been reached since the end of last centenary, with the assistance of advancing technologies, like High Resolution Transmission Electron Microscopy (HRTEM) [5–7,10–13], Atom Probe Tomography (APT) [1,13–17], Differential Scanning Calorimetry (DSC) [18–20], electrical resistivity measurements [21–26], Phase Field Crystal (PFC) modeling [4,24], etc. The precipitation sequence has been established and gradually accepted as: supersaturated solid solution (SSSS)→clusters/GP zones→β″→β′→β [27]. The clusters/GP zones are the early stage precipitates with size of several nanometers, usually formed during natural aging or first minutes of artificial aging process. β″ and β′ phases are formed in the artificial aging process before over ageing. β is the final stable phase. The AlMgSi alloys are usually strengthened by the meta-stable phases. However, debates in this field still remain. For example, detailed crystal structure and composition of the meta-stable phases, precipitation kinetics that is quite related to the natural aging effect, quantitative characterization of the precipitates is under debate [13–16]. The atomic structure and evolution of the early clusters are ambiguous yet [5–7]. Characterization accuracy and effectiveness are still dependent on the methods. Therefore, the investigation on the details of the precipitation process is still in progress.

Recently reported investigations proved that the electrical resistivity measurement method is feasible and convenient for quantitative analysis of the reaction kinetics and precipitate volume fraction in AlMgSi alloys [21,25,26]. When compared to the TEM, APT, and DSC techniques, its merits are: (i) it is sensitive due to the high sensitivity of resistivity to even atomic-scale changes of the microstructure, such as solution atoms, defects, precipitates; (ii) there are no need of complicated sample preparation or sophisticated equipments like TEM system; and, (iii) samples used are much larger than those for HRTEM, APT or DSC, so that the accuracy would be better [26]. The existing resistivity investigations of AlMgSi alloys were mostly conducted after aging and samples were immersed in liquid nitrogen to keep a constant temperature during testing [21–26]. This would cause time interval between the precipitation and the measuring, including the heating and cooling process. Some information about the evolution of the precipitates would be missing, due to the high time and temperature sensitivity of the precipitation in AlMgSi alloys. Transients experienced by the sample also require serious attention, especially when the heating and cooling rates fluctuates.

In the present study, we proposed an in-situ electrical resistivity measuring method, which is much easier and of higher efficiency, for the real-time monitoring of the precipitation behavior of AlMgSi alloys. The sample temperature and electrical resistivity was continuously measured during the whole thermal course from quenching to natural or artificial aging. By this method, details of the precipitation process were analyzed. The relationship between the precipitation-induced resistivity variation during the artificial aging process, mechanical properties after aging, microstructure evolution, and the precipitation behavior was investigated. The industrial and scientific application feasibility of the proposed in-situ electrical resistivity measurement method for describing the precipitation behavior of the AlMgSi alloys was evaluated.

2. Materials and Methods

6063 aluminum profiles with composition of 0.55Mg-0.43Si-0.20Fe-0.001Cu-0.001Mn-0.004Cr-0.01Ti-banance Al (wt %) were used as raw materials. After extrusion, the profiles were immediately quenched from 530 °C to room temperature by the water spray system on the profile extruding machine. Then, they were stored for one month before conducting artificial aging. 150 °C, 175 °C, and 195 °C were selected as low, medium, and high artificial aging temperatures, which can also be considered as representative of the temperatures in industrial production.

Electrical resistance was measured during the whole artificial aging process, on 4 × 35 × 150 mm sheet samples cut from the profiles, by a four-point probe system, as shown in Figure 1. Four pure aluminum wires with 0.6 mm in diameter and 1000 mm in length were used as the probes. The probes were spot welded on the sample surface. For consistency, the welding voltage was kept constant to obtain solder points with the same size. A constant current of 100 mA was applied to the sample by the outside two probes. The resulting potential (in the scale of nano-volt) was measured by the inner two probes and recorded by a counting computer. Then, the sample resistance R under the testing distance was calculated by the relationship of $R = U/I$, where U and I stands of voltage and current respectively. Sample temperature was measured by a thermocouple inserted into a hole with diameter of 1.2 mm drilled on the sample surface. Microstructure and precipitate morphology of samples after aging were investigated through transmission electron microscopy (TEM) and high resolution TEM (HRTEM) observations. Specimens for TEM were first mechanically grinded to about 100 µm in thickness and then further thinned by twin-jet electro-polishing with a solution of 30 vol % nitric acid and 70 vol % methanol at −20 °C, 24 V and 80 mA. TEM observations were performed on a Tecnai F20 TEM system (FEI, Hillsboro, OR, USA) operated at 200 KV. To verify the consistency of the results, at least ten spots in different crystal grains were analyzed for each specimen. Mechanical strength and elongation of samples after aging were measured by uniaxial tensile testing, according to ASMT standards E8M-09. Standard dumbbell-type tensile specimens with gauge length of 60 mm and width of 10 mm were used. Tests were performed on an INSTRON-3382 tensile testing machine (Instron, Norwood, MA, USA) with a stretching speed of 1 mm/min. For each aging situation, three specimens were tested in order to verify the tensile results were in consistence.

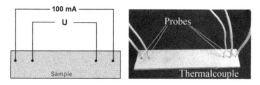

Figure 1. A sketch of the experimental set up for resistance measuring of the sample.

3. Results

3.1. Resistance and Resistivity

The real-time measured electrical resistance and sample temperature during the whole artificial aging process are shown in Figure 2. As shown in Figure 2a, the resistance sharply increased upon heating, and drastically decreased during cooling, a behavior that was caused by the temperature effect on the aluminum alloy matrix. After the aging temperature reached the pre-set value, the resistance gradually decreased as aging time was prolonged. During the aging process, the sample temperature was kept stable, as shown by Figure 2b.

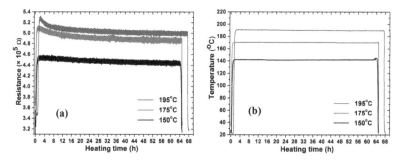

Figure 2. Resistance and sample temperature measurement results: (a) variation of the resistance during the whole aging process; and (b) sample temperatures during the whole aging process.

The real resistivity of the samples was calculated from the measured resistance and sample size values. The recorded resistance R comprises the system resistance of the instrument R_0, contact resistance between the probes and sample R_c, and the real resistance of the sample R_s, as $R = R_0 + R_c + R_s$. When the temperature reaches the pre-set value, the R_0 and R_c can be considered to be constant. At a fixed temperature, the sample resistance R_s is proportional to the testing distance: $R_s = \rho_s L/A$, where ρ_s is the resistivity, L is the testing distance (namely the distance between the inner probes), A is the cross-sectional area of the sample. By varying the distance at the end of the aging process, series of total resistances R were obtained which should be proportional to the testing distances. Therefore, the fixed R_0 and R_c value can be extracted from the intercept of linear fitting of the resistance values measured on different distances, as shown in Figure 3. By subtracting the R_0 and R_c from R, the R_s was obtained. Then, the resistivity value was calculated using the expression $\rho_s = R_s \times A/L$.

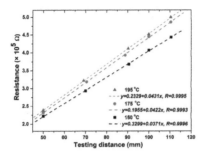

Figure 3. Resistance value of the samples R at the end of aging process measured on different distances and their linear fittings.

The variation of the resistivity with the aging time (in this case, the aging time is set as zero when the sample temperature reached the pre-set value) under different aging temperatures is shown in Figure 4. The electrical resistivity values are around several dozens of $n\Omega \cdot m$, which is in the same order of magnitude of those reported in the literature [24,25]. The resistivity of the samples continuously decreases with aging time, as in the case of the resistance value. By plotting the resistivity versus the logarithmic scale of the aging time, the curve was divided into two or three parts. In each part, the resistivity decreases linearly with the logarithmic scale of aging time, which has also been found in other studies [23,26]. In this way, the whole precipitation process was separated into two or three stages. It reflects that two or three different precipitation reaction types occurred. It can also be seen that these stages end at a different aging time. When the aging temperature is 195 °C and 175 °C, the first stage ends within 0.5 h and 1 h, respectively, while for 150 °C, it is prolonged to about 5 h. When the aging temperatures are 195 °C and 175 °C, the end of the second stage is after 4 h and 11 h, respectively. The end of the third stage, which seems to occur after a long time, is not clearly observed in the present study since the experimental aging time was limited to 60 h. Nevertheless, it can be easily concluded that, as expected, the sequence of reactions is faster when the temperature is higher.

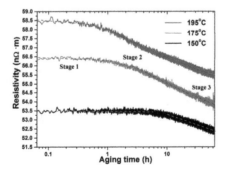

Figure 4. Variation of the resistivity of the samples with aging time under different temperature.

The derivative of the resistivity to the aging time, which would reflect the precipitation reaction rate, is shown in Figure 5. The figure shows that the reaction rate continuously decreases. When the aging temperature is 195 °C or 175 °C, the initial fast decrease in resistivity during the first stage, which comes at completion within several hours, is followed by a slower decrease in the second stage.

The inflection point in the curves distinguishes the transition between the first and the second stages (4–6 h for 195 and 12–14 h for 175 °C, respectively). Such an obvious inflection point is not observed in the curve for 150 °C. The higher the aging temperature, the faster the initial decrease in resistivity. These phenomena should all be related to the precipitation type and reaction kinetics.

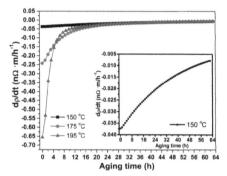

Figure 5. Differentiation of resistivity to aging time under different aging temperature. The inset figure is a magnified picture of the curve for samples aged at 150 °C.

3.2. Mechanical Properties

The relationship between the mechanical properties of the samples and aging temperature and time is shown in Figure 6. The tensile and yield strength firstly increase sharply with the aging time, and then reach a plateau at a maximum value, which is called peak aging strength. A slight decrease of the strength is observed for samples aged at 195 °C when aging time exceeds 4 h, which is called over aging phenomenon [28]. The time required for reaching the peak aging status, at the time when the strength reaches the maximum [28], is shorter when the aging temperature is higher (4 h for 195 °C, 11 h for 175 °C, and 36 h for 150 °C, designated as the peak aging time). The elongation values of the samples show continuous decrease in all the cases. It is also observed that the peak strength is higher when the aging temperature is lower, for example, the maximum tensile strength for samples aged at 150 °C reaches 231 MPa, while it is 209 MPa for those aged at 195 °C. These phenomena should all be related to the precipitation rate, phase type, number density, and size of the precipitates in the matrix [14,16,17], which have obvious different strengthening effects. This would be discussed together with the evolution of the microstructure of the precipitates.

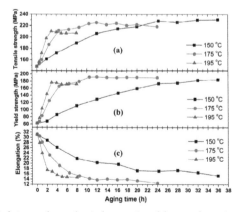

Figure 6. Relationship between the mechanical properties of the samples, aging temperature and aging time (**a**) tensile strength; (**b**) yield strength; and, (**c**) elongation.

3.3. Microstructure of Precipitates

The morphology of the precipitates in the samples after different stages of artificial aging were investigated by TEM. Representative bright field TEM and HRTEM images taken along the $<100>_{Al}$ zone axis are shown in Figure 7. Precipitates were all found to grow along <100> direction of Al matrix. For naturally aged samples without artificial aging, no evidence of large precipitates was found, as shown in Figure 7a. For artificially aged samples under different aging conditions, three kinds of precipitates were observed: spherical GP zones, needle-like β'' phase, and rod-like β' phase, as shown in Figure 7b–f. In the sample not artificially aged, very small clusters is believed to exist in the matrix since the raw materials have been stored for one month, during which the materials must have undergone natural aging [12,21–24]. But, their total morphology is hard to be observed by ordinary TEM images in samples with such low alloy composition, and also due to their small size (around 1–3 m) and highly coherence with the matrix [1,2,10,11]. They are identified by their white contrast in the HRTEM images, as shown in Figure 7f. GP zones were observed to have spherical morphology and size of 3–5 nm in short time artificially aged samples. They can be seen also only in the HRTEM images. But, it was much easier since they have a little larger size and number density. Examples were shown by Figure 7b,g for the sample which has been aged at 175 °C for 1 h. The β'' phase and β' phase were distinguished by their needle-like and rod-like morphology, respectively, and different contrast surroundings in the matrix, as described in the literature [3,11]. This was further verified by their monoclinic and hexagonal crystal structure, respectively [11], identified by HRTEM images, as shown in Figure 7h, taken from the sample aged at 175 °C for 24 h.

The phase type, size and number density evolution of different precipitates with artificial aging time at different temperatures were clearly reflected by the TEM images. After very short time of artificial aging at each selected temperature, also described as stage 1 in Figure 4, the precipitation process starts with the formation of GP zones and the beginning of transformation from GP zones to β'' phase. A typical image is presented in Figure 7b for the sample aged at 175 °C for 1 h. A large amount of GP zones and a little trace of β'' phase with length of about 20 nm were observed. As aging proceeds, the GP zones totally transformed to β'' phase. When the aging time prolongs to the peak aging status, also in the time range described as stage 2 in Figure 4, the β'' phase gradually coarsened. The length of the β'' phase particles in peak aged samples increased to 40–120nm, as shown in Figure 7c, for the sample aged at 195 °C for 3 h. The total number density of the precipitates also increased with aging time. It was verified by comparing TEM images for samples aged at the same temperature but for different aging times, the number density of β'' phase is obviously increased as the aging time increases. After the peak aging time at 175 °C and 195 °C, described as stage 3 in Figure 4, the length of some precipitates abnormally reached several hundreds of nanometers, and the morphology of precipitates became inhomogeneous, as shown in Figure 7d, for the sample aged at 175 °C for 24 h. β' phase also starts to form. However, it is interesting to find that when aging temperature is 150 °C, the length of β'' phase still stays at the level of 15–30 nm, even when the aging time reaches 36 h, as shown in Figure 7e. The small β'' phase is homogenously distributed, and no β' phase is found. This indicates that aging at lower temperature, 150 °C for example, has much lower coarsening and β' phase formation kinetics.

It can be summarized from the TEM results: (i) GP zones form quickly at the early stage of artificial aging process; (ii) for peak-aged samples, the β'' phase is the main precipitate; (iii) as the aging processes, the number density of precipitates increases obviously; (iv) raising the artificial temperature promotes growth of the precipitates and formation of β' phase; and, (v) the transformation from β'' phase to β' phase is harder to occur at low temperature (150 °C) than at higher temperature (175 °C or 195 °C).

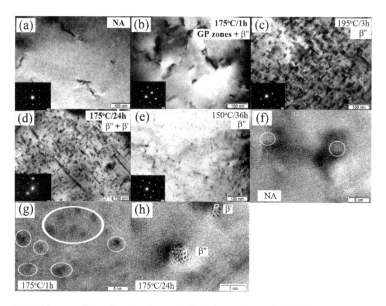

Figure 7. TEM images of the microstructures of selected samples aged at different temperatures and aging times: (**a**) naturally aged (NA) sample without artificial aging; (**b**) 175 °C for 1 h; (**c**) 195 °C for 3 h; (**d**) 175 °C for 24 h; (**e**) 150 °C for 36 h; and, (**f–h**) HRTEM images taken from (**a,b,d**), respectively.

4. Discussion

All of the experimental results in present study confirmed the same precipitation sequence of AlMgSi alloys, as reported in the literature [27], which was either indirectly reflected in the resistivity measurement results and mechanical properties, or directly observed during the TEM analysis. This sequence is obviously related to the precipitate phase type, particle size, and reaction kinetics, since different precipitates contribute to the electrical resistivity and strengthening effect in a markedly different manner [21,25]. Therefore, the evolution of these precipitates during the whole artificial aging process, and the consistency of the experimental results will be discussed.

4.1. Precipitation and Phase Transformation Process

The precipitation and phase transformation process was well revealed by the resistivity variation curves, mechanical strength, and microstructure evolution. The whole artificial aging process is clearly divided into three stages, as reflected by Figure 4. Precipitate type and morphology, size, density etc. evolve in sequence, with typical characters as shown in Figure 7.

Formation and transition of the precipitates accounts for the variation of electrical resistivity. At the quenched and naturally aged state, the solute atoms exist in the matrix or in the form of clusters (consisted of several atoms), which have different electronic surroundings to the Al atoms, act as serious scattering centers for electron conducting. But, when precipitation occurs, the solute atoms were accumulated to form much larger precipitates. In this way, their electron scattering effect was weakened. Larger precipitates also affect the electron conduction, but it is not so obvious as clusters or solute atoms. As the precipitates grow up, their nearby distance becomes larger. In this way, it provides more convenient pathway for electron conduction and further decreases the resistivity [21–25]. Therefore, the sample has the highest electrical resistivity in the original state. As artificial aging begins, GP zones form quickly from solute atoms or clusters (as shown in Figure 7b), which leads to obvious decline of the resistivity. Therefore, the first stage in Figure 4 is characterized by the formation of GP zones. As aging further proceeds, GP zones, which are also considered as the pre-phase of β″ [1,8,13,27], transforms to the needle-like β″ phase. Since the size of β″ phase (at least 20 nm in length) is much larger than that of GP zones (2–5 nm in diameter), the total number of the precipitates decreases. This process further reduces the electrical resistivity. This process continues to the peak aging status, since the main precipitate is β″ phase for peak aging samples, as shown in Figure 7c. Thus, the second stage in Figure 4 is assigned to transition from GP zones to β″ phase. As the aging time is prolonged to the third stage, the rod-like β′ phase starts to precipitate by consuming the β″ phase, and both of the β″ and β′ phase start to coarsen, as shown by Figure 7d. The distance between neighboring precipitates is further increased, which would facilitate the passage of conducting electrons [23,25]. In this way, the coarsening of the precipitates also contributes to decline of the resistivity. Therefore, the resistivity continuously decreased during the whole artificial aging process, either by formation or phase transformation and coarsening of the precipitates. But, it is worth mentioning that the formation of β″ phase contribute to most decrease of the resistivity, while other transitions have not so obvious influence.

The evolution of the mechanical properties of samples after artificial aging is also related to the formation and phase transition of the precipitates. In the first two stages, the base material is mainly strengthened by small GP zone and/or β″ phase. But, in last stage, the precipitates greatly coarsened. It weakens the dislocation-particle interaction, and thus reduces the effectiveness of their strengthening effect [29,30]. At the early aging status, the precipitate sizes remain very small. The movement of dislocations when encountering precipitate obstacles is mainly in the shearing mode, which has excellent strengthening effect. The larger the amount of fine β″ phase the sample contains, the higher its strength. It shown consistently by the mechanical strength curves and TEM images, as the amount of fine β″ phase increased (before reaching peak-aging), the strength gradually increases. This accounts for the fact that the samples aged at 150 °C, i.e., in a condition that results in large amount of fine β″ phase, exhibit the highest peak strength. If the precipitates size (namely the aspect ratio) reaches and exceeds the critical particle radius, the dislocation movement transforms from the shearing to the bypassing mode, which has not so strong strengthening effect. Although the total number of precipitates increases, coarsening of the precipitates makes the dislocation by-pass easier. Therefore, for samples at the plateau or over-aged status, their strength increases not so obviously, or even decreases.

The evolution of the resistivity, strength, precipitate type and size shows an extremely good consistence. The fast increase in the strength-time curves corresponds to stage 1 and 2 in resistivity variation curves, during which GP zones form and then transform to β″ phase. The plateau in strength curves implies the aging process is in stage 3, in which transition of β″ phase to β′ phase and coarsening of both phases take place.

4.2. Precipitation Kinetics

The precipitation and phase transformation in AlMgSi alloys are atomic diffusion-controlled reaction. The atom diffusion is largely assisted by the spare vacancies, either reserved from quenching or intentionally designed [1,6,12–15,17]. Thus it is obviously driven by vacancy diffusion [1,6,12–15,17]. Activity of the vacancies greatly influences their diffusion and precipitation reaction kinetics, which is also related to temperature and the particular reaction. The resistivity curves in the present study reflect the precipitation kinetics well. Since the stages in the electrical resistivity and mechanical strength curves correspond to precipitation types, their variation trends stand for different reaction kinetics. The reaction kinetics is discussed in two situations: (i) for different reaction types at the same aging temperature; and, (ii) for individual reaction type at different aging temperatures.

The measured resistivity can be expressed by the following equation: $\rho = \rho^* + \sum_i \rho_i C_i + \sum_j P_j f_j$, where the ρ^* is the resistivity of the Al matrix at the measuring temperature; ρ_i and C_i are the specific resistivity and concentration of the ith solution atoms; P and f_j is the scattering power and concentration of the jth precipitate with an average size in the microstructure of pure host metal [26]. The relationship between C_i and f_j can be expressed by $C_i = C_{oi} - N_i f_j$, where the C_{oi} and N_i are the initial concentration of the ith solute atom and its equivalent number in jth precipitate. The precipitation process during artificial aging is considered to follow the Johnson-Mehl, Avrami, Kolmogorov (JMAK) kinetics model [11–13]: $f_r = 1 - \exp(-kt^n)$, where f_r is the relative fraction of the precipitate; k is temperature dependent rate constant; t is the aging time; and, n a numerical exponent for JMAK relationship. For the solution treated and T4 temper AlMgSi alloys, n is always assigned to be 1 [31–34]. In the present study, the original materials were quenched and then stored in room temperature, which equaled to the T4 temper, therefore, n is also considered to be 1. The constant k represents the reaction kinetics, which follows the Arrhenius model: $k = k_o \exp(-Q/RT)$, where k_o, Q, R and T are the pre-exponential constant, apparent activation energy related to the precipitation process, the universal gas constant and temperature, respectively [25,26,31,32]. Therefore, by re-arranging the above relations, the measured resistivity of the sample is found to be in linear relationship to the logarithmic scale of the aging time for an individual reaction, and the linear fitting slope containing the origination of k reflects the reaction kinetics.

As shown in Figure 4, the linear decreasing relationship of the resistivity to aging time represents three types of reactions. The slope of each linear fitting thus reflects the reaction kinetics. It can be seen that the slope for the second stage, which is attributed to formation of β'' phase, is the largest slope for all the curves. This implies that the transformation from GP zones to β'' phase has higher kinetics than other reactions. In the early state, formation of GP zones is determined by the response of clusters that contain large amount of solution atoms. Distance between the small clusters is small, and diffusion of these atoms is rather active, with the assistance of quenched in vacancies. However, it can be either dissolved into the matrix or acts as the nucleation site for GP zones growth [9,18]. In other words, the competition between the dissolution and nucleation processes slows down the formation of GP zones. Once there exists GP zones with size large enough, formation of β'' phase occurs fast. Other studies by DSC have also reported such phenomenon [18]. Therefore, for the first stage, even though the concentration of the clusters at this moment is pretty high, kinetics for formation of GP zones is lower than formation of β'' phase in the second stage. As the reaction comes to the third stage, the solution atoms in the matrix may have been all diffused into the precipitates, phase transition from β'' to β' and size growing of the precipitates all depends on the atom diffusion rate without the assistance of vacancies. Therefore, this speed of this process is lowered down when compared to the formation of β'' phase. It is also observed in Figure 5. The differential of the resistivity to aging time becomes to be much lower for the last stage. Therefore, the formation of β'' phase have the highest kinetics that is facilitated by both vacancies and existing nucleation sites.

For the same kind of stages aged at different temperatures, the higher the temperature is, the higher kinetics it has. This is seen by the higher slope of the linear fitting of resistivity in Figure 4 and sharper variation rate of the derivation of the resistivity in Figure 5. It is much easier to understand that the atoms or vacancies have higher diffusion capacity under elevated temperatures. Therefore, for each individual reaction stage, it finishes in much shorter time at higher temperature. As a example, it can be seen in Figure 4, the formation of GP zones under 195 °C only takes about 0.5 h, but it requires 5 h when the temperature is 150 °C.

4.3. Accuracy and Application of the In-Situ Resistivity Measuring Method

The accuracy of the in-situ resistivity measurement method is evaluated by its data error level. As shown in Figure 3, when testing distance is 110 nm, the R_0 and R_c comprises 4.66%, 4.41%, and 7.44% of R for 195 °C, 175 °C, and 150 °C, respectively. The proportion R_c in R should be some lower than these values. The testing error is mainly caused by the fluctuation of the welded contact surface between the probe and the sample (R_c). It can be further controlled to be even lower by increasing the testing distance. The linear fitting error stays at extremely low level below 0.1% (with fitting level higher than 99.9%). Therefore, the final testing data are confirmed to be highly trustable. The results also showed better accuracy and stability than those reported in the literature [21–26], since the data were in-situ and continuously recorded.

Further applications of the in-situ resistivity measuring method can be both industrial or scientific. On the one hand, it can be employed as a monitoring technique for aluminum profiles manufacturing enterprises. Insufficient aging in partial products that are located in some blind corners in the big industrial aging furnace is often encountered during practical experience. It may be caused by inhomogeneous temperature field, uneven heating, fluctuation of power, or other unknown reasons. It is troublesome because it is unpredictable before ending the aging process and after testing. The usual approach for compensating this problem is by conducting supplementary aging or re-aging, which is time and energy consuming. The method presented in this study would be an effective solution. By calibration of the resistivity of aged samples for any given alloy composition, their target mechanical properties can be marked. Therefore, using the in-situ resistivity measuring method, the aging status, and properties of the products can be monitored in real time. This way, the insufficient aging problem is eliminated. On the other hand, the measured resistivity data may also be useful for further quantitative analysis of the precipitation behavior or modeling of mechanical strength. For example, the relative volume fraction of the precipitates can be simply calculated by its initial, real-time, and final resistivity values (ρ_0, ρ_t, ρ_f) through the expression: $f_r = (\rho_0 - \rho_t)/(\rho_0 - \rho_f)$ [29,31]. It is similar to those using the data obtained by DSC and isothermal calorimetry experiments [19,20]. Moreover, it is suggested that this method would be also useful for analysis of the cluster and GP zones evolution, since the variation of resistivity is highly related to them [24]. Together with other characterization techniques, APT, for example, quantitative investigations would be feasible.

Based on the above discussion, advantages of the in-situ resistivity measurement method are: (i) simplicity, because the sample is easy to prepared; (ii) convenience, no professional skills or sophisticated equipment is required for conducting the measurements, with trustable results in low testing error level; (iii) effectiveness, variation of the obtained resistivity data well reveals the trend of mechanical and microstructure evolution under different aging conditions; and, (iv) high efficiency, for a specific aging situation, only one sample is sufficient for different aging times, while in the reported method, a serial of samples is desirable [21,25].

5. Conclusions

In summary, the present work illustrates a detailed study of the precipitation process of AlMgSi alloys during artificial aging by proposing a simple, convenient, and effective real-time resistivity characterization method. The precipitation processes of AlMgSi alloys during artificial aging, namely formation of GP zones, transition from GP zones to β'' phase, transformation from β'' phase to β' phase, and coarsening of both phases, were clearly separated into three stages based on the variation of the resistivity. Feasibility of monitoring the phase transformation and reaction kinetics of the precipitates by this technique has been proved. It has been revealed that the formation of β'' phase has the highest reaction kinetics. Formation of fine needle-like β'' phase before coarsening and transition to larger β' phase contributes most to the mechanical strength. The proposed method is also recommended for both industrial and scientific applications like real-time monitoring technique for checking the aging completion degree in industrial production, characterization of the cluster and GP zones, and quantitative analysis or modeling of the precipitation process.

Acknowledgments: This work was supported by the National Key Research and Development Program of China [grant number: 2016YFB0101700], the National Natural Science Foundation of China [grant numbers: U1664252, 51475156].

Author Contributions: H.H. and L.L. conceived and designed the experiments; H.H., L.Z. and S.L. performed the experiments; H.H. and S.L. analyzed the data; X.W. contributed to discussion of the data and paper writing; H.Z. and L.L. contributed reagents/materials/analysis tools; H.H. and L.L. wrote the paper together.

Conflicts of Interest: The authors declare no conflict of interest.

References

1. Zandbergen, M.W.; Xu, Q.; Cerezo, A.; Smith, G.D.W. Study of precipitation in Al-Mg-Si alloys by Atom Probe Tomography, I. Microstructural changes as a function of ageing temperature. *Acta Mater.* **2015**, *101*, 136–148. [CrossRef]
2. Chen, J.H.; Costan, E.; Van Huis, M.A.; Xu, Q.; Zandbergen, H.W. Atomic pillar-based nanoprecipitates strengthen AlMgSi alloys. *Science* **2006**, *312*, 416–419. [CrossRef] [PubMed]
3. Pogatscher, S.; Antrekowitsch, H.; Leitner, H.; Ebner, T.; Uggowitzer, P.J. Mechanisms controlling the artificial aging of Al-Mg-Si Alloys. *Acta Mater.* **2011**, *59*, 3352–3363. [CrossRef]
4. Esmaeili, S.; Lloyd, D.J. Modeling of precipitation hardening in pre-aged AlMgSi(Cu) alloys. *Acta Mater.* **2005**, *53*, 5257–5271. [CrossRef]
5. Saito, T.; Ehlers, F.J.H.; Lefebvre, W.; Hernandez-Maldonado, D.; Bjørge, R.; Marioara, C.D.; Andersen, S.J.; Mørtsell, E.A.; Holmestad, R. Cu atoms suppress misfit dislocations at the β″ Al interface in Al-Mg-Si alloys. *Scr. Mater.* **2016**, *110*, 6–9. [CrossRef]
6. Werinos, M.; Antrekowitsch, H.; Kozeschnik, E.; Ebner, T.; Moszner, F.; Löffler, J.F.; Uggowitzer, P.J.; Pogatscher, S. Ultrafast artificial aging of Al-Mg-Si alloys. *Scr. Mater.* **2016**, *112*, 148–151. [CrossRef]
7. Li, K.; Idrissi, H.; Sha, G.; Song, M.; Lu, J.B.; Shi, H.; Wang, W.L.; Ringer, S.P.; Du, Y.; Schryvers, D. Quantitative measurement for the microstructural parameters of nano-precipitates in Al-Mg-Si-Cu alloys. *Mater. Charact.* **2016**, *118*, 352–362. [CrossRef]
8. Buchanan, K.; Colas, K.; Ribis, J.; Lopez, A.; Garnier, J. Analysis of the metastable precipitates in peak-hardness aged Al-Mg-Si(-Cu) alloys with differing Si contents. *Acta Mater.* **2017**, *132*, 209–221. [CrossRef]
9. Marioara, C.D.; Andersen, S.J.; Jansen, J.; Zandbergen, H.W. The influence of temperature and storage time at RT on nucleation of the β″ phase in a 6082 Al-Mg-Si alloy. *Acta Mater.* **2003**, *51*, 789–796. [CrossRef]
10. Wenner, S.; Marioara, C.D.; Ramasse, Q.M.; Kepaptsoglou, D.M.; Hagec, F.S.; Holmestada, R. Atomic-resolution electron energy loss studies of precipitates in an Al-Mg-Si-Cu-Ag alloy. *Scr. Mater.* **2014**, *74*, 92–95. [CrossRef]
11. Li, K.; Béché, A.; Song, M.; Sha, G.; Lu, X.; Zhang, K.; Du, Y.; Ringer, S.P.; Schryvers, D. Atomistic structure of Cu-containing β″ precipitates in an Al-Mg-Si-Cu alloy. *Scr. Mater.* **2014**, *75*, 86–89. [CrossRef]
12. Valiev, R.Z.; Murashkina, M.Y.; Sabirov, I. A nanostructural design to produce high-strength Al alloys with enhanced electrical conductivity. *Scr. Mater.* **2014**, *76*, 13–16. [CrossRef]
13. Liu, C.H.; Lai, Y.X.; Chen, J.H.; Tao, G.H.; Liu, L.M.; Ma, P.P.; Wu, C.L. Natural-aging-induced reversal of the precipitation pathways in an Al-Mg-Si alloy. *Scr. Mater.* **2016**, *115*, 150–154. [CrossRef]
14. Pogatscher, S.; Kozeschnik, E.; Antrekowitsch, H.; Werinos, M.; Gerstl, S.S.A.; Löffler, J.F.; Uggowitzer, P.J. Process-controlled suppression of natural aging in an Al-Mg-Si alloy. *Scr. Mater.* **2014**, *89*, 53–56. [CrossRef]
15. Aruga, Y.; Kozuka, M.; Takaki, Y.; Sato, T. Effects of natural aging after pre-aging on clustering and bake-hardening behavior in an Al-Mg-Si alloy. *Scr. Mater.* **2016**, *116*, 82–86. [CrossRef]
16. Li, H.; Liu, W. Nanoprecipitates and Their Strengthening Behavior in Al-Mg-Si Alloy during the Aging Process. *Metall. Mater. Trans. A* **2017**, *48*, 1990–1998. [CrossRef]
17. Guo, M.X.; Sha, G.; Cao, L.Y.; Liu, W.Q.; Zhang, J.S.; Zhuang, L.Z. Enhanced bake-hardening response of an Al-Mg-Si-Cu alloy with Zn addition. *Mater. Chem. Phys.* **2015**, *162*, 15–19. [CrossRef]
18. Esmaeili, S.; Lloyd, D.J. Characterization of the evolution of the volume fraction of precipitates in aged AlMgSiCu alloys using DSC technique. *Mater. Charact.* **2005**, *55*, 307–319. [CrossRef]
19. Aouabdia, Y.; Boubertakh, A.; Hamamda, S. Precipitation kinetics of the hardening phase in two 6061 aluminium alloys. *Mater. Lett.* **2010**, *64*, 353–356. [CrossRef]
20. Giersberg, L.; Milkereit, B.; Schick, C.; Kessler, O. In Situ Isothermal Calorimetric Measurement of Precipitation Behaviour in Al-Mg-Si Alloys. *Mater. Sci. Forum* **2014**, *794*, 939–944. [CrossRef]
21. Esmaeili, S.; Lloyd, D.J.; Poole, W.J. Effect of natural aging on the resistivity evolution during artificial aging of the aluminum alloy AA6111. *Mater. Lett.* **2005**, *59*, 575–577. [CrossRef]
22. Raeisinia, B.; Poole, W.J.; Lloyd, D.J. Examination of precipitation in the aluminum alloy AA6111 using electrical resistivity measurements. *Mater. Sci. Eng. A* **2006**, *420*, 245–249. [CrossRef]
23. Seyedrezai, H.; Grebennikov, D.; Mascher, P.; Zurob, H.S. Study of the early stages of clustering in Al-Mg-Si alloys using the electrical resistivity measurements. *Mater. Sci. Eng. A* **2009**, *525*, 186–191. [CrossRef]

24. Fallah, V.; Langelier, B.; Ofori-Opoku, N.; Raeisinia, B.; Provatas, N.; Esmaeili, S. Cluster evolution mechanisms during aging in Al-Mg-Si alloys. *Acta Mater.* **2016**, *103*, 290–300. [CrossRef]
25. Esmaeili, S.; Poole, W.J.; Lloyd, D.J. Electrical Resistivity Studies on the Precipitation Behaviour of AA6111. *Mater. Sci. Forum* **2000**, *331–337*, 995–1000. [CrossRef]
26. Esmaeili, S.; Vaumousse, D.; Zandbergen, M.W.; Poole, W.J.; Cerezo, A.; Lloyd, D.J. A study on the early-stage decomposition in the Al-Mg-Si-Cu alloy AA6111 by electrical resistivity and three-dimensional atom probe. *Philos. Mag.* **2007**, *87*, 3797–3816. [CrossRef]
27. Edwards, G.A.; Stiller, K.; Dunlop, G.L.; Couper, M.J. The precipitation sequence in Al-Mg-Si alloys. *Acta Mater.* **1998**, *46*, 3893–3904. [CrossRef]
28. Siddiqui, A.R.; Abdullah, H.A.; Al-Belushi, K.R. Influence of aging parameters on the mechanical properties of 6063 aluminium alloy. *J. Mater. Process. Technol.* **2000**, *102*, 234–240. [CrossRef]
29. Liu, G.; Zhang, G.J.; Ding, X.D.; Sun, J.; Chen, K.H. Modeling the strengthening response to aging process of heat-treatable aluminum alloys containing plate/disc- or rod/needle-shaped precipitates. *Mater. Sci. Eng. A* **2003**, *344*, 113–124. [CrossRef]
30. Myhr, O.R.; Grong, Ø.; Andersen, S.J. Modelling of the age hardening behaviour of Al-Mg-Si alloys. *Acta Mater.* **2001**, *49*, 65–75. [CrossRef]
31. Luo, A.; Lloyd, D.J.; Gupta, A.; Youdelis, W.V. Precipitation and dissolution kinetics in Al-Li-Cu-Mg alloy 8090. *Acta Mater.* **1993**, *41*, 769–776. [CrossRef]
32. Esmaeili, S.; Lloyd, D.J.; Poole, W.J. Modeling of precipitation hardening for the naturally aged Al-Mg-Si-Cu alloy AA6111. *Acta Mater.* **2003**, *51*, 3467–3481. [CrossRef]
33. Cluff, D.R.A.; Esmaeili, S. Prediction of the effect of artificial aging heat treatment on the yield strength of an open-cell aluminum foam. *J. Mater. Sci.* **2008**, *43*, 1121–1127. [CrossRef]
34. Raeisinia, B. A Study of Precipitation in the Aluminum Alloy AA6111. Master's Thesis, Tehran University, Tehran, Iran, 2000.

 *metals*

Article

Dissolution of $M_{23}C_6$ and New Phase Re-Precipitation in Fe Ion-Irradiated RAFM Steel

Zheng Yang [1], Shuoxue Jin [2,*], Ligang Song [2], Weiping Zhang [1], Li You [3] and Liping Guo [1,*]

1   Hubei Key Laboratory of Nuclear Solid Physics, Key Laboratory of Artificial Micro- and Nano-Structures of Ministry of Education and School of Physics and Technology, Wuhan University, Wuhan 430072, China; yangzheng34@msn.com (Z.Y.); zhangweiping@whu.edu.cn (W.Z.)
2   Multi-Discipline Research Center, Institute of High Energy Physics, Chinese Academy of Sciences, Beijing 100049, China; songlg@ihep.ac.cn
3   Key Laboratory for Advanced Metals and Materials, University of Science and Technology Beijing, Beijing 100083, China; youli-0104321@163.com
*   Correspondence: jinshuoxue@ihep.ac.cn (S.J.); guolp@whu.edu.cn (L.G.); Tel.: +86-10-8823-5971 (S.J.); +86-27-6875-2481 (ext. 2223) (L.G.)

Received: 16 April 2018; Accepted: 12 May 2018; Published: 14 May 2018

Abstract: The $M_{23}C_6$ precipitate plays a major role in preventing the sliding of the grain boundary and strengthens the matrix in the reduced-activation ferritic/martensic (RAFM) steel. However, its stability might be reduced under irradiation. The microstructural instability of the $M_{23}C_6$ precipitates in the RAFM steels irradiated at 300 °C with Fe ions up to a peak dose of 40 dpa was investigated by transmission electron microscopy. A "Core/Shell" morphology was found for the pre-existing $M_{23}C_6$ and a large number of new small phases appeared in parallel near the periphery of the precipitates after irradiation. The loss of crystallinity of the $M_{23}C_6$ periphery due to the dissolution of carbon atoms into the interface (C-rich "Shell") actually decreased the size of the Cr-rich "Core". The new phase that formed around the pre-existing precipitates was M_6C (Fe_3W_3C), which was formed through the carbide transformation of $M_{23}C_6$ to M_6C.

Keywords: $M_{23}C_6$; ion irradiation; M_6C; amorphization; RAFM steels

1. Introduction

Reduced-activation ferritic/martensic (RAFM) steels have excellent mechanical strength, low thermal expansion coefficient, high thermal conductivity and good resistance to radiation-induced swelling and helium embrittlement [1–4]. Thus, they have potential uses in new-generation fission reactors and future fusion reactors [5]. It is well known that several typical RAFM steels can be used for test blanket modules in the International Thermonuclear Experimental Reactor [6], which includes Eurofer97, 9Cr2WVTa and China Low Activation Martensitic (CLAM) steels. These RAFM steels undergo various transformation processes, including transformation of the ferritic steel structure (body centered cubic) to a martensitic steel structure (body centered tetragonal) after quenching. The $M_{23}C_6$ (Cr-rich carbide) precipitate preferentially forms at the martensitic lath and grain boundaries after thermal treatment. The $M_{23}C_6$ phase is very stable at temperatures below 820 °C [7], which plays a critical role in retaining the physical and mechanical properties of RAFM steels at elevated temperatures [6]. This prevents the grain boundary sliding [8] and strengthens the matrix by exerting a large Zener pinning force on the martensitic lath boundaries compared to the MX phase in RAFM steels [7,9]. However, under thermal, stress and irradiation conditions, the precipitate stability may be reduced [10–12]. Irradiation can increase the diffusion rates due to the production of point defects. The interface between the second-phase particle and the matrix was shown to function as a sink that traps point defects. It is reported that the interface is an important factor in element segregation and swelling behaviors due to the trapping of gas atoms and point defects [9,13,14]. The size of the $M_{23}C_6$ precipitates increases in the 9Cr-2WV steels during neutron irradiation [15] and CLAM steel after H and D ion irradiation [16]. The irradiation-induced amorphous transformation of $M_{23}C_6$ precipitates was observed in 800-MeV proton irradiated martensitic steels [17–19] and in 9Cr-1Mo ferritic/martensitic steel irradiated with 0.5 dpa at low temperatures [20]. The changes in the specimen microstructure, such as the secondary phases/precipitates induced by irradiation, will affect mechanical properties, including fracture toughness [21]. Therefore, examining and understanding the mechanism of $M_{23}C_6$ phase instabilities under irradiation is essential in evaluating the response of $M_{23}C_6$ against irradiation, which is also an essential part of developing radiation-resistant RAFM steels.

In this work, we investigate the microstructural instability of $M_{23}C_6$ precipitates in the RAFM steels due to irradiation. The mechanism of dissolution and re-precipitation will be discussed.

2. Experimental

2.1. Material Preparation

The chemical compositions of the RAFM steel test materials are shown in Table 1. The bulk specimens underwent heat treatment as follows: quenching at 1000 °C for 40 min and water cooled, before being tempered at 740 °C for 2 h and air cooled. A disc-like specimen with a diameter of 3 mm was punched out from a ~0.1 mm thick sample cut from the bulk RAFM steel and thinned to its final thickness. The standard TEM specimens were prepared by the conventional jet electropolishing methods using 5% perchloric acid and 95% ethanol polishing solution at −30 °C.

Table 1. Chemical compositions of Reduced-activation ferritic/martensic (RAFM) steel in wt. %.

Fe	C	Cr	W	V	Mn	Si	P	S
Bal.	0.088	9.24	2.29	0.25	0.49	0.25	0.0059	0.001

2.2. Irradiation Experiments

The irradiation experiments were carried out in the Accelerator Laboratory of Wuhan University and the samples were irradiated to a peak dose of 40 dpa with 100-keV Fe ion at 300 °C. The damage profile calculated by the Stopping and Range of Ions in Materials (SRIM) code simulation is shown in Figure 1, where the displacement energy was E_d = 40 eV. A uniform ion beam current (~1 μA) was created for holding by scanning in both the horizontal and vertical directions. The sample temperature was monitored with a thermocouple throughout the irradiation process, which touched the implantation surface of the sample. The microstructure analysis and composition measurements were conducted with a field-emission-gun analytical Tecnai G2 F30 microscope (Beijing, China), which had a nominal spot size of 0.5 nm, equipped with an energy dispersive X-ray spectroscopy (EDX) system.

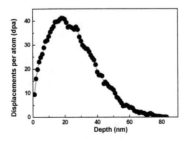

Figure 1. Depth profiles of the displacement damage (dpa) for the RAFM steel irradiated to a peak dose of 40 dpa with 100-keV Fe ions.

3. Results and Discussion

Figure 2 shows the unirradiated microstructure of a RAFM steel sample containing many pre-existing carbides with sizes of approximately 100–300 nm along the long axis in the matrix. A bright-field micrograph after 40 dpa of Fe ion irradiation at 300 °C is shown in Figure 3. A systematic selected-area electron diffraction (SAED) pattern of the pre-existing precipitates obtained from the unirradiated and irradiated specimens are shown in the upper right of the panels in Figures 2 and 3a, respectively. The calculated d-spacing of the (111), (200), (331), (422) and (511) spots were 0.6145 nm, 0.532 nm, 0.244 nm, 0.217 nm and 0.205 nm, respectively, which is parallel with the Pcpdfwin file (No. 781500) that originated from the $M_{23}C_6$ consisting mainly of Cr and Fe. It should be noted that the $(11\bar{1})$ and (200) reflections make a [011] zone axis (Figure 2), which paralleled the $[0\bar{1}1]$ zone axis by (422), $(1\bar{3}\bar{3})$ and $(5\bar{1}\bar{1})$ reflections in Figure 3a.

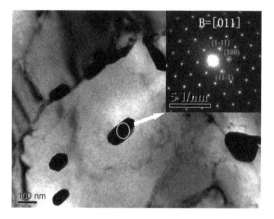

Figure 2. Bright-field micrograph of the unirradiated RAFM steel, in which many precipitates ware located at the martensitic lath and grain boundaries. The microstructure of pre-existing precipitates was identified by the selected-area electron diffraction (SAED) patterns that are shown in the upper right of panels. The diffraction patterns indicate that it is a $M_{23}C_6$ precipitate.

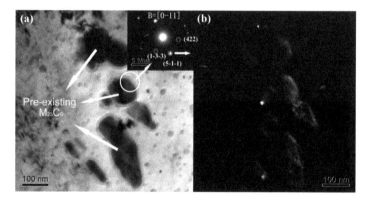

Figure 3. (**a**) Micrograph of the RAFM steel after 40 dpa Fe ion irradiation at 300 °C. An obvious change takes place at the periphery of the pre-existing precipitates. Simultaneously, a large number of small precipitates appear in the matrix; (**b**) is the dark-field image of (**a**), which is taken from the $(5\bar{1}\bar{1})$ diffraction spot indicated by the white arrow.

Remarkable changes in the original pre-existing $M_{23}C_6$ precipitates were found in the 40 dpa Fe ion irradiated specimen, which is shown in Figure 3a. The contrast at the periphery of the $M_{23}C_6$ precipitates was reduced or weakened. A large number of new small phases appeared in parallel near the pre-existing $M_{23}C_6$ precipitate after the irradiation. The original pre-existing $M_{23}C_6$ precipitate periphery (the matrix/carbide interface) seemed to grow, with the width of the extended region being about 10–30 nm. In the right inset of Figure 3a, the SAED patterns show the amorphous diffraction halos accompanied by diffraction spots. The patterns contain three parts: the $M_{23}C_6$ precipitate, matrix/carbide interface and the matrix, which is marked by a white circle. The $(5\overline{1}\overline{1})$, $(1\overline{3}3)$ and (422) diffraction spots originate from the $M_{23}C_6$ precipitate as discussed above. The dark-field image of the $M_{23}C_6$ precipitate shown in Figure 3b was from the $(5\overline{1}\overline{1})$ diffraction spot marked by the white arrow. On the one hand, all the small phases were completely black in Figure 3b, which further indicated that the structure of the small new phase is not parallel to $M_{23}C_6$. On the other hand, Figure 3b shows that all $M_{23}C_6$ carbides were bright and the peripheral region was brighter than the center of carbide. The electron transmittance in the periphery region was higher than that in the $M_{23}C_6$ carbides. The appearance of diffraction halos indicated that the brighter periphery region (i.e., the matrix/carbide interface) is amorphous. From Figure 3, one can conclude that the pre-existing $M_{23}C_6$ precipitate periphery was unstable under Fe ion irradiation. The loss of crystallinity of the $M_{23}C_6$ particles also appeared in our previous studies [22,23]. The seemingly extended region indicates the dissolution of the pre-existing $M_{23}C_6$ precipitates, which is similar to the heavy ion-irradiated or neutron-irradiated ferritic steels at 300–500 °C [24,25]. They also indicate that the pre-existing $M_{23}C_6$ precipitates were partially dissolved after irradiation, especially at elevated irradiation temperatures. Kai et al. [24] also indicated that the modification of pre-existing $M_{23}C_6$ precipitates due to 14-MeV Ni ion irradiation was very similar to the results of the thermal annealed ferritic steel. A vague explanation was given that irradiation would accelerate the thermal diffusion and in turn, enhance the thermal/aging effect, which is consistent with the mechanism of radiation-enhanced diffusion [26]. However, a more detailed reason for the dissolution of pre-existing $M_{23}C_6$ should also be investigated from the view point of elemental segregation.

The compositions of the precipitate, peripheral amorphous region and matrix in the 40 dpa irradiated RAFM steel at 300 °C were investigated with STEM/EDX microanalysis (Figure 4). Fe, Cr, W and C were analyzed using EDX mapping, with no other elements having appeared in the spectra. The EDX spectra and the elemental compositions (in wt. %) of the different regions marked by white arrays (Figure 4a) are shown in Figure 4b–d. Compared to the chemical composition of the matrix (Figure 4d), the precipitate includes a greater content of C and Cr elements (Figure 4a). This also indicates that the precipitate is Cr-rich $M_{23}C_6$, which is consistent with the SAED result. The M mainly represents the Cr element and also contains Fe, W and other elements [16,27,28]. The EDX spectra collected from the interface between the precipitate and the matrix has a higher C amount. In addition, an elemental map was constructed in the same region as that in Figure 4a, while Figure 4e shows the High-Angle Annular Dark Field (HAADF) STEM images of the same precipitate. Figure 4f–i shows the EDX-Kα mapping of Fe, Cr, W and C in the pre-existing precipitates after irradiation. The bright contrast in EDX elemental map represents a high elemental concentration. Therefore, Cr is mainly present in the precipitate. They also highlight the increased Fe and W contents in the pre-existing precipitates. However, considering the uneven thickness between the precipitates and the matrix (the precipitate region is thicker than that of the matrix in Figure 4), Fe and W elements should be distributed homogeneously. It should be noted that a high C element concentration appears at the interface between the precipitate and the matrix (i.e., the amorphous region). The reasons for this are not clear but one possibility is that the $M_{23}C_6$ phases decompose during irradiation because of the loss of carbon atoms into the interface through irradiation-enhanced diffusion. The STEM/EDX map reveals a "Core/Shell" morphology for the $M_{23}C_6$ precipitates after irradiation with 40 dpa Fe ions and a percolated structure of C-enrichment extending from the precipitate. It has a C-rich "Shell" and a Cr-rich "Core", with C atoms segregated in the peripheral amorphous region. As discussed above, the C-rich "Shell" region is brighter than the $M_{23}C_6$ carbides in the dark-field image. The electron transmittance in the C-rich "Shell" region is higher than that in the precipitate and matrix. As presented in the literature, the irradiation increases the diffusion rates and sustains the point defect flux from the precipitate/matrix to the sinks (i.e., the interface between carbide and matrix). The defect concentration gradient near the interface makes the undersize atoms (i.e., C atoms) segregate to the sinks assisted by the formation of the solute–self-interstitial complexes [29]. The oversized Cr atoms become depleted along the interface between the $M_{23}C_6$ precipitate and the matrix according to the solute-drag self-interstitial driven mechanisms [28–30]. Thus, the ion irradiation enhances the loss of crystallinity of the $M_{23}C_6$ phase and decreases the size of the Cr-rich "core" through the irradiation-enhanced diffusion.

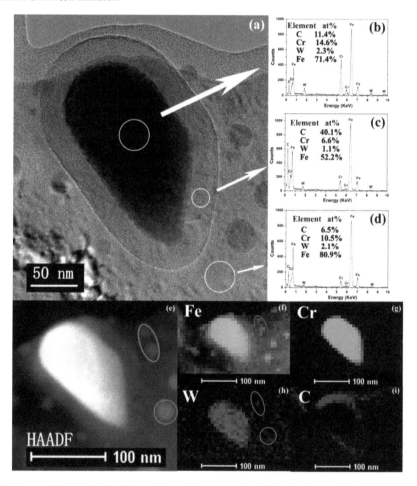

Figure 4. (a) Micrograph of the $M_{23}C_6$ precipitates after Fe ion irradiation; (**b–d**) Chemical compositions in the different regions marked by the white arrows are analyzed by EDX; (**b**) EDX energy spectrum of the precipitate region; (**c**) EDX results at the matrix/carbide interface (amorphous region); (**d**) EDX spectrum of the matrix; (**e**) HAADF–STEM image and (**f–i**) corresponding EDX elemental mappings of Fe, Cr, W and C elements for the pre-existing $M_{23}C_6$ precipitate and the new phase, respectively.

A large number of small new precipitates gathered around the pre-existing $M_{23}C_6$ precipitate in Figure 5, which are not $M_{23}C_6$ as indicated by Figure 3b. Thus, the EDX elemental maps were also measured for the new phase (see Figure 4f–i). The Fe and W elements were enriched in the new small precipitates (see the red circles in Figure 4). Tanigawa et al. [31] predicted the possibility of the formation of M_6C in the F82H steel after irradiation at 300 °C, but no detailed microstructure examination was performed. In order to clarify the structure of the new phase, a high-resolution TEM (HRTEM) image of the small precipitate was obtained as shown in Figure 5b. The calculated d-spacing of the (220) and (422) diffraction spots were 0.226 nm and 0.401 nm, respectively (Pcpdfwin file No. 781990). This confirms that the small precipitate has the M_6C (i.e., Fe_3W_3C) structure. Irradiation can increase the diffusion rates through the production of point defects, which can be trapped by the interface between $M_{23}C_6$ precipitates and matrices. In-situ observations revealed that the small $M_{23}C_6$ fragment can separate from the pre-existing $M_{23}C_6$ precipitates in 2-MV electron irradiated F82H steel [32]. The irradiation-induced phase transformations are possible because the free energies of the different phases are changed by the excess energy introduced into the lattice. The carbide transformations of $M_{23}C_6$ to M_6C have been observed at 700 °C by Inoue et al. [33], which was determined based on the following relationships of crystal orientations between $M_{23}C_6$ and M_6C: $(\bar{1}10)$ $M_{23}C_6$//$(\bar{1}10)$ M_6C, $(\bar{1}11)$ $M_{23}C_6$//$(11\bar{1})$ M_6C, and (112) $M_{23}C_6$//(112) M_6C. Therefore, the formation of a new M_6C phase near the pre-existing $M_{23}C_6$ precipitate may be transformed from a small $M_{23}C_6$ fragment separated from the pre-existing precipitates, with irradiation at elevated temperature promoting these processes. As discussed above, the $M_{23}C_6$ carbides can prevent sliding of the grain boundary [8] and strengthen the mechanical properties of the RAFM steels [6]. The dissolution of the pre-existing $M_{23}C_6$ carbides was detrimental to the mechanical properties, especially the ductile-to-brittle transition temperature [34]. Simultaneously, the high energy precipitate–matrix interfaces due to Cr and C segregation favored fracture paths [5]. The precipitation of M_6C caused a reduction in material toughness [35]. Thus, the phase transformation ($M_{23}C_6 \rightarrow M_6C$) resulted in degraded mechanical properties, especially the ductile-to-brittle transition temperature. The effect of the $M_{23}C_6$ dissolution and re-precipitation of M_6C on the fracture toughness requires further research. Ghidelli et al. [21] suggested a possible approach to study the micro-scale fracture toughness by obtaining pillar splitting measurements.

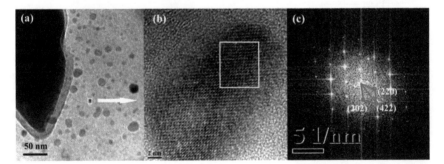

Figure 5. (a) TEM micrograph of the small precipitates in the matrix of RAFM steel irradiated with 40 dpa Fe ion at 300 °C; (b) Corresponding HRTEM image of the small precipitate; and (c) Fast Fourier transformation (FFT) pattern of the white square region in (b).

4. Conclusions

In the present work, we investigated the instability of $M_{23}C_6$ precipitates in ion-damaged RAFM steel via STEM/EDX. A "Core/Shell" structure (C-rich "Shell" and Cr-rich "Core") was formed in the pre-existing $M_{23}C_6$ precipitates after the irradiation with 40 dpa Fe ion at 300 °C. The SAED measurements combined with the bright/dark-field contrast images showed that the extended C-rich "Shell" was amorphous with carbon atoms percolated into the interface. Ion irradiation actually enhances the loss of crystallinity of the $M_{23}C_6$ phase and decreases the size of the Cr-rich "Core" through irradiation-enhanced diffusion. The small new phase gathering around the pre-existing $M_{23}C_6$ precipitate has a M_6C (Fe_3W_3C) structure, which was confirmed by SAED and EDX analyses. The irradiation at elevated temperatures promoted the separation of small $M_{23}C_6$ fragments from the pre-existing precipitates and transformation into a new M_6C phase.

Author Contributions: Z.Y., S.J. and L.G. conceived and designed the experiments; W.Z. and L.Y. performed the experiments; S.J. and L.S. analyzed the data; Z.Y. and S.J. wrote the paper.

Funding: This research was funded by the [National Natural Science Foundation of China] grant number [11505192, 11775162 and 11775236].

Conflicts of Interest: The authors declare no conflict of interest.

References

1. Huang, Q.Y. Status and improvement of CLAM for nuclear application. *Nucl. Fusion* **2017**, *57*, 086042. [CrossRef]
2. Yu, J.N.; Huang, Q.Y.; Wan, F.R. Research and development on the China low activation martensitic steel (CLAM). *J. Nucl. Mater.* **2007**, *367*, 97–101. [CrossRef]
3. Yin, F.S.; Jung, W.S.; Chung, S.H. Microstructure and creep rupture characteristics of an ultra-low carbon ferritic/martensitic heat-resistant steel. *Scr. Mater.* **2007**, *57*, 469–472. [CrossRef]
4. Li, Q.S.; Shen, Y.Z.; Zhu, J.; Huang, X.; Shang, Z.X. Evaluation of Irradiation Hardening of P92 Steel under Ar Ion Irradiation. *Metals* **2018**, *8*, 94. [CrossRef]
5. Di Martino, S.F.; Riddle, N.B.; Faulkner, R.G. Controlling the ductile to brittle transition in Fe-9%Cr ODS steels. *J. Nucl. Mater.* **2013**, *442*, S124–S132. [CrossRef]
6. Yan, B.Y.; Liu, Y.C.; Wang, Z.J.; Liu, C.X.; Si, Y.H.; Li, H.J.; Yu, J.X. The Effect of Precipitate Evolution on Austenite Grain Growth in RAFM Steel. *Materials* **2017**, *10*, 1017. [CrossRef] [PubMed]
7. Xiao, X.; Liu, G.Q.; Hu, B.F.; Wang, J.S.; Ma, W.B. Microstructure Stability of V and Ta Microalloyed 12%Cr Reduced Activation Ferrite/Martensite Steel during Long-term Aging at 650 °C. *J. Mater. Sci. Technol.* **2015**, *31*, 311–319. [CrossRef]
8. Duan, Z.X.; Pei, W.; Gong, X.B.; Chen, H. Superplasticity of Annealed H13 Steel. *Materials* **2017**, *10*, 870. [CrossRef] [PubMed]
9. Shen, T.L.; Wang, Z.G.; Yao, C.F.; Sun, J.R.; Li, Y.F.; Wei, K.F.; Zhu, Y.B.; Pang, L.L.; Cui, M.H.; Wang, J.; et al. The sink effect of the second-phase particle on the cavity swelling in RAFM steel under Ar-ion irradiation at 773 K. *Nucl. Instrum. Methods Phys. Res. Sect. B* **2013**, *307*, 512–515. [CrossRef]
10. Tan, L.; Katoh, Y.; Snead, L.L. Stability of the strengthening nanoprecipitates in reduced activation ferritic steels under Fe^{2+} ion irradiation. *J. Nucl. Mater.* **2014**, *445*, 104–110. [CrossRef]
11. Tanigawa, H.; Sakasegawa, H.; Ogiwara, H.; Kishimoto, H.; Kohyama, A. Radiation induced phase instability of precipitates in reduced-activation ferritic/martensitic steels. *J. Nucl. Mater.* **2007**, *367*, 132–136. [CrossRef]
12. Dong, Q.S.; Yao, Z.W.; Wang, Q.; Yu, H.B.; Kirk, M.A.; Daymond, M.R. Precipitate Stability in a Zr-2.5Nb-0.5Cu Alloy under Heavy Ion Irradiation. *Metals* **2017**, *7*, 287. [CrossRef]
13. Rowcliffe, A.F.; Lee, E.H. High-Temperature Radiation-Damage Phenomena in Complex Alloys. *J. Nucl. Mater.* **1982**, *108*, 306–318. [CrossRef]
14. Zhang, C.H.; Chen, K.Q.; Wang, Y.S.; Sun, J.G.; Hu, B.F.; Jin, Y.F.; Hou, M.D.; Liu, C.L.; Sun, Y.M.; Han, J.; et al. Microstructural changes in a low-activation Fe-Cr-Mn alloy irradiated with 92 MeV Ar ions at 450 °C. *J. Nucl. Mater.* **2000**, *283*, 259–262. [CrossRef]
15. Klueh, R.L.; Alexander, D.J.; Rieth, M. The effect of tantalum on the mechanical properties of a 9Cr-2W-0.25V-0.07Ta-0.1C steel. *J. Nucl. Mater.* **1999**, *273*, 146–154. [CrossRef]
16. Zhao, M.Z.; Liu, P.P.; Zhu, Y.M.; Wan, F.R.; He, Z.B.; Zhan, Q. Effects of hydrogen isotopes in the irradiation damage of CLAM steel. *J. Nucl. Mater.* **2015**, *466*, 491–495. [CrossRef]
17. Dai, Y.; Bauer, G.S.; Carsughi, F.; Ullmaier, H.; Maloy, S.A.; Sommer, W.F. Microstructure in Martensitic Steel DIN 1.4926 after 800 MeV proton irradiation. *J. Nucl. Mater.* **1999**, *265*, 203–207. [CrossRef]
18. Dai, Y.; Carsughi, F.; Sommer, W.F.; Bauer, G.S.; Ullmaier, H. Tensile properties and microstructure of martensitic steel DIN 1.4926 after 800 MeV proton irradiation. *J. Nucl. Mater.* **2000**, *276*, 289–294. [CrossRef]

19. Dai, Y.; Maloy, S.A.; Bauer, G.S.; Sommer, W.F. Mechanical properties and microstructure in low-activation martensitic steels F82H and Optimax after 800-MeV proton irradiation. *J. Nucl. Mater.* **2000**, *283*, 513–517. [CrossRef]

20. Sencer, B.H.; Garner, F.A.; Gelles, D.S.; Bond, G.M.; Maloy, S.A. Structural evolution in modified 9Cr-1Mo ferritic/martensitic steel irradiated with mixed high-energy proton and neutron spectra at low temperatures. *J. Nucl. Mater.* **2002**, *307*, 266–271. [CrossRef]

21. Ghidelli, M.; Sebastiani, M.; Johanns, K.E.; Pharr, G.M. Effects of indenter angle on micro-scale fracture toughness measurement by pillar splitting. *J. Am. Ceram. Soc.* **2017**, *100*, 5731–5738. [CrossRef]

22. Jin, S.X.; Guo, L.P.; Yang, Z.; Fu, D.J.; Liu, C.S.; Tang, R.; Liu, F.H.; Qiao, Y.X.; Zhang, H.D. Microstructural evolution of P92 ferritic/martensitic steel under argon ion irradiation. *Mater. Charact.* **2011**, *62*, 136–142. [CrossRef]

23. Jin, S.X.; Guo, L.P.; Li, T.C.; Chen, J.H.; Yang, Z.; Luo, F.F.; Tang, R.; Qiao, Y.X.; Liu, F.H. Microstructural evolution of P92 ferritic/martensitic steel under Ar^+ ion irradiation at elevated temperature. *Mater. Charact.* **2012**, *68*, 63–70. [CrossRef]

24. Kai, J.J.; Kulcinski, G.L. 14 MeV nickel-ion irradiated HT-9 ferritic steel with and without helium pre-implantation. *J. Nucl. Mater.* **1990**, *175*, 227–236. [CrossRef]

25. Maziasz, P.J.; Klueh, R.L.; Vitek, J.M. Helium Effects on Void Formation in 9Cr-1MoVNB and 12Cr-1MoVW Irradiated in HFIR. *J. Nucl. Mater.* **1986**, *141*, 929–937. [CrossRef]

26. Chen, J.H.; Guo, L.P.; Liu, C.X.; Luo, F.F.; Li, T.C.; Zheng, Z.C.; Jin, S.X.; Yang, Z. Enhancement of room temperature ferromagnetism in Mn-implanted Si by He implantation. *Appl. Phys. Lett.* **2012**, *101*, 132413. [CrossRef]

27. Fang, C.M.; van Huis, M.A.; Sluiter, M.H.F. Formation, structure and magnetism of the γ-(Fe,M)$_{23}$C$_6$ (M = Cr, Ni) phases: A first-principles study. *Acta Mater.* **2016**, *103*, 273–279. [CrossRef]

28. Klimiankou, M.; Lindau, R.; Moslang, A. Direct correlation between morphology of (Fe,Cr)$_{23}$C$_6$precipitates and impact behavior of ODS steels. *J. Nucl. Mater.* **2007**, *367*, 173–178. [CrossRef]

29. Lu, Z.; Faulkner, R.G.; Was, G.; Wirth, B.D. Irradiation-induced grain boundary chromium microchemistry in high alloy ferritic steels. *Scr. Mater.* **2008**, *58*, 878–881. [CrossRef]

30. Jiang, Z.H.; Feng, H.; Li, H.B.; Zhu, H.C.; Zhang, S.C.; Zhang, B.B.; Han, Y.; Zhang, T.; Xu, D.K. Relationship between Microstructure and Corrosion Behavior of Martensitic High Nitrogen Stainless Steel 30Cr15Mo1N at Different Austenitizing Temperatures. *Materials* **2017**, *10*, 861. [CrossRef] [PubMed]

31. Tanigawa, H.; Sakasegawa, H.; Klueh, R.L. Irradiation effects on precipitation in reduced-activation ferritic/martensitic steels. *Mater. Trans.* **2005**, *46*, 469–474. [CrossRef]

32. Kano, S.; Yang, F.; Shen, J.; Zhao, Z.; McGrady, J.; Hamaguchi, D.; Ando, M.; Tanigawa, H.; Hiroaki, A. Investigation of instability of M$_{23}$C$_6$ particles in F82H steel under electron and ion irradiation conditions. *J. Nucl. Mater.* **2018**, *502*, 263–269. [CrossRef]

33. Inoue, A.; Masumoto, T. Carbide Reactions (M$_3$C-M$_7$C$_3$-M$_{23}$C$_6$-M$_6$C) during Tempering of Rapidly Solidified High-Carbon Cr-W and Cr-Mo Steels. *Metall. Trans. A* **1980**, *11*, 739–747. [CrossRef]

34. Wang, W.; Mao, X.D.; Liu, S.J.; Xu, G.; Wang, B. Microstructure evolution and toughness degeneration of 9Cr martensitic steel after aging at 550 A degrees C for 20000 h. *J. Mater. Sci.* **2018**, *53*, 4574–4581. [CrossRef]

35. Shiba, K.; Tanigawa, H.; Hirose, T.; Sakasegawa, H.; Jitsukawa, S. Long-term properties of reduced activation ferritic/martensitic steels for fusion reactor blanket system. *Fusion Eng. Des.* **2011**, *86*, 2895–2899. [CrossRef]

 *metals*

Article

Effect of the Martensitic Transformation on the Stamping Force and Cycle Time of Hot Stamping Parts

Maider Muro [1], Garikoitz Artola [1], Anton Gorriño [2] and Carlos Angulo [2],*

[1] Metallurgy Research Centre IK4 AZTERLAN, Aliendalde Auzunea 6, 48200 Durango, Spain; mmuro@azterlan.es (M.M.); gartola@azterlan.es (G.A.)

[2] Department of Mechanical Engineering, University of the Basque Country (UPV/EHU), Plaza Ingeniero Torres Quevedo 1, 48013 Bilbao, Spain; antonio.gorrino@ehu.es

* Correspondence: carlos.angulo@ehu.es; Tel.: +34-94-601-4217

Received: 26 April 2018; Accepted: 24 May 2018; Published: 26 May 2018

Abstract: Stamping dies perform two functions in the hot stamping process of body-in-white components. Firstly, they form the steel sheet into the desired shape and, secondly, they quench the steel at a cooling rate that leads to hardening by means of the austenite-γ to martensite transformation. This microstructural change implies a volume expansion that should lead to a force peak in the press, which has yet to be detected in industrial practice. In this study, a set of hot stamping laboratory tests were performed on instrumented Al–Si-coated 22MnB5 steel flat formats to analyze the effect of the stamping pressure on the detection of the expected peak. Plotting the sheet temperature and pressure curves against time allowed us to identify and understand the conditions in which the force peak can be detected. These conditions occurred most favorably when the stamping pressure is below 5 MPa. It is thus possible to determine the exact moment at which the complete hardening transformation occurs by monitoring the local pressing force of the tool in areas where the pressure exerted on the metal format is below 5 MPa. This information can be applied to optimize the time needed to open the dies in terms of the complete martensitic transformation.

Keywords: hot stamping; press hardening; martensitic expansion; force peak; cycle time

---

1. Introduction

The basis of hot stamping technology is a combination of the forming and subsequent quenching of steel sheet formats in a single die stroke as described by Reference [1–3]. The cooling sequence during hot stamping is a major technological factor since it drives the final mechanical properties of the parts. The cooling of the steel sheet during hot stamping consists of four consecutive steps. First, the austenitized steel sheet is transferred from the furnace to the press (radiation + convection); second, the die approaches the sheet format until first contact is achieved (radiation + convection + conduction); third, sheet forming is carried out in the press stroke that starts from the first die–sheet contact to the bottom dead center position (BDC) (conduction + convection); and, finally, the press is kept in the stroke's BDC until the steel sheet undergoes a martensitic transformation (conduction).

The last step takes most of the takt time since it is intended to ensure that quenching is completed during the stamping procedure [4]. During the quenching stage, heat is conducted from the stamped part towards the forming tool across the contact surface between both elements. Both the bulk and contact thermal conductivities and tool cooling systems are of the utmost importance for this step. An improvement in any of the aforementioned four steps leads to higher production rates in the press and inclines the production bottleneck upstream towards the austenitizing furnace. R&D efforts have sought to exploit this process by exploring high conductivity steels [5], applying conformal cooling strategies [6], and studying the interfacial heat transfer coefficient (IHTC) between the sheet and the die [7–11]. The relationship between the IHTC and the nominal contact pressure during the quenching stage has awakened interest in the field due to the force and position control capabilities of the stamping presses. Pressing force is one of the few process variables than can be monitored and controlled straight away in an industrial hot stamping line.

Nevertheless, this force is not distributed uniformly on the part's surface during the quenching step. In a first approach, the actual local pressure on each tool-sheet contact point is affected by the angle between the part surface and the moving direction of the press (Figure 1). This leads to IHTC differences along the stamped part that increase the cooling rate variability already caused by the tool's shape, contact sequence, and cooling channel performance. As a result, the martensitic transformation is not performed in unison but sequentially, where areas with the lowest IHTCs are the last to transform. The total quenching time that the press must be kept in the BDC for a total martensitic transformation to be achieved must thus account for the lowest local pressure sites, as they drive the lowest IHTC values.

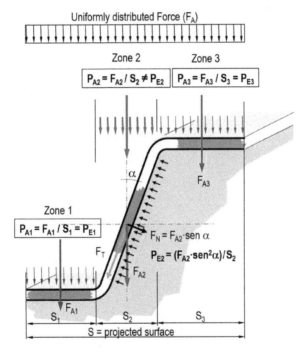

Figure 1. Schematic pressure distribution on flanges and flat areas in a hot stamped profile. "A" subscripts stand for "Apparent" and "E" subscripts stand for "Effective".

The quenching of these sites implies a volume expansion related to the austenite-to-martensite transformation. Provided the press position is fixed in the BDC and the sheet fills the clearance between the upper and lower tools, this expansion should theoretically lead to a local pressure peak [12,13]. However, this phenomenon has seldom been reported in the literature by authors seeking to experimentally reproduce hot stamping conditions [8]. The thermography is actually used in the hot stamping industry to detect cooling issues in the part and/or tool normally caused by cooling channel clogging, leakage, water supply shortage, etc. [11], and in other applications [14]. This technique is useful to understand the process without contact with the material. This work explores the simultaneous evolution of the actual pressure and sheet temperature in a range of nominal stamping pressures from 12.5 to 2.5 MPa to identify the conditions under which the martensite expansion can be properly measured. The detection of the force caused by quenching expansion has direct applications in terms of defining the minimum achievable takt time for a given set of pressing setups and tool designs.

The interest in controlling this cycle time is evidenced by the modeling efforts presented in the literature [15,16]. Previous approaches to this issue have been based both on numerical modelling of the process or hardness/microstructure surveys after a part has been extracted from the press [17], and searching for online dwell time optimization methods. Monitoring of the martensitic expansion force on the other hand, leads to a real-time monitoring of the quenching completion, the simultaneousness of which takes into consideration the process of obtaining measurements and decision making.

2. Materials and Methods

The sheet steel material employed for our experimental work was the Al–Si pre-coated quenchable 22MnB5 steel, commonly known under the commercial name USIBOR 1500P® (ArcelorMittal, Luxembourg City, Luxembourg), whose chemical composition is given in Table 1.

Table 1. Chemical composition of USIBOR 1500P® (in wt % with balanced Fe).

C	Si	Mn	P	S	Al	B	N_2	Ti	Cr	Ni
0.22	0.27	1.23	0.014	0.003	0.041	0.003	0.006	0.036	0.21	0.05

The dimensions of the samples were 80 × 90 mm, with a thickness of 3 mm. In each sample, a 1 mm diameter hole was machined by electrical discharge machining, right in the middle of the thickness, to a 30 mm depth (Figure 2a). Inside the hole, an AISI 316L sheathed type K thermocouple of 1 mm diameter was fixed (Figure 2b).

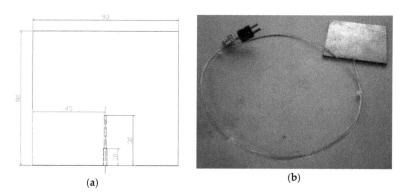

Figure 2. (a) Test sample sketch; (b) samples with the K thermocouple inserted.

A flat tool with an active surface of 150 × 150 mm was used. The tool was manufactured in ductile iron grade GJS-700-2 with a hardness of 275 HB. The tool design is shown in Figures 3 and 4. The cooling channels in both halves of the tool were fabricated with a nominal diameter of 10 mm. The cooling water for the tests was taken from the line at a temperature of 12.5 °C and flow rate of 20 L/min.

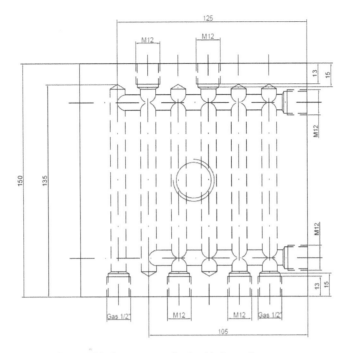

Figure 3. Die demonstrator sketch with the cooling system.

Figure 4. The die demonstrator placed on the hydraulic press.

The tests were carried out on a MTS 810 hydraulic press (Material Testing Systems, Eden Prairie, MN, USA) that was programed in a closed-loop pressure control until the working pressure was achieved, and then immediately shifted to a blockage position for 10 s before the load was released.

Before carrying out the tests, the pressure homogeneity of the setup was verified by employing a Pressurex-Micro® (Sensor Products Inc., Madison, NJ, USA) SPF-ER indication film. This film reveals a contact pressure map between the two surfaces by employing the interaction between the two foils of the film that leads to different degrees of magenta coloring depending on the pressure in each contact spot, see Reference [18]. Subsequently, using the Topaq® imaging system (Sensor Products Inc., Madison, NJ, USA), the magenta color map was translated to a high resolution, full-color representation of the pressure distribution. A 0.5 mm thickness Pressurex® (Sensor Products Inc., Madison, NJ, USA) film was employed in this study.

Initially, the alignment of the tool and the press was verified using the pressure sensing films. The film was placed directly on the bottom of the tool and several loads were checked. Figure 5 shows the magenta-colored profile for 54 kN. Once the indication film was marked with the Topaq® image system (Sensor Products Inc., Madison, NJ, USA) and measurements of the color density were performed, the pressure was calculated and plotted in a 3D image (Figure 6). This way it was verified that the assembly was set to contact primarily in the center of the tool, where the samples were placed during test.

The same test was repeated using USIBOR 1500P® (ArcelorMittal, Luxembourg City, Luxembourg) sheet blanks of 80 × 90 mm by placing the film between the sheet and both the upper and lower tools. Figure 7 shows the result for an applied pressure force of 85 kN (11.8 MPa pressure). It is relevant that, despite the test being performed in a flat configuration, there are significant pressure gradients in the samples. A nominal pressure of 12 MPa turns out to be translated in a pressure map of 12 ± 4 MPa. The pressure force distribution analyses indicate that the press applied higher pressure on the right side than on the front, back, and left side. So, it is clear that despite industrial process parameters being taken on average pressures values, the actual pressure distributions are randomly distributed around the nominal value.

Contact pressure

Color density of the pressed film

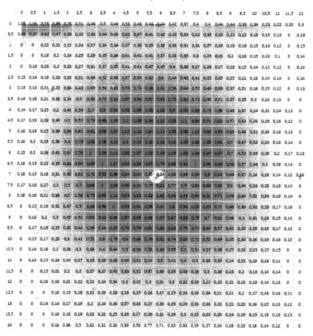

Front side of the hydraulic press

Figure 5. The force press track over the 120 × 140 mm film.

Contact pressure

Between die demonstrator – Film (120 x 140 mm) – 54 kN – 3.2 MPa)

Average measured pressure (MPa)

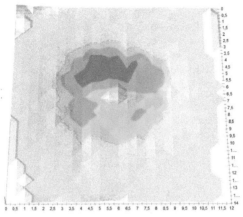

Die demonstrator width (cm)

Front side of the hydraulic press

Figure 6. Pressure distribution in a 3D image.

Contact Pressure

Pressure map – 12 MPa set-point

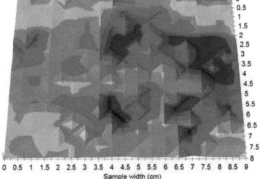

Sample width (cm)

Front side of the hydraulic press

Figure 7. Distribution of the applied pressure over a 80 × 90 mm USIBOR 1500P® steel sheet blank (ArcelorMittal, Luxembourg City, Luxembourg).

The austenitizing of the samples for the press hardening was performed without a protective atmosphere in an electric resistance furnace (LAC LH 39/13) whose dimensions were 0.31 m × 0.31 m × 0.31 m (width × length × height). Samples were heated one by one. Each sample was heated for approximately 300 s to an austenitizing temperature of 900 °C and they were extracted from the furnace and placed inside the tooling in less than 4 s. The austenitizing temperature and time were selected in order to reproduce industrial hot stamping conditions. The temperature setpoint of 900 °C was selected to ensure full austenitization of the 22MnB5 while avoiding grain growth. Furthermore, the 300 s heating time not only ensured austenitization but also Al–Si coating enrichment in Fe, avoiding liquid aluminum dropping (eutectic Al–Si melting point is at 577 °C in the absence of Fe).

Press hardening tests were then conducted at the corresponding pressure set-points (2.5, 5, 7.5, 10, and 12.5 MPa).

After the pressing tests, and with the aim of ensuring that press hardening was carried out correctly during the tests, all samples were metallurgically analyzed by means of metallographic inspections and hardness measurements. Three HP (High Pressure), T (Thermocouple), and LP (Low Pressure) zones in the sample were checked for each specimen, as shown in Figure 8. The hardness variations observed in the three zones were insignificant.

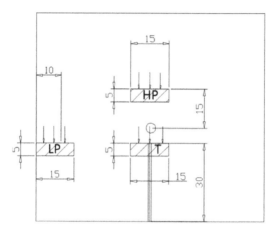

Figure 8. Identification of the zones subjected to metallurgical testing, T (Thermocouple), HP (High Pressure), and LP (Low Pressure).

The hardness was measured in a cross section of the Low Pressure zone of each sample via Vickers hardness control measurements, in accordance with EN ISO 6507-1: 2006, on the core (half thickness) and periphery (0.3 mm from the surface) of the cross section of the metallographic specimens. The microstructures were analyzed using a Leica MEF4 optical microscope (Leica Microsystems, Wetzlar, Germany) and etching of the metallographic samples was carried out with Nital 4.

3. Results

The recorded pressure–temperature–time curves are presented below, starting from the maximum employed pressure. Figure 9 shows the 12.5 MPa set-point curves. For this case, the pressure graph is smooth without any overshot. The hardening was successful, as shown by the microstructure in Figure 10 and corroborated by the HV10 hardness measurements in Table 2.

Thus, most of the martensitic transformation in the sample must have occurred while the press was still in closed-loop pressure control. Some parts of the sheet may have been undergoing residual martensitic transformation, as indicated by the thermocouple records, which reached the martensite start (Ms) temperature about a second after the pressure signal arrived to the set-point.

As the pressure set-point was reduced to 10 MPa, martensitic expansion was detected as an overshot in the pressure curve (Figure 11). The start of the overshot was 0.1 s after the target pressure was reached, and the pressure ramp-up lasted 2 s. In agreement with the observations in Figure 9, the Ms in the thermocouple location was reached about a second later than the overshot, indicating that the hardening did not proceed in a synchronized manner, but occurred over a period of seconds.

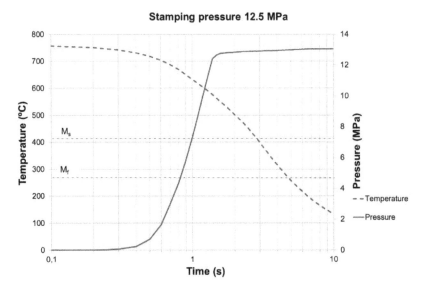

Figure 9. Temperature–pressure–time plot for a pressure set-point of 12.5 MPa.

Figure 10. Martensitic microstructure of the material stamped at the 12.5 MPa set-point and analyzed in LP zone.

Table 2. Vickers hardness measurement sampling for each pressure set-point. Average of five values and confidence interval of 95% of the mean.

Pressure Set-Point (MPa)	HV10 Hardness
12.5	487 ± 7
10.0	477 ± 18
7.5	478 ± 9
5.0	488 ± 10
2.5	481 ± 10

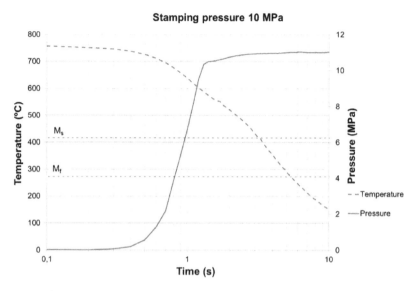

Figure 11. Temperature–pressure–time plot for a pressure set-point of 10 MPa.

Overshot becomes easier to identify as the pressure set-point decreases. The lower the pressure, the later the start of the overshot and the longer its duration. For a 7.5 MPa stamping pressure, the start of the martensitic expansion signal was delayed by 0.2 s after the pressure set-point was reached, which lasted 2.2 s (Figure 12).

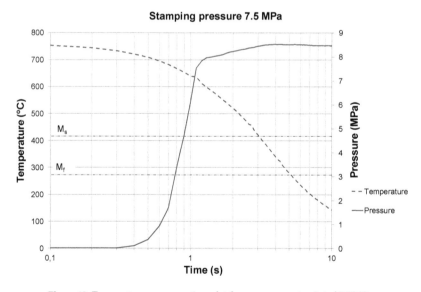

Figure 12. Temperature–pressure–time plot for a pressure set-point of 7.5 MPa.

For the two lowest pressure set-points on the testing schedule, 5 and 2.5 MPa, thermal shrinkage of the sample was detected as a relaxation of the exerted pressure (Figures 13 and 14). In these two cases, there is an initial peak of stress, followed by a reduction of the pressure that is attributed to the thermal expansion coefficient. In terms of the delay of the onset of the overshot, the measured values are 0.3 s and 0.4 s, while the duration of the expansion is 2.5 s and 2.9 s, for 5 MPa and 2.5 MPa, respectively. In these two cases, the Ms in the thermocouple position is reached during the overshot ramp.

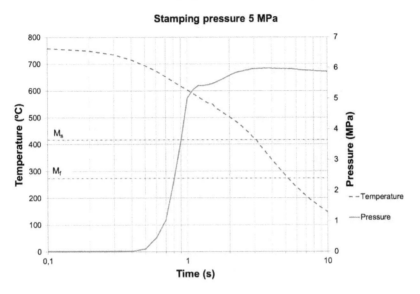

Figure 13. Temperature–pressure–time plot for a pressure set-point of 5 MPa.

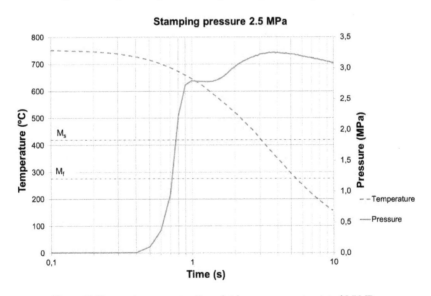

Figure 14. Temperature–pressure–time plot for a pressure set-point of 2.5 MPa.

Figure 15 shows microstructural confirmation of the correct quenching of the samples even for the case of the lowest working pressure. This result is backed-up by the hardness measurements shown earlier in Table 2.

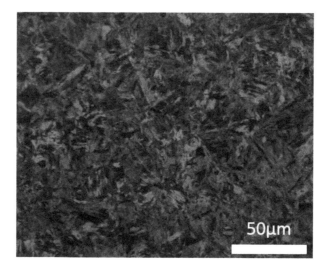

Figure 15. Martensitic microstructure of the material stamped at 2.5 MPa set-point and analyzed in LP zone.

According to the foregoing records, the lower the working pressure, the better the resolution of the martensitic expansion that can be obtained in terms of pressure overshot, as the hardening is carried out at a slower pace along the sample.

4. Discussion

The results reported above show a phenomenon that allows us to detect when the martensitic transformation was complete during hot stamping, and under certain conditions:

1. Force monitoring must exist on the lowest quenching pressure site of the stamped part.
2. The press force set-point must be achieved before quenching starts.
3. The site of the stamped part must have a flat sheet-to-tool contact surface.
4. It must be possible to program the press stroke so that it maintains its BDC during quenching.

Provided these conditions are met, the martensitic expansion of the quenched steel can be measured in terms of a force increase and stabilization. This feature can be potentially applied to industrial hot stamping processes as a means of reducing the quenching time. Press opening can be commanded immediately after the quenching finishes, thus reducing the over-holding times that are employed to account for any uncertainties in the precise instant that full-part quenching is achieved.

It is understandable that this behavior has not been observed during industrial production. All in all, industrial presses monitor average forces and manufacture complex part shapes, because of this the final quenching expansion cannot be resolved. Regarding laboratory hot stamping tests, some authors have supplied pressure (or force) and/or temperature versus time curves for the hot stamping of both flat and shaped samples.

The same overshot was observed in Reference [8], but corresponded to an omega-shaped sample, meaning that there was no way to distinguish any expansion effect on the flanges, which is the most conflictive point for quenching hot stamped parts.

Regarding the lack of martensitic expansion reporting, also shown by Reference [7], two alternatives related to the press programming are proposed as an explanation. On the one hand, if the pressure ramp-up is too slow, the martensitic transformation can happen while the press is in load control and the proportional integral derivative program is in a position that compensates for any expansion, thus hiding any overshot. On the other hand, if closed-loop pressure control is set and the control reaction time is very quick, the quenching might be carried-out unnoticed and smooth force variations would only be registered.

5. Conclusions

It is worth pointing out that despite the low pressures that have been used in this work, martensitic transformation has been achieved over a very narrow timeline, which points to potential future work. Considering all the industrial effort involved in reducing cycle times and costs, the next research steps should be focused on developing tool-design strategies and taking advantage of the achieved results by takt-time reduction.

This work shows that it is possible to develop a system/device that is capable of monitoring quenching in real time during hot stamping. If force is monitored in the slowest cooling spots of the part, detection of the martensitic expansion makes it possible to control the production equipment and optimize the process.

Author Contributions: All the four authors conceived, designed, performed the press hardening experiments and interpreted the results together; M.M. and G.A. prepared the experimental setup (tool, cooling system, press programming, sample instrumentation and data acquisition devices); M.M., G.A. and A.G. analyzed the data; A.G. performed the pressure homogeneity distribution measurements and M.M., G.A. and C.A. wrote the paper.

Funding: The authors gratefully acknowledge the funding provided by the Department of Research and Universities of the Basque Government under Grant No. IT947-16 and the University of the Basque Country UPV/EHU under Program No. UFI 11/29.

Conflicts of Interest: The authors declare no conflict of interest.

References

1. Karbasian, H.; Tekkaya, A.E. A review on hot stamping. *J. Mater. Process. Technol.* **2010**, *210*, 2103–2118. [CrossRef]
2. Neugebauer, R.; Schieck, F.; Polster, S.; Mosel, A.; Rautenstrauch, A.; Schönherr, J.; Pierschel, N. Press hardening—An innovative and challenging technology. *Arch. Civil Mech. Eng.* **2012**, *12*, 113–118. [CrossRef]
3. Mori, K.; Bariani, P.F.; Behrens, B.-A.; Brosius, A.; Bruschi, S.; Maeno, T.; Merklein, M.; Yanagimoto, J. Hot stamping of ultra-high strength steel parts. *CIRP Ann. Manuf. Technol.* **2017**, *66*, 755–777. [CrossRef]
4. Palm, C.; Vollmer, R.; Aspacher, J.; Gharbi, M. Increasing performance of hot stamping systems. *Procedia Eng.* **2017**, *207*, 765–770. [CrossRef]
5. Ghiotti, A.; Bruschi, S.; Medea, F.; Hamasaiid, A. Tribological behavior of high thermal conductivity steels for hot stamping tools. *Tribol. Int.* **2016**, *97*, 412–422. [CrossRef]
6. He, B.; Ying, L.; Li, X.; Hu, P. Optimal design of longitudinal conformal cooling channels in hot stamping tools. *Appl. Therm. Eng.* **2016**, *106*, 1176–1189. [CrossRef]
7. Caron, E.J.F.R.; Daun, K.J.; Wells, M.A. Experimental heat transfer coefficient measurements during hot forming die quenching of boron steel at high temperatures. *Int. J. Heat Mass Transf.* **2014**, *71*, 396–404. [CrossRef]
8. Abdulhay, B.; Bourouga, B.; Dessain, C. Thermal contact resistance estimation: Influence of the pressure contact and the coating layer during a hot forming process. *Int. J. Mater. Form.* **2012**, *5*, 183–197. [CrossRef]
9. Mendiguren, J.; Ortubay, R.; Saenz de Argandoña, E.; Galdos, L. Experimental characterization of the heat transfer coefficient under different close loop controlled pressures and die temperatures. *Appl. Therm. Eng.* **2016**, *99*, 813–824. [CrossRef]
10. Salomonsson, P.; Oldenburg, M.; Akerström, P.; Bergman, G. Experimental and numerical evaluation of the heat transfer coefficient in press hardening. *Steel Res. Int.* **2009**, *80*, 841–845. [CrossRef]
11. Gorriño, A.; Angulo, C.; Muro, M.; Izaga, J. Investigation of thermal and mechanical properties of quenchable high-strength steels in hot stamping. *Metall. Mater. Trans. B* **2016**, *47*, 1527–1531. [CrossRef]
12. Naderi, M.; Saeed-Akbari, A.; Bleck, W. The effects of non-isothermal deformation on martensitic transformation in 22MnB5 steel. *Mater. Sci. Eng. A* **2008**, *487*, 445–455. [CrossRef]

13. Chang, Y.; Tang, X.; Zhao, K.; Hu, P.; Wu, Y. Investigation of the factors influencing the interfacial heat transfer coefficient in hot stamping. *J. Mater. Process. Technol.* **2016**, *228*, 25–33. [CrossRef]
14. Ancona, F.; Palumbo, D.; De Finis, R.; Demelio, G.P.; Galietti, U. Automatic procedure for evaluating the Paris Law of martensitic and austenitic stainless steels by means of thermal methods. *Eng. Fract. Mech.* **2016**, *163*, 206–219. [CrossRef]
15. Wang, L.; Zhu, B.; Wang, Y.; An, X.; Wang, Q.; Zhang, Y. An online dwell time optimization method based on parts performance for hot stamping. *Procedia Eng.* **2017**, *207*, 759–764. [CrossRef]
16. Muvunzi, R.; Dimitrov, D.M.; Matope, S.; Harms, T.M. Development of a model for predicting cycle time in hot stamping. *Procedia Manuf.* **2018**, *21*, 84–91. [CrossRef]
17. Gu, Z.; Lu, M.; Lu, G.; Li, X.; Xu, H. Effect of contact pressure during quenching on microstructure and mechanical properties of hot-stamping parts. *J. Iron. Steel Res. Int.* **2015**, *22*, 1138–1143. [CrossRef]
18. Rider, C. Pressure Distribution Image Analysis Process. U.S. Patent 2002/0129658 A1, 19 September 2002.

Article

Microstructure Evolution in Super Duplex Stainless Steels Containing σ-Phase Investigated at Low-Temperature Using In Situ SEM/EBSD Tensile Testing

Christian Oen Paulsen [1,2,*,†], Runar Larsen Broks [1,†], Morten Karlsen [1,3,†], Jarle Hjelen [1,4,†] and Ida Westermann [1,2,†]

[1]   Department of Materials Science and Engineering, Norwegian University of Science and Technology (NTNU), NO-7491 Trondheim, Norway; runarbro@stud.ntnu.no (R.L.B.); mortenka@statoil.com (M.K.); jarle.hjelen@ntnu.no (J.H.); ida.westermann@ntnu.no (I.W.)
[2]   Centre for Advanced Structural Analysis (CASA), Norwegian University of Science and Technology (NTNU), NO-7491 Trondheim, Norway
[3]   Equinor ASA, NO-7053 Trondheim, Norway
[4]   Department of Geoscience and Petroleum, Norwegian University of Science and Technology (NTNU), NO-7491 Trondheim, Norway
*   Correspondence: christian.o.paulsen@ntnu.no; Tel.: +47-73-59-49-21
†   These authors contributed equally to this work.

Received: 25 May 2018; Accepted: 19 June 2018; Published: 22 June 2018

Abstract: An in situ scanning electron microscope (SEM) study was conducted on a super duplex stainless steel (SDSS) containing 0%, 5% and 10% σ-phase. The material was heat treated at 850 °C for 12 min and 15 min, respectively, to achieve the different amounts of σ-phase. The specimens were investigated at room temperature and at −40 °C. The microstructure evolution during the deformation process was recorded using electron backscatter diffraction (EBSD) at different strain levels. Both σ-phase and χ-phase were observed along the grain boundaries in the microstructure in all heat treated specimens. Cracks started to form after 3–4% strain and were always oriented perpendicular to the tensile direction. After the cracks formed, they were initially arrested by the matrix. At later stages of the deformation process, cracks in larger σ-phase constituents started to coalesce. When the tensile test was conducted at −40 °C, the ductility increased for the specimen without σ-phase, but with σ-phase present, the ductility was slightly reduced. With larger amounts of σ-phase present, however, an increase in tensile strength was also observed. With χ-phase present along the grain boundaries, a reduction of tensile strength was observed. This reduction seems to be related to χ-phase precipitating at the grain boundaries, creating imperfections, but not contributing towards the increase in strength. Compared to the effect of σ-phase, the low temperature is not as influential on the materials performance.

Keywords: in situ tensile testing; super duplex stainless steel; SDSS; low-temperature; σ-phase; SEM; EBSD; microstructure analysis

---

1. Introduction

Duplex stainless steels (DSS) consist of two phases: austenite and ferrite. The two phases, in combination with the alloying elements, result in a steel with superior mechanical properties and corrosion resistance compared to steels with similar cost. DSS was first developed by the oil and gas industry for use in the North Sea. Here, it is typically used in process pipe systems and fittings exposed to corrosive environments at elevated temperature (up to 150 °C in H_2S atmosphere) [1].

DSS typically contains 22% Cr, 5% Ni and 0.18% N, to achieve the desired phase composition and corrosion properties. If better corrosion properties are required, super duplex stainless steel (SDSS) can be used instead. This alloy contains a higher amount of Cr, Ni and N typically 25%, 7% and 0.3%, respectively. In order to achieve the desired phase composition, it is annealed at 1050 °C, and left there until a 50-50 phase balance between ferrite and austenite is obtained. During cooling after the heat treatment, precipitation of intermetallic phases (σ, χ, π and R) may occur. These intermetallics have been found to considerably reduce the mechanical properties and corrosion resistance of the material [2–13]. The most common of these phases is the σ-phase. Even small amounts (<0.5%) of σ-phase will significantly reduce the fracture toughness [2,14]. This reduction, combined with the short time it takes for the phase to form and the deteriorating effect on corrosion properties, is what makes σ-phase a dangerous and strongly unwanted intermetallic.

The χ precipitates on ferrite/ferrite grain boundaries and occurs before the σ-phase [13,15]. In addition, the χ-phase is a metastable phase, consumed during σ-phase precipitation [13]. σ-phase typically forms on austenite/ferrite boundaries, but can also form on ferrite/ferrite boundaries. The σ-phase forms in the temperature range 675 °C–900 °C. After 10 min at 850 °C, small amounts of σ-phase will start to precipitate [6,16,17]. σ-phase has a body centered tetragonal crystal structure with the lattice parameters a = 8.8 Å and c = 4.55 Å, while the χ-phase has a body centered cubic crystal structure with lattice parameter a = 8.8 Å [18]. The lattice parameters of both are significantly larger than the 2.87 Å and 3.65 Å for ferrites and austenite, respectively [19,20]. The chemical composition of σ-phase includes, in addition to Fe, approximately 30–60% Cr and 4–10% Mo. The χ-phase differs from σ-phase with a higher Mo content and a lower content of Cr [15]. As a result, since χ-phase has a higher atomic weight, it is possible to distinguish it from σ-phase in a scanning electron microscope (SEM) using Z-contrast. In such an image, the χ-phase will appear brighter. Since Cr and Mo are stabilizing elements for ferrite, the σ and χ-phase will form at the expense of ferrite. Following the eutectoid reaction $\alpha \rightarrow \sigma + \gamma$ or $\alpha \rightarrow \chi + \gamma$, an increase in the austenite phase will also occur [15,17]. The surrounding area will be depleted of Cr and Mo, which are important elements for corrosion protection and, as a consequence, leaving the material exposed. This is especially troublesome in SDSS since these are mostly selected to operate in areas requiring a corrosion resistance superior to DSS.

A study by Børvik et al. [2] looked into the low-temperature effect on σ-phase in DSS. It was found that the temperature had a minor effect on the tensile ductility, while increasing amounts of σ-phase in the structure considerably reduced the ductility. Another study by Kim et al. [21] investigated the low-temperature mechanical behavior of SDSS containing σ-phase. Here, the material was tested in a universal tensile test machine, equipped with a sub-zero chamber. After the specimens were tested, the microstructure was investigated, comparing the amount of σ-phase present with microcrack length. Microcracks were found to have propagated through the entire σ-phase, relating the crack length to the size of σ-phase inclusions. As in Børvik et al. [2], the influence of temperature was observed to be minor. In addition, in the tensile tests performed at −50 °C, no strain-induced martensite was produced.

In the present study, in situ SEM tensile tests have been conducted on an SDSS with 0%, 5% and 10% σ-phase present in the microstructure. The tensile tests were carried out at both room temperature and sub-zero temperature (−40 °C). The microstructure was monitored with secondary electron imaging and electron backscatter diffraction (EBSD). Images were acquired at different loading steps. From these results, it is possible to observe the microstructure evolution and study the effects of χ-phase and σ-phase on the microstructure during the deformation process.

2. Materials and Methods

2.1. Material and Heat Treatments

The investigated material in this study was a 2507 SDSS, with the chemical composition listed in Table 1. This pipe was manufactured by welding a rolled plate along the length of the pipe. The microstructure of the steel investigated here contained more ferrite than austenite, 56.3% and

43.7%, respectively. The grain size in the two phases is also different, with ferrite having larger grains compared with austenite grains. These results have been summarized in Table 2. In addition, the grains have different morphology in different directions. Figure 1 gives a phase map of the pipe in three different directions. Here, LD, RD, and TD are abbreviations for longitudinal direction, radial direction, and transverse directions, respectively. The meaning of these are illustrated in Figure 2a.

Table 1. Chemical composition of 2507 SDSS.

Element	C	Si	Mn	P	S	Cr	Mo	Ni	Cu	W	N
wt%	0.018	0.42	0.52	0.017	0.001	25.55	3.46	8.28	0.72	0.52	0.25

Table 2. Microstructure statistics summarized. The data was collected from EBSD scans.

Average	Ferrite	Austenite	Overall
Composition	56.3%	43.7%	100%
Grain size	9 µm	6.5 µm	7.9 µm

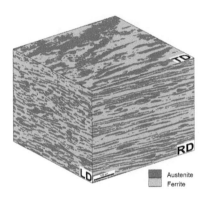

Figure 1. An illustration of the microstructure with the phases illustrated. The dimensions of the cube are 500 µm × 500 µm × 500 µm, green representing ferrite and red representing austenite. In the bottom right corner of each side, the plane normal is given. LD, RD, and TD are illustrated in Figure 2a.

Specimens being used for EBSD analysis need a completely smooth surface, where the deformation layer at the surface has to be removed. For SDSS this was done by grinding and polishing down to 1 µm, followed by electropolishing. The settings used are summarized in Table 3. Specimens were spark eroded from a 10 mm thick pipe to the dimensions in Figure 2b. All specimens were parallel to the length of the pipe, as illustrated in Figure 2a. The observed plane in the specimen during an in situ experiment has TD as plane normal.

Table 3. Parameters used during the electropolishing.

Electrolyte	Struers A2
Voltage [V]	20
Time [s]	15
Temperature [°C]	22

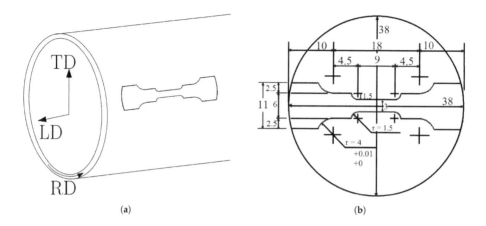

(a)   (b)

Figure 2. (a) an illustration on how the specimens were taken from the SDSS pipe and gives the definition of LD, RD and TD. The specimen dimensions is magnified compared to the pipe for illustration purpose; (b) specimen geometry, with all measurements in mm. The specimen had an original thickness of 2 mm before grinding and polishing.

The material was heat treated to achieve different amounts of σ-phase in the structure. Specimens were placed in a pre-heated oven at 850 °C for 12 min, 15 min, 20 min and 25 min. Cooling was performed by quenching in a water bath at room temperature. The heat treatment and the resulting amount of σ-phase achieved are summarized in Table 4. Phase maps from EBSD scans were used to quantify amounts of σ-phase present. These will not be exact measurements since they were only taken from the surface. The results from Elstad [16] was used to determine the heat treatment procedures used in this work. However, σ-phase precipitation was not constant with the same heat treatment being performed. Resulting in significant variation in σ-phase content during the heat treatment. All specimens in this work are heat treated as described in Table 4, but only specimens with amounts roughly in the region indicated in the third column were used for the in situ tests. However, the deviation was less than 1% for the 5% specimen, measured by EBSD. For the specimens with larger amounts of σ-phase present, the deviation was 1–2%.

Table 4. Heat treatment and resulting amount of σ-phase in the specimens tested.

Temperature [°C]	Time [min]	Amount σ-Phase [%]
-	-	0
850	12	5
850	15	10
850	20	15
850	25	20

2.2. Materials Characterization

During this experiment, the microstructure was monitored using secondary electron imaging and EBSD. Images were acquired at different loading steps. At each step, the same area (350 μm × 350 μm) was recorded with EBSD, using a step size of 1 μm. From these results, it is possible to observe the microstructure evolution and study the effects of σ-phase and χ-phase on the microstructure during the deformation process. The microscope used was a Field Emission SEM Zeiss Ultra 55 Limited Edition (Jena, Germany) with a NORDIF UF-1100 EBSD detector (Trondheim, Norway), with the microscope settings given in Table 5. Z-contrast imaging mode was used in order to distinguish σ-phase from χ-phase during the experiments.

Table 5. SEM parameters used during EBSD acquisition.

Acceleration Voltage [kV]	20
Working distance [mm]	24.6–25.4
Tilt angle [°]	70
Aperture size [μm]	300
Probe current [nA]	65–70

2.3. Tensile Testing

The specimen was deformed using a spindle-driven in situ tensile device. This device was placed inside the vacuum chamber of the SEM to monitor the microstructure. In situ tensile tests were carried out at both room temperature and at −40 °C for specimens containing 0%, 5% and 10% σ-phase. Tensile tests were also performed on specimens containing 15% and 20% of σ-phase, however, no in situ investigation or low-temperature testing was carried out on these specimens, due to their purely brittle behavior. The in situ tensile tests were carried out with a constant ramp speed of 1 μm/s. This corresponds to a strain rate of 1.11×10^{-4} s^{-1}. For further reading and previous use of the in situ device, the reader is referred to [22,23]. When performing the sub-zero experiments, a cold finger was attached to the specimen as shown in Figure 3. This cold finger is made from 99.99% Cu. It goes from the specimen, through the microscope door, into a dewar filled with liquid N. The blue and white wire seen in Figure 3 is a thermocouple. It was placed between the screw-head and specimen throughout the experiment. The temperature was measured to be in the interval −35 °C and −45 °C for all specimens. However, the fluctuations in temperature are assumed to be due to the variable thermal resistance between the thermocouple and specimen. This variation is a result of the thermocouple shifting position during straining. The temperature is assumed constant and reported as −40 °C in this paper.

Figure 3. Cold finger attached to the specimen with a thermocouple placed between the screw-head and specimen.

3. Results

3.1. Tensile Properties and Fracture Surfaces

The tensile test curves for the specimens tested in this work are shown in Figure 4. As seen from these curves, the specimens with more than 10% σ-phase present exhibit a purely brittle behavior at room temperature and do not deform plastically before fracture. For that reason, these specimens are not suitable for in situ and low-temperature investigations. Hence, only specimens containing roughly 5% and 10% σ-phase are further investigated. The specimens containing 0% σ-phase are included as a reference.

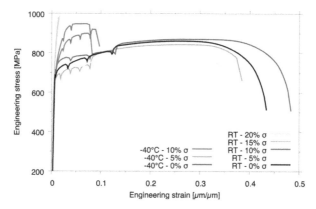

Figure 4. Tensile test curves obtained during the in situ tensile tests. The drops in the curves are when the test is paused for acquiring EBSD data.

The stress–strain curves in Figure 4 show that small amounts of σ-phase greatly affect the tensile properties of the material. Another observation is the short time at the critical temperature it takes before the material is completely brittle (cf. Table 4). Specimens containing 15% and 20% σ-phase only deforms elastically before fracture. A general remark is that the yield strength increases at low temperature and the strain at fracture decrease with an increasing amount of intermetallic phases. Conversely, for the material not heat treated, there is an increase of fracture strain. In addition, the drops in the curves are from when the tensile test is paused for EBSD acquisition. A curious observation from the tensile test curve is that the tests containing 5% σ-phase have a lower yield strength and ultimate tensile strength (UTS). In Figure 5, the microstructure of one of these specimens can be seen. Along the grain boundaries, the χ-phase has precipitated as a thin continuous layer of approximately 200 nm thickness. This image was acquired during the test at room temperature, after 4% strain. In the center of the image is a σ-phase island, with two cracks marked with white circles. The χ-phase also contains numerous small cracks, seen in the black circles in Figure 5, which seem to contribute towards a reduction in strength. When the amount of σ-phase increases, it also adds towards increased tensile strength.

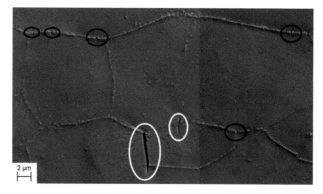

Figure 5. A close up micrograph from the specimen containing 5% σ-phase, after 4% strain, tested at room temperature. Along the grain boundaries, the χ-phase can be found, and, in the center, a larger island of σ-Phase is seen. The white circles show cracks in the σ-phase and the black circles show the cracks in the χ-phase.

In Figure 6, the fracture surfaces of the specimens tested at room temperature with 0%, 5% and 10% σ-phase are presented. To the left is an overview of the total surface area and to the right is a close-up image showing the fracture surface at a higher magnification. The reference sample exhibits classic ductile fracture features, with a large reduction of area and the typical cup and cone dimpled structure at the surface. This is also expected when compared to the tensile test curve (Figure 4). In the specimen with 5% σ-phase present, Figure 6b, some reduction in area is observed—however, not as great as in the test with 0% σ-phase present. In addition, here the fracture surface appears to be mixed between a ductile dimpled structure and a brittle faceted structure. Conversely, the specimen containing 10% σ-phase, Figure 6c, has all the characteristics of a brittle fracture. There is little to no reduction in area and completely faceted fracture surface, despite having a 10% fracture strain.

Figure 6. Fracture surfaces for the tensile test specimens with (**a**) 0%, (**b**) 5% and (**c**) 10% σ-phase, tested at room temperature. To the left is the total fracture area and to the right is a close-up of the fracture surface.

3.2. Microstructure Evolution

During the tensile tests, specimens containing different amounts of σ-phase were recorded using secondary electron imaging and EBSD to observe the microstructure throughout the deformation process. EBSD scans were obtained at the same area at approximately 0%, 2% and 6% strain of all tested specimens. Each of the tensile test curves in Figure 4 showed a drop when the test was paused for the acquisition of EBSD scans and secondary electron imaging. An observation is that the specimen with 0% σ-phase, tested at −40 °C, has a greater fracture strain than the specimen tested at room temperature.

In all specimens containing σ-phase, cracks were observed throughout the microstructure. These were observed to form after 3–4% strain in all the specimens, initiating in the σ-phase. Typical size of the cracks is seen in Figure 7. In Figure 8, two micrographs acquired at 6% and 10% strain show several micro-sized cracks in the σ-phase. During further straining, these cracks widen and appears to propagate deeper into the specimen. The ferrite and austenite grain boundaries act as a barrier for the cracks to propagate further. However, the larger constituents of σ-phase in the matrix contain large cracks, which eventually will propagate through the matrix. This is seen in the center of both frames in Figure 8. The microcracks in Figure 8a grow, and, in Figure 8b (black circle), they have coalesced, forming one large crack. A close-up of this crack is shown in Figure 9. This is a phase map superimposed on to an image quality (IQ) map from EBSD, acquired with a step size of 50 nm. From this map, it can be seen that the crack propagates along grain boundaries when it is moving through the matrix. When the cracks start to coalesce, the material is close to fracturing, as the volume fraction of cracks is increasing fast. The micrograph in Figure 8b was acquired after 10% strain. The specimen fractured after being strained less than 1% further. It is possible to see how the cracks in the white circles widen from Figure 8a to Figure 8b. Presumably, they are propagating through the thickness of the material.

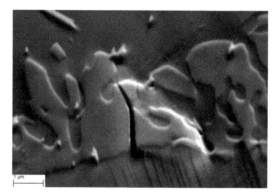

Figure 7. Micro-crack formed in the σ-phase during the initial stages of deformation. This frame is acquired after 6% strain, during the low temperature test with 5% σ-phase.

In Figure 10, the grain orientation spread (GOS) in the different tests are shown. All curves are obtained after 6% global strain. The GOS gives a quantitative description of the crystallographic orientation gradients in individual grains [24,25]. It is found by calculating the average orientation deviation of all points in a grain from the average grain orientation. A higher spread would indicate that those grains are accommodating a larger deformation compared to a lower spread. However, as seen from the graphs, there is, in general, a low spread, with peaks for all tests around 1–2°. One notable deviation is the curves from the experiment at −40 °C with 0% σ-phase present in Figure 10b. These grains seem to accommodate more deformation, with a larger GOS distribution compared to other curves in Figure 10. Another observation is that the ferrite and austenite phases have nearly identical curves in the low-temperature test in Figure 10b, while the phases are behaving differently at room temperature (Figure 10a). During the room temperature tests, all curves for ferrite grains have a taller peak compared to austenite grains. In addition, the specimen with 5% σ has a higher GOS peak-value compared to the specimen containing 10% σ-phase when tested at −40 °C.

At room temperature, the austenite for both tests is fairly similar, while the ferrite is accommodating more deformation in the specimen with 5% σ-phase.

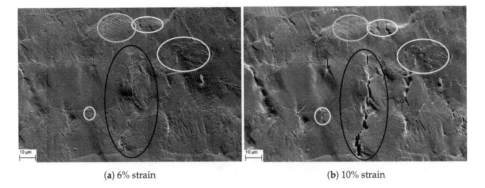

(a) 6% strain   (b) 10% strain

Figure 8. Micrographs of cracks formed in the σ-phase, taken from the test carried out at room temperature with 10% σ-phase present. Some cracks are restricted by the matrix while some propagate and coalesce. In the green circle, a heavily deformed austenite grain with cross-slip is seen. The white circles show microcracks restrained by the matrix and the large crack in the black circle was formed when many smaller cracks coalesce. A close-up of this crack is shown in Figure 9.

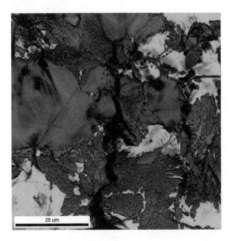

Figure 9. A close-up of the crack shown in the black circle in Figure 8b. This is a phase map with an IQ map overlay, acquired by EBSD. The red is austenite, green is ferrite and mustard yellow is σ-phase. The EBSD-scan of this area was acquired with a step size of 50 nm.

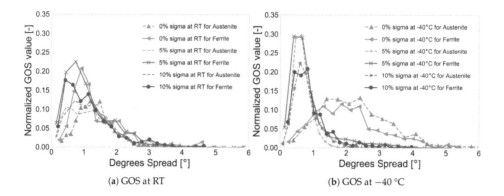

Figure 10. The grain orientation spread curves for the different specimens. (**a**) are the specimens tested at room temperature and (**b**) are the specimens tested at −40 °C. All curves were taken after 6% strain. The solid lines are ferrite and the dashed lines are austenite.

4. Discussion

During this work, different specimens of super duplex stainless steel, containing varying amounts of σ-phase have been investigated, during an in situ SEM tensile test. Each specimen was taken from a pipe segment and heat treated to get different amounts of σ-phase present. In general, it took roughly 10 min for the intermetallic phases to start forming at 850 °C. During the next 10 min, approximately 15% of σ-phase had precipitated, and the material had changed to an utterly brittle behavior, as seen in Figure 4. It proved hard to meet our targets of 5% and 10% σ-phase, sometimes achieving 0% after 13 min and other times 15% after 15 min at 850 °C. However, when the σ-phase starts to precipitate, it forms fast. Since duplex steels are being heat treated, typically at 1050 °C, to achieve its final microstructure and often goes through other heat treatment, e.g., welding, a thorough control of the cooling rate is crucial. In addition, no ductile-to-brittle transition was observed in this work. This was also the case in the work of Børvik et al. [2] and Kim et al. [21]. In these works, DSS and SDSS, respectively, were tested at −50 °C and no transition was observed. This means that, if the material has a ductile-to-brittle transition temperature, it is lower than −50 °C.

As seen from Figure 4, additions of σ-phase significantly reduce the ductility. This phenomenon is also well documented by others in previous studies [2,3,5,7,9]. However, in this study, the microstructure has been closely monitored during the tensile test to elucidate how it is accommodating the σ-phase in relation to deformation. The GOS in grains from the austenite and ferrite (shown in Figure 10) suggests that the presence of σ-phase and low temperature (−40 °C) is influencing the deformation behavior of the matrix. A consequence of presence of σ-phase is a lower fraction of ferrite. This altered phase balance, in combination with much harder particles containing numerous cracks explains this difference in behavior between specimens with and without σ-phase present. However, the primary concern is the brittle nature of σ-phase. Cracks were observed in the σ-phase at 3–4% strain in all specimens, and all cracks were oriented perpendicular to the tensile direction. During the initial stages, the surrounding matrix restricts the growth of the crack. As the material is strained further, the cracks continues to widen. Eventually, the cracks start to propagate and coalesce. In specimens with higher amounts of σ-phase, the propagation occurs earlier, following the shorter distance to the nearest σ-phase inclusion. In addition, the σ-phase particles are larger and the cracks, therefore, grow to a larger size.

The influence of temperature seems to make the σ-phase somewhat more brittle, resulting in a higher UTS and lower ductility. Austenite and ferrite grains seem to behave similarly during the low-temperature tests with σ-phase present when studying Figure 10b. However, during the test at room temperature, the ferrite accommodates more deformation compared to the austenite. This is seen from the curves in Figure 10a. The reason for the ferrite being more active is believed to be due to the fact that ferrite has 48 active slip systems at room temperature. Conversely, austenite has 12 slip systems and they are not dependent on temperature. With more slip systems available, there are more ways for the dislocations to propagate. In addition, the specimens without any σ-phase present have a larger GOS compared to the specimens containing σ-phase. This indicates that the presence of σ-phase in the structure is retarding the deformation of ferrite and austenite. This is also observed through visual inspection of micrographs. There are more slip lines present, at equal strain level, in specimens without σ-phase present.

An observation of a specimen with 0% σ-phase, tested at −40 °C, has a greater fracture strain than the specimen tested at room temperature. It could be expected that the ferrite would have a brittle behavior at this temperature. A reason for this behavior might be due to the fact that SDSS is a highly alloyed material, containing elements improving the low-temperature performance of ferrite. In addition, the presence of austenite will improve low-temperature performance. It has been reported in several studies that austenitic steels have increased ductility at −50 °C in static uniaxial tensile tests [26–28].

Looking at the tensile test curve in Figure 4 for the tests with 5% σ-phase, a lower tensile strength compared to the curve without any σ-phase present is observed. Conversely, a greater amount of σ-phase gives a contribution towards increased strength. An explanation for this can be the relative amount of χ-phase present. As seen from the black circles in Figure 5, the χ-phase precipitates along grain boundaries and is very brittle containing many cracks. These cracks result in the observed reduction of tensile strength. However, the size of the cracks in χ-phase are subcritical and does not contribute towards a large reduction in ductility. The specimen containing 5% σ-phase is still a very ductile material, with a fracture strain of 35%–38%. This is in contrast to previously reported literature. As mentioned in the Introduction, it has been reported that specimens with only 0.5% σ-phase have significantly reduced fracture toughness. However, as discussed in Børvik et al. [2] and Børvik et al. [3], DSS are more sensitive towards σ-phase with respect to fracture toughness than to tensile ductility. In this work, all specimens were tested strain rate of 1.11×10^{-4} s^{-1}. In addition, the tensile tests were paused at certain intervals to acquire images and EBSD scans. In Børvik et al. [2], an increase in flow stress of about 30% was found for DSS when the strain rate was increased from 5×10^{-4} s^{-1} to 50 s^{-1} based on tensile tests.

No strain-induced martensite was observed in any of the specimens investigated in this work. This indicates a very stable austenitic phase. However, this is not unexpected, since the σ-phase is formed at the expense of ferrite, not austenite. The alloying elements added to stabilize the austenitic phase are still present in the matrix. In the work by Kim et al. [21], there was also no martensite observed.

5. Conclusions

- The cracks in χ-phase contribute towards a lower flow stress but were not of critical size concerning a large reduction in tensile ductility. The specimens with small amounts of χ-phase and σ-phase still retained a ductility of 35%.
- Visible cracks start to form after 3–4% strain, regardless of σ-phase content and they all form perpendicular to the tensile direction.
- During the initial stages of deformation, the cracks are constrained by the ferrite/austenite matrix. However, during the later stages, these cracks start to propagate through the material and coalesce. This occurs moments before fracture.
- The ferrite accommodates more deformation than austenite at room temperature tests; however, during low-temperature tests, both phases have a more equal behavior during deformation.
- At low temperature, with σ-present, the material had slightly higher flow stress and lower ductility. However, the amount of σ-phase present is the most important aspect when it

comes to duplex steels. It alters the phase balance of ferrite and austenite and deteriorates the mechanical properties.

Author Contributions: C.O.P. is the first author and analyzed the data and wrote the paper. The experiments were performed by C.O.P. and R.L.B. I.W., M.K., and J.H. conceived, designed and supervised the experiments. In addition, they contributed to the interpretation of data and editing the paper.

Funding: This research received no external funding

Conflicts of Interest: The authors declare no conflict of interest.

Abbreviations

The following abbreviations are used in this manuscript:

CASA	Centre for Advanced Structural Analysis
DSS	Duplex Stainless Steel
EBSD	Electron Backscatter Diffraction
GOS	Grain Orientation Spread
IQ	Image Quality
LD	Longitudinal Direction
NTNU	Norwegian University of Science and Technology
RD	Radial Direction
RT	Room Temperature
SDSS	Super Duplex Stainless Steel
SEM	Scanning Electron Microscope
TD	Transverse Direction
UTS	Ultimate Tensile Strength

References

1. NORSOK Standard. Materials Selection. 2004. Available online: http://www.standard.no/pagefiles/1174/m-dp-001r1.pdf (accessed on 11 June 2018).
2. Børvik, T.; Lange, H.; Marken, L.A.; Langseth, M.; Hopperstad, O.S.; Aursand, M.; Rørvik, G. Pipe fittings in duplex stainless steel with deviation in quality caused by sigma phase precipitation. *Mater. Sci. Eng. A* **2010**, *527*, 6945–6955, doi:10.1016/j.msea.2010.06.087. [CrossRef]
3. Børvik, T.; Marken, L.A.; Langseth, M.; Rørvik, G.; Hopperstad, O.S. Influence of sigma-phase precipitation on the impact behaviour of duplex stainless steel pipe fittings. *Ships Offshore Struct.* **2016**, *11*, 25–37, doi:10.1080/17445302.2014.954303. [CrossRef]
4. Lee, Y.H.; Kim, K.T.; Lee, Y.D.; Kim, K.Y. Effects of W substitution on ς and χ phase precipitation and toughness in duplex stainless steels. *Mater. Sci. Technol.* **1998**, *14*, 757–764, doi:10.1179/mst.1998.14.8.757. [CrossRef]
5. Kim, S.B.; Paik, K.W.; Kim, Y.G. Effect of Mo substitution by W on high temperature embrittlement characteristics in duplex stainless steels. *Mater. Sci. Eng. A* **1998**, *247*, 67–74, doi:10.1016/S0921-5093(98)00473-0. [CrossRef]
6. Lopez, N.; Cid, M.; Puiggali, M. Influence of σ-phase on mechanical properties and corrosion resistance of duplex stainless steels. *Corros. Sci.* **1999**, *41*, 1615–1631, doi:10.1016/S0010-938X(99)00009-8. [CrossRef]
7. Chen, T.H.; Yang, J.R. Effects of solution treatment and continuous cooling on σ-phase precipitation in a 2205 duplex stainless steel. *Mater. Sci. Eng. A* **2001**, *311*, 28–41, doi:10.1016/S0921-5093(01)00911-X. [CrossRef]
8. Chen, T.H.; Weng, K.L.; Yang, J.R. The effect of high-temperature exposure on the microstructural stability and toughness property in a 2205 duplex stainless steel. *Mater. Sci. Eng. A* **2002**, *338*, 259–270, doi:10.1016/S0921-5093(02)00093-X. [CrossRef]
9. Zucato, I.; Moreira, M.C.; Machado, I.F.; Lebrão, S.M.G. Microstructural characterization and the effect of phase transformations on toughness of the UNS S31803 duplex stainless steel aged treated at 850 °C. *Mater. Res.* **2002**, *5*, 385–389, doi:10.1590/S1516-14392002000300026. [CrossRef]

10. Cvijović, Z.; Radenković, G. Microstructure and pitting corrosion resistance of annealed duplex stainless steel. *Corros. Sci.* **2006**, *48*, 3887–3906, doi:10.1016/j.corsci.2006.04.003. [CrossRef]

11. Michalska, J.; Sozańska, M. Qualitative and quantitative analysis of σ and χ phases in 2205 duplex stainless steel. *Mater. Charact.* **2006**, *56*, 355–362, doi:10.1016/j.matchar.2005.11.003. [CrossRef]

12. Souza, C.M.; Abreu, H.F.G.; Tavares, S.S.M.; Rebello, J.M.A. The σ phase formation in annealed UNS S31803 duplex stainless steel: Texture aspects. *Mater. Charact.* **2008**, *59*, 1301–1306, doi:10.1016/j.matchar.2007.11.005. [CrossRef]

13. Pohl, M.; Storz, O.; Glogowski, T. Effect of intermetallic precipitations on the properties of duplex stainless steel. *Mater. Charact.* **2007**, *58*, 65–71, doi:10.1016/j.matchar.2006.03.015. [CrossRef]

14. Calliari, I.; Zanesco, M.; Ramous, E. Influence of isothermal aging on secondary phases precipitation and toughness of a duplex stainless steel SAF 2205. *J. Mater. Sci.* **2006**, *41*, 7643–7649, doi:10.1007/s10853-006-0857-2. [CrossRef]

15. Escriba, D.; Materna-Morris, E.; Plaut, R.; Padilha, A. Chi-phase precipitation in a duplex stainless steel. *Mater. Charact.* **2009**, *60*, 1214–1219, doi:10.1016/J.MATCHAR.2009.04.013. [CrossRef]

16. Elstad, K.R. In Situ Tensile Testing During Continuous EBSD Mapping of Super Duplex Stainless Steel Containing Sigma Phase. 2016. Available online: http://hdl.handle.net/11250/2418016 (accessed on 25 May 2018).

17. Stradomski, Z.; Dyja, D. Sigma Phase Precipitation in Duplex Phase Stainless Steels. 2009. Available online: http://www.ysesm.ing.unibo.it/Abstract/57Dyja.pdf (accessed on 25 May 2018).

18. Padilha, A.F.; Rios, P.R. Decomposition of Austenite in Austenitic Stainless Steels. *ISIJ Int.* **2002**, *42*, 325–327, doi:10.2355/isijinternational.42.325. [CrossRef]

19. Cahn, R.W.; Haasen, P.; Kramer, E.J. *Materials Science and Technology: A Comprehensive Treatment—Volume 1: Structure of Solids*; Wiley-VCH: Weinheim, Germnay, 2005.

20. Cahn, R.W.; Haasen, P.; Kramer, E.J. *Materials Science and Technology: A Comprehensive Treatment—Volume 7: Constitution and Properties of Steel*; Wiley-VCH: Weinheim, Germnay, 2005.

21. Kim, S.K.; Kang, K.Y.; Kim, M.S.; Lee, J.M. Low-temperature mechanical behavior of super duplex stainless steel with sigma precipitation. *Metals* **2015**, *5*, 1732–1745, doi:10.3390/met5031732. [CrossRef]

22. Karlsen, M.; Hjelen, J.; Grong, Ø.; Rørvik, G.; Chiron, R.; Schubert, U.; Nilsen, E. SEM/EBSD based in situ studies of deformation induced phase transformations in supermartensitic stainless steels. *Mater. Sci. Technol.* **2008**, *24*, 64–72, doi:10.1179/174328407X245797. [CrossRef]

23. Karlsen, M.; Grong, Ø.; Søfferud, M.; Hjelen, J.; Rørvik, G.; Chiron, R. Scanning Electron Microscopy/Electron Backscatter Diffraction—Based Observations of Martensite Variant Selection and Slip Plane Activity in Supermartensitic Stainless Steels during Plastic Deformation at Elevated, Ambient, and Subzero Temperatures. *Metall. Mater. Trans. A* **2009**, *40*, 310–320, doi:10.1007/s11661-008-9729-5. [CrossRef]

24. Jorge-Badiola, D.; Iza-Mendia, A.; Gutiérrez, I. Study by EBSD of the development of the substructure in a hot deformed 304 stainless steel. *Mater. Sci. Eng. A* **2005**, *394*, 445–454, doi:10.1016/j.msea.2004.11.049. [CrossRef]

25. Mitsche, S.; Poelt, P.; Sommitsch, C. Recrystallization behaviour of the nickel-based alloy 80 A during hot forming. *J. Microsc.* **2007**, *227*, 267–274, doi:10.1111/j.1365-2818.2007.01810.x. [CrossRef] [PubMed]

26. Byun, T.; Hashimoto, N.; Farrell, K. Temperature dependence of strain hardening and plastic instability behaviors in austenitic stainless steels. *Acta Mater.* **2004**, *52*, 3889–3899, doi:10.1016/J.ACTAMAT.2004.05.003. [CrossRef]

27. Lee, K.J.; Chun, M.S.; Kim, M.H.; Lee, J.M. A new constitutive model of austenitic stainless steel for cryogenic applications. *Comput. Mater. Sci.* **2009**, *46*, 1152–1162, doi:10.1016/J.COMMATSCI.2009.06.003. [CrossRef]

28. Park, W.S.; Yoo, S.W.; Kim, M.H.; Lee, J.M. Strain-rate effects on the mechanical behavior of the AISI 300 series of austenitic stainless steel under cryogenic environments. *Mater. Des.* **2010**, *31*, 3630–3640, doi:10.1016/j.matdes.2010.02.041. [CrossRef]

 *metals*

Article

Evaluation of Irradiation Hardening of P92 Steel under Ar Ion Irradiation

Qingshan Li [1], Yinzhong Shen [1,*], Jun Zhu [2], Xi Huang [1] and Zhongxia Shang [3]

[1] School of Mechanical Engineering, Shanghai Jiao Tong University, No. 800 Dongchuan Road, Shanghai 200240, China; liqingshan999@163.com (Q.L.); huangxi060231@126.com (X.H.)
[2] Jiangsu Nuclear Power Co., Ltd., CNNP, No. 28 Haitang Mid-Road, Lianyungang 222042, China; zhujun0345@126.com
[3] School of Materials Science and Engineering, Shanghai Jiao Tong University, 800 Dongchuan Road, Shanghai 200240, China; shangzx_568@126.com
* Correspondence: shenyz@sjtu.edu.cn; Tel.: +86-21-3420-7117

Received: 25 December 2017; Accepted: 25 January 2018; Published: 27 January 2018

Abstract: P92 steel was irradiated with Ar ion up to 10 dpa at 200, 400, and 700 °C. The effect of Ar ion irradiation on hardness was investigated with nanoindentation tests and microstructure analyses. It was observed that irradiation-induced hardening occurred in the steel after Ar ion irradiation at all three temperatures to 10 dpa. The steel exhibited significant hardening at 200 and 700 °C, and slight hardening at 400 °C under Ar ion irradiation. Difference in the magnitude of irradiation-induced hardening at different temperature in the steel is attributed to different changes in the microstructure of the steel that arose from the irradiation. Irradiation-induced hardening in the P92 steel irradiated at 200 °C is attributed to the occurrence of both dislocation loops and other fine irradiation defects during irradiation. Slight hardening in the steel irradiated at 400 °C mainly arises from the annihilation of defect clusters at this temperature. The occurrence of fine Ar bubbles with high number density during the Ar ion irradiation at 700 °C resulted in the significant hardening in the steel.

Keywords: ferritic steel; irradiation; nanoindentation; hardness; transmission electron microscopy (TEM); microstructure

1. Introduction

High chromium ferritic/martensitic (F/M) steels have been considered as candidate materials for the structural components in Generation IV nuclear reactors—such as gas-cooled fast reactor (GFR), lead-cooled reactor (LFR), sodium-cooled fast reactor (SFR), and super critical water-cooled reactor (SCWR)—due to their good swelling resistance, excellent thermal properties, and low thermal expansion [1–3]. However, these components will be subjected to extreme conditions such as higher temperatures, higher neutron doses, and extremely corrosive environment, which are beyond the experience of the current reactors [3]. Irradiation damage can lead to a series of changes in microstructure, probably giving rise to a hardening of structural materials. Irradiation-induced hardening results in the degradation of fracture properties, such as ductile-to-brittle transition temperature (DBTT) shift, at low irradiation temperature [4–7]. Therefore, it is necessary to investigate the influence of irradiation on the microstructure and mechanical properties of high chromium F/M steels.

Ion irradiation technique has been recognized as a means to simulate neutron damage due to many advantages in investigating the irradiation effects on the mechanical properties of reactor structural materials, such as high damage rate, little or no residual radioactivity, and a lower cost. Additionally, irradiation conditions can be easily controlled [7,8]. A large number of studies have reported the mechanical properties of F/M steels after irradiation with neutrons [6], Ne ions [9], Kr ions [10,11], H ions, He ions, and electron beams [12,13]. Inert gas, especially He, created by nuclear reaction has a significant influence on the microstructural evolution of structural materials exposed to neutrons in nuclear reactors. Although simulations recently indicated bubble formation and loss of structural integrity induced by Ar ion and Xe ion were much less compared to the case of He ion in bcc Fe [14], inert gas is still widely applied to ion irradiation experiments. Therefore, in the present work, the Ar ion was selected to create similar irradiation damage as He created in reactors. Irradiation damage depth is limited up to a few microns, so nanoindentation testing combined with ion irradiation is usually applied to investigate the influence of irradiation conditions on irradiation-induced hardening of structural materials for reactors [15–17].

Attributed to solution strengthening by W and Mo additions and precipitation strengthening by V and Nb additions on the basis of P91 steel, P92 steel exhibits excellent high temperature mechanical properties, which makes it a promising candidate material for in-core applications in sodium fast reactor (SFR) and fusion systems [11,18]. The irradiation response of P92 steel has received much attention in the past few years [10,11,19–22]. Most research focused on the microstructural evolution of P92 steel irradiated with Kr ions at 200 and 400 °C [10,11]; with Ar ions at room temperature, 290, 390, and 550 °C [19,20]; and with protons at 500 °C [21]. When F/M steels are used as fuel cladding in a fast reactor, the operating temperature is expected to approach 650–700 °C [18]. Although microstructural evolution in P92 steel during irradiation has been investigated systematically in their previous work, there is a lack of available data associated with the mechanical properties of irradiated P92 steel. In addition, there appears to be few previous reports of hardening behavior in F/M steels under ion irradiation at high temperatures (especially higher than 600 °C). In this paper, irradiation-induced change in the hardness of P92 steel was evaluated through Ar ion irradiation at 200, 400, and 700 °C to 10 dpa in combination with nanoindentation tests and microstructure analyses. The mechanism of irradiation-induced hardening in P92 steel under Ar ion irradiation at different irradiation temperatures to 10 dpa was also discussed.

2. Materials and Methods

The initial material used in this study was commercial P92 steel. The chemical composition of the P92 steel is showed in Table 1. The P92 steel was normalized at 1050 °C for 30 min then cooled by air, and then tempered at 765 °C for 60 min followed by air cooling. The normalized and tempered plate was cut into specimens of 20 mm × 6 mm × 1 mm with spark erosion, then ground and polished to a mirror finish. Irradiation experiment was carried out on an implantation facility (Institute of Modern Physics of the Chinese Academy of Sciences, Lanzhou, China) using 250 keV Ar^{2+} ions at 200, 400, and 700 °C under high vacuum less than 1×10^{-4} Pa. Polished samples were irradiated up to 10 dpa and the dose rate of Ar ion irradiation was about 1.56 dpa/h for P92 steel samples. During irradiation experimentation, two small pieces were taken to suppress the samples to prevent the samples from falling from the sample plate. Correspondingly, the irradiated and unirradiated regions were acquired due to the suppressed regions not being irradiated by Ar ions. The irradiation temperature of sample was controlled using thermocouples with a fluctuation within ±5 °C. The distribution of the damage level and the implanted Ar ions in the irradiated samples after irradiation were simulated through the Stopping and Range of Ions in Matter (SRIM) output vacancy file with quick Kinchin and Pease method [23], as shown in Figure 1. Depth profile of the irradiation damage has a peak at a depth of about 100 nm.

Table 1. Chemical composition of P92 steel (in wt %).

C	Si	Cr	Mn	W	Mo	Nb	Ni	V	N	B	Fe
0.093	0.14	8.75	0.41	1.62	0.505	0.052	0.207	0.183	0.063	0.003	Bal.

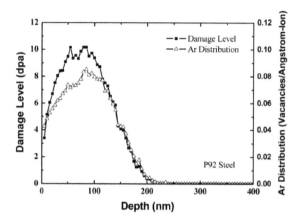

Figure 1. The depth profiles of displacement damage and Ar ions distribution in irradiated P92 steel simulated by the Stopping and Range of Ions in Matter (SRIM).

Nanoindentation hardness was measured by a MTS Nano Indenter equipped with a Berkovich type indentation tip (MTS Cooperation, Nano Instruments Innovation Center, TN, USA). The calibration of the bluntness of the indentation tip is based on the Oliver–Pharr method [24]. The hardness as a function of indenter depth was acquired by the continuous stiffness measurement (CSM) method. In order to obtain reliable results, more than six measurements were carried out for each irradiation condition.

Thin-foil TEM (transmission electron microscope, JEOL Ltd., Tokyo, Japan) specimens with a diameter of 3 mm were prepared from the unirradiated side of the irradiated steel samples by grinding and thinning process, and then followed by an ion milling in Gatan-691 precision ion polishing system. The thin-foil specimens were examined in a JEOL-2100 TEM at 200 kV.

3. Results

Figure 2 presents the average hardness of the P92 steel before and after Ar ion irradiation at 200, 400, and 700 °C to 10 dpa. For all sheet samples, the phenomenon that hardness decreases with indent depth increasing was observed at indentation depth larger than 100 nm, which was caused by the indentation size effect (ISE) [25]. Conversely, a reverse ISE that hardness increases with depth increasing close to surface. It is clearly noted that irradiation-induced hardening was revealed for all irradiated samples. The difference in magnitudes of irradiation-induced hardening at different temperatures were shown in Figure 3, $\Delta H = H_{irr} - H_{uni}$, where ΔH is defined as variable quantity of hardness after Ar ion irradiation, H_{irr} is the hardness after Ar ions irradiation and H_{uni} is the corresponding original hardness at the same depth. Commonly, the hardness data near the surface is not adapted to investigate the irradiation-induced hardening due to testing artifacts [26]. Besides, there are some challenges for the measurement of mechanical properties of ion implanted layers, including indentation size effect, pile-up effect, sink-in effect, and residual stresses. The pile-up and sink-in effect can be revised by scanning electron microscope (SEM) on the residual indents [27]. Many methods have been developed to estimate surface residual stress [28–30]. Compared with the hardness of the unirradiated region, the hardness of regions irradiated with Ar ion irradiation at 200 and 700 °C to 10 dpa increased to about 1 and 1.5 GPa, respectively. The hardness after Ar ions irradiation at 400 °C was slightly higher than the unirradiated region. The most severe irradiation-induced hardening was observed at 700 °C, while slight irradiation-induced hardening behavior was found at 400 °C.

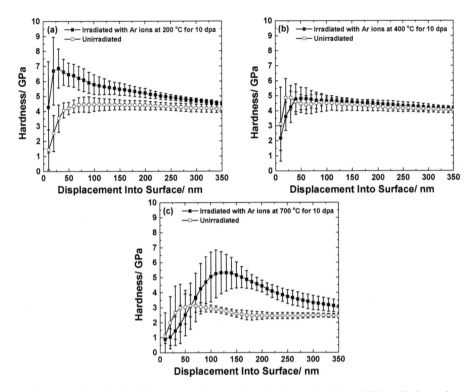

Figure 2. Indentation depth dependence of averaged nanoindentation hardness of P92 steel before and after Ar ion irradiation to 10 dpa at (**a**) 200 °C, (**b**) 400 °C, and (**c**) 700 °C.

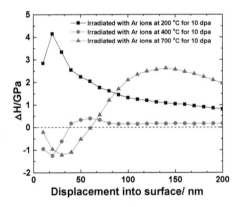

Figure 3. Indentation depth dependence of averaged nanoindentatiion hardness variation of P92 steel after Ar ion irradiation to 10 dpa at 200, 400, and 700 °C.

4. Discussion

In recent years, many studies on the irradiation hardening behavior of F/M steel have been done through tensile, Vickers hardness, or nanoindentation tests [26,31–34]. Significant hardening was observed in HCM12A and T91 steels after irradiation with 2.0 MeV protons at 400 °C, while both steels irradiated at 500 °C showed a slight increase in their hardness [31]. Kimura et al. [32]

compared the yield stress increment of 9Cr-2W F/M steel before and after irradiation in Fast Flux Test Facility (FFTF)/Materials Open Test Assembly (MOTA) and found the largest irradiation hardening was observed at 373 °C (646 K) at 10–15 dpa, while softening was observed in the steel specimens irradiated at temperatures above 430 °C (703 K) to doses of 40 and 59 dpa. Xin et al. [33] found that irradiation-induced hardening was confirmed in CLAM steel irradiated with 140 keV He ions with fluences up to 1×10^{16}/cm^2 at 27,200, and 400 °C. Their results indicated that CLAM steel exhibited significant hardening at 200 °C and hardening effect decreased at 400 °C, nevertheless, negligible hardening was observed when irradiation temperature was increased up to 600 °C. However, there are few reports of significant irradiation-induced hardening in high chromium F/M steels at a high irradiation temperature of 700 °C. Besides, slightly irradiation-induced hardening at 400 °C observed in this work is another concern.

Figure 4 presents the TEM micrographs of the unirradiated and irradiated P92 steel taken from thin-foil samples. As shown in Figure 4a, clear martensite lath structure and high density of dislocations and precipitates were observed in the unirradiated P92 steel sample. Figure 4b–e illustrates the microstructural evolution of the P92 steel irradiated at 200, 400, and 700 °C to 10 dpa, suggesting that the micrographs of the irradiated samples were quite different from those of the unirradiated samples. Dislocation loops were observed in all irradiated samples, as indicated with dashed circles in Figure 4b–d. The average diameter and the density of dislocation loops were investigated statistically in different regions of the irradiated samples. The diameter of dislocation loops in the steel samples irradiated to 10 dpa at 200, 400, and 700 °C were determined by measuring the diameters of 271, 226, and 187 dislocation loops respectively from more than 25 TEM images using $\mathbf{g} = \{200\}$ for each irradiated sample. The dislocation loops density was estimated by n/V, which was used to calculate bubbles density [35], where n is the amount of dislocation loops, and V is the volume of the selected regions which was estimated with a thin-foil thickness of about 50 nm. For the samples irradiated at 200, 400, and 700 °C, the average dislocation loops diameter was determined to be about 7, 13, and 17 nm, and the number density of dislocation loops was calculated to be about 2.43×10^{21}, 1.83×10^{21}, and 1.15×10^{21} m^{-3}, respectively. These results indicated that the size of dislocation loops increased, on the contrary, the density of dislocation loops decreased with the increase of irradiation temperature from 200 to 700 °C. In addition, fine bubbles were observed to homogeneously distribute in the sample irradiated at 700 °C, as showed in Figure 4e. These bubbles have an average diameter of about 3 nm and a very high number density of about 1.87×10^{25} m^{-3} [36].

Irradiation-induced hardening is a consequence of interactions of dislocations with irradiation-induced defects including point defects, vacancy and interstitial clusters, dislocation loops and lines, voids and bubbles, and precipitates [37]. The theories of irradiation hardening are based on the assumption that these irradiation defects can act as obstacles to the glide of dislocations and different hardening models have been proposed by Azevedo [38]. Combined with TEM analysis regarding the number density and size of dislocation loops in the CLAM steel irradiated with single-(He$^+$) and sequential-(He$^+$ plus H$^+$ subsequently), the magnitude of irradiation-induced hardening was well expounded in terms of a dispersed barrier hardening model [39]. The model can be written as

$$\Delta\sigma_y = M \cdot \alpha \cdot \mu \cdot b(Nd)^{0.5} \tag{1}$$

where $\Delta\sigma_y$ is the increment in yield strength, M is the Taylor factor, α is the strength of barrier, μ is the shear modulus, b is the magnitude of the Burger's vector for moving dislocations, N is the number density of obstacles, and d is the obstacle diameter [14]. Therefore, the number density and size of dislocation loops would determine the magnitude of irradiation hardening in the irradiated CLAM steel. In the case of P92 steel, the product of density and size of dislocation loops in the P92 steel irradiated at 400 °C was larger than that irradiated at 200 °C according to a dispersed barrier hardening model. Meanwhile, a smaller magnitude of irradiation-induced hardening was observed in the P92 steel irradiated at 400 °C. Therefore, point defects and other defect clusters should be taken into consideration to understand the magnitude of irradiation-induced hardening in P92 steel irradiated at 200 and 400 °C with the exception of dislocation loops. It was suggested that irradiation-induced defects with a small size contributing to irradiation hardening could not be observed due to the resolution limit of TEM [40]. It can be speculated that irradiation hardening in P92 steel irradiated at 200 °C is attributed to the production of both dislocation loops and other fine irradiation defects. The annihilation between vacancy and interstitial atoms occurs during irradiation independent of irradiation temperature, while the annihilation will become more obvious with increasing irradiation temperature [33]. Thus, it can be inferred that defect clusters decreased in the irradiated P92 steel due to the occurrence of annihilation of vacancy and interstitial atoms partly at 400 °C. The annihilation of irradiation defects during irradiation at 400 °C is the main factor leading to slight hardening in the irradiated P92 steel at this temperature.

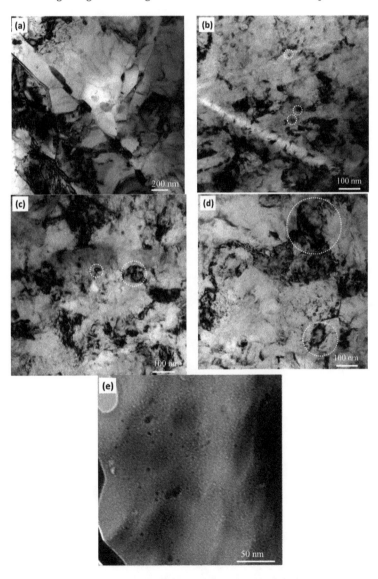

Figure 4. TEM micrographs of P92 steel (**a**) before and after Ar ion irradiation to 10 dpa at (**b**) 200 °C, (**c**) 400 °C, and (**d,e**) 700 °C.

With the increasing of irradiation temperature up to 700 °C, significant irradiation hardening was observed in the irradiated P92 steel. At high temperatures, irradiation-induced defect clusters will become unstable, meanwhile the absorption of point defects to dislocations makes the hardening decrease. However, fine bubbles were observed to homogeneously distribute in the sample irradiated at 700 °C, as shown in Figure 4e. It has been confirmed that high-density helium bubbles could act as obstacles to dislocation motion leading to hardening [41,42]. Therefore, the irradiation-induced hardening in the irradiated P92 steel at 700 °C can result from the formation of uniformly distributed Ar bubbles with a high number density and fine size. These uniformly distributed Ar bubbles can be considered as a stronger barrier to the motion of dislocation resulting in the hardening in the irradiated P92 steel. In addition, it can be inferred that the hardness increase caused by the formation of Ar bubbles is stronger than the hardness decrease resulting from annihilation of other irradiation defects. The formation of cavities at a higher irradiation temperatures was also observed in other studies. It has been reported that cavities with a high number density were observed in Fe-Cr-Mn alloy irradiated with Ar ions at 450 °C [43]. Recently, irradiation-induced cavity was found in P92 steel irradiated with Ar ions at 550 °C to 7 dpa, while no cavity was observed in the steel irradiated at temperatures below 550 °C up to a high dose of 34.5 dpa [19,20]. In the present study, Ar bubbles contributing to hardening were also merely observed in the steel irradiated at 700 °C, which might be due to a sufficient mobility for Ar atoms to nucleate and form Ar bubbles at the high irradiation temperature. Thus, the irradiation-induced hardening in the P92 steel occurred at 700 °C is mainly attributed to the production of these high-density and fine Ar bubbles.

5. Conclusions

Irradiation-induced hardening of P92 steel irradiated with Ar ions has been investigated with nanoindentation tests and microstructure analyses. Hardening behavior occurs in the steel after Ar ion irradiation at temperatures of 200, 400, and 700 °C to 10 dpa. P92 steel exhibits significant hardening at 200 and 700 °C, and slight hardening at 400 °C. Irradiation hardening in steel is sensitive to irradiation temperatures. Dominant irradiation-induced defects are different at different temperatures. Irradiation hardening observed in the P92 steel irradiated at 200 °C is considered to be attributed to the production of defects cluster and dislocation loops. Significant hardening in the P92 steel irradiated at 700 °C is due to the generation of high-density and fine Ar bubbles. The annihilation of irradiation defects during irradiation at 400 °C is a major factor leading to slight hardening in the irradiated steel.

Acknowledgments: This investigation was supported by the Key Program of National Natural Science Foundation of China (51034011), ITER-National Magnetic Confinement Fusion Program (2011GB113001), and Shanghai Pujiang Program. The authors thank Bo Ji and Aidang Shan from Shanghai Jiao Tong University for polishing steel samples and providing heat-treated steel plates, respectively.

Author Contributions: Qingshan Li analyzed the experimental data and completed this paper. Yinzhong Shen designed this experiment and revised the manuscript. Jun Zhu performed the nanoindentation experiments. Xi Huang and Zhongxia Shang helped to complete the microstructure analysis.

Conflicts of Interest: The authors declare no conflict of interest.

References

1. Kimura, A.; Cho, H.S.; Toda, N.; Kasada, R.; Yutani, K.; Kishimoto, H.; Iwata, N.; Ukai, S.; Fujiwara, M. High burn up fuel cladding materials R & D for advanced nuclear systems nano-sized oxide dispersion strengthening steels. *J. Nucl. Sci. Technol.* **2007**, *44*, 323–328. [CrossRef]
2. Boutard, J.L.; Alamo, A.; Lindau, R.; Rieth, M. Fissile core and Tritium-Breeding Blanket: Structural materials and their requirements. *C. R. Phys.* **2008**, *9*, 287–302. [CrossRef]
3. Murty, K.L.; Charit, I. Structural materials for Gen-IV nuclear reactors: Challenges and opportunities. *J. Nucl. Mater.* **2008**, *383*, 189–195. [CrossRef]
4. Calder, A.F.; Bacon, D.J.; Barashev, A.V.; Osetsky, Y.N. On the origin of large interstitial clusters in displacement cascades. *Philos. Mag.* **2010**, *90*, 863–884. [CrossRef]

5. Bloom, E.E.; Zinkle, S.J.; Wiffen, F.W. Materials to deliver the promise of fusion power—Progress and challenges. *J. Nucl. Mater.* **2004**, *329–333*, 12–19. [CrossRef]
6. Matijasevic, M.; Van Renterghem, W.; Almazouzi, A. Characterization of irradiated single crystals of Fe and Fe–15Cr. *Acta Mater.* **2009**, *57*, 1577–1585. [CrossRef]
7. Kohyama, A.; Katoh, Y.; Ando, M.; Jimbo, K. A new Multiple Beams—Material Interaction Research Facility for radiation damage studies in fusion materials. *Fusion Eng. Des.* **2000**, *51–52*, 789–795. [CrossRef]
8. Was, G.S. *Fundamentals of Radiation Materials Science*; Springer: New York, NY, USA, 2007; pp. 545–565, ISBN 978-3-540-49471-3.
9. Zhang, C.H.; Jang, J.; Kim, M.C.; Cho, H.D.; Yang, Y.; Sun, Y.M. Void swelling in a 9Cr ferritic/martensitic steel irradiated with energetic Ne-ions at elevated temperatures. *J. Nucl. Mater.* **2008**, *375*, 185–191. [CrossRef]
10. Topbasi, C.; Kaoumi, D.; Motta, A.T.; Kirk, M.A. Microstructural evolution in NF616 (P92) and Fe–9Cr–0.1C model alloy under heavy ion irradiation. *J. Nucl. Mater.* **2015**, *466*, 179–186. [CrossRef]
11. Topbasi, C.; Motta, A.T.; Kirk, M.A. In situ study of heavy ion induced radiation damage in NF616 (P92) alloy. *J. Nucl. Mater.* **2012**, *425*, 48–53. [CrossRef]
12. Zhao, F.; Qiao, J.; Huang, Y.; Wan, F.; Ohnuki, S. Effect of irradiation temperature on void swelling of China Low Activation Martensitic steel (CLAM). *Mater. Charact.* **2008**, *59*, 344–347. [CrossRef]
13. Peng, L.; Huang, Q.; Ohnuki, S.; Yu, C. Swelling of CLAM steel irradiated by electron/helium to 17.5 dpa with 10 appm He/dpa. *Fusion Eng. Des.* **2011**, *86*, 2624–2626. [CrossRef]
14. Gai, X.; Lazauskas, T.; Smith, R.; Kenny, S.D. Helium bubbles in bcc Fe and their interactions with irradiation. *J. Nucl. Mater.* **2015**, *464*, 382–390. [CrossRef]
15. Fischer-Cripps, A.C. *Nanoindentation*, 2nd ed.; Springer: New York, NY, USA, 2004; pp. 147–161, ISBN 978-1-4419-9871-2.
16. Takayama, Y.; Kasada, R.; Sakamoto, Y.; Yabuuchi, K.; Kimura, A.; Ando, M.; Hamaguchi, D.; Tanigawa, H. Nanoindentation hardness and its extrapolation to bulk-equivalent hardness of F82H steels after single- and dual-ion beam irradiation. *J. Nucl. Mater.* **2013**, *442*, S23–S27. [CrossRef]
17. Kasada, R.; Konishi, S.; Yabuuchi, K.; Nogami, S.; Ando, M.; Hamaguchi, D.; Tanigawa, H. Depth-dependent nanoindentation hardness of reduced-activation ferritic steels after MeV Fe-ion irradiation. *Fusion Eng. Des.* **2014**, *89*, 1637–1641. [CrossRef]
18. Klueh, R.L.; Nelson, A.T. Ferritic/martensitic steels for next-generation reactors. *J. Nucl. Mater.* **2007**, *371*, 37–52. [CrossRef]
19. Jin, S.X.; Guo, L.P.; Yang, Z.; Fu, D.J.; Liu, C.S.; Tang, R.; Liu, F.H.; Qiao, Y.X.; Zhang, H.D. Microstructural evolution of P92 ferritic/martensitic steel under argon ion irradiation. *Mater. Charact.* **2011**, *62*, 136–142. [CrossRef]
20. Jin, S.; Guo, L.; Li, T.; Chen, J.; Yang, Z.; Luo, F.; Tang, R.; Qiao, Y.; Liu, F. Microstructural evolution of P92 ferritic/martensitic steel under Ar+ ion irradiation at elevated temperature. *Mater. Charact.* **2012**, *68*, 63–70. [CrossRef]
21. Huang, Y.; Wharry, J.P.; Jiao, Z.; Parish, C.M.; Ukai, S.; Allen, T.R. Microstructural evolution in proton irradiated NF616 at 773 K to 3 dpa. *J. Nucl. Mater.* **2013**, *442*, S800–S804. [CrossRef]
22. Huang, X.; Shen, Y.; Zhu, J. Influence of Ar-ions irradiation on the oxidation behavior of ferritic—martensitic steel P92 in supercritical water. *J. Nucl. Mater.* **2015**, *457*, 18–28. [CrossRef]
23. Ziegler, J.F. SRIM-2008 Program. 2008. Available online: http://www.srim.org (accessed on 12 January 2017).
24. Oliver, W.C.; Pharr, G.M. An improved technique for determining hardness and elastic modulus using load and displacement sensing indentation experiments. *J. Mater. Res.* **1992**, *7*, 1564–1583. [CrossRef]
25. Nix, W.D.; Gao, H. Indentation size effects in crystalline materials: A law for strain gradient plasticity. *J. Mech. Phys. Solids* **1998**, *46*, 411–425. [CrossRef]
26. Kasada, R.; Takayama, Y.; Yabuuchi, K.; Kimura, A. A new approach to evaluate irradiation hardening of ion-irradiated ferritic alloys by nano-indentation techniques. *Fusion Eng. Des.* **2011**, *86*, 2658–2661. [CrossRef]
27. Hardie, C.D.; Roberts, S.G. Nanoindentation of model Fe-Cr alloys with self-ion irradiation. *J. Nucl. Mater.* **2013**, *433*, 174–179. [CrossRef]
28. Suresh, S.; Giannakopoulos, A.E. A new method for estimating residual stresses by instrumented sharp indentation. *Acta Mater.* **1998**, *46*, 5755–5767. [CrossRef]

29. Janssen, G.C.A.M.; Abdalla, M.M.; van Keulen, F.; Pujada, B.R.; van Venrooy, B. Celebrating the 100th anniversary of the Stoney equation for film stress: Developments from polycrystalline steel strips to single crystal silicon wafers. *Thin Solid Films* **2009**, *517*, 1858–1867. [CrossRef]

30. Ghidelli, M.; Sebastiani, M.; Collet, C.; Guillemet, R. Determination of the elastic moduli and residual stresses of freestanding Au–TiW bilayer thin films by nanoindentation. *Mater. Des.* **2016**, *106*, 436–445. [CrossRef]

31. Allen, T.R.; Tan, L.; Gan, J.; Gupta, G.; Was, G.S.; Kenik, E.A.; Shutthanandan, S.; Thevuthasan, S. Microstructural development in advanced ferritic—martensitic steel HCM12A. *J. Nucl. Mater.* **2006**, *351*, 174–186. [CrossRef]

32. Kimura, A.; Morimura, T.; Narui, M.; Matsui, H. Irradiation hardening of reduced activation martensitic steels. *J. Nucl. Mater.* **1996**, *233–237*, 319–325. [CrossRef]

33. Xin, Y.; Ju, X.; Qiu, J.; Guo, L.; Chen, J.; Yang, Z.; Zhang, P.; Cao, X.; Wang, B. Vacancy-type defects and hardness of helium implanted CLAM steel studied by positron-annihilation spectroscopy and nano-indentation technique. *Fusion Eng. Des.* **2012**, *87*, 432–436. [CrossRef]

34. Tanigawa, H.; Klueh, R.L.; Hashimoto, N.; Sokolov, M.A. Hardening mechanisms of reduced activation ferritic/martensitic steels irradiated at 300 °C. *J. Nucl. Mater.* **2009**, *386–388*, 231–235. [CrossRef]

35. Hu, W.; Luo, F.; Shen, Z.; Guo, L.; Zheng, Z.; Wen, Y.; Ren, Y. Hydrogen bubble formation and evolution in tungsten under different hydrogen irradiation conditions. *Fusion Eng. Des.* **2015**, *90*, 23–28. [CrossRef]

36. Shen, Y.; Zhu, J.; Huang, X. Ar ion irradiation hardening of high-Cr ferritic/martensitic steels at 700 °C. *Metals Mater. Int.* **2016**, *22*, 181–186. [CrossRef]

37. Anderoglu, O.; Byun, T.S.; Toloczko, M.; Maloy, S.A. Mechanical performance of ferritic martensitic steels for high dose applications in advanced nuclear reactors. *Metall. Mater. Trans. A* **2013**, *44*, 70–83. [CrossRef]

38. Azevedo, C.R.F. A review on neutron-irradiation-induced hardening of metallic components. *Eng. Fail. Anal.* **2011**, *18*, 1921–1942. [CrossRef]

39. Fu, Z.Y.; Liu, P.P.; Wan, F.P.; Zhan, Q. Helium and hydrogen irradiation induced hardening in CLAM steel. *Fusion Eng. Des.* **2015**, *91*, 73–78. [CrossRef]

40. Lei, J.; Ding, H.; Shu, G.G.; Wan, Q.M. Study on the mechanical properties evolution of A508-3 steel under proton irradiation. *Nucl. Instrum. Meth. Phys. Res. Sec. B Beam Interact. Mater. Atoms* **2014**, *338*, 13–18. [CrossRef]

41. Osetsky, Y.N.; Stoller, R.E. Atomic-scale mechanisms of helium bubble hardening in iron. *J. Nucl. Mater.* **2015**, *465*, 448–454. [CrossRef]

42. Zhang, H.; Zhang, C.; Yang, Y.; Meng, Y.; Jang, J.; Kimura, A. Irradiation hardening of ODS ferritic steels under helium implantation and heavy-ion irradiation. *J. Nucl. Mater.* **2014**, *455*, 349–353. [CrossRef]

43. Zhang, C.; Chen, K.; Wang, Y.; Sun, J.; Hu, B.; Jin, Y.; Hou, M.; Liu, C.; Sun, Y.; Han, J.; et al. Microstructural changes in a low-activation Fe-Cr-Mn alloy irradiated with 92 MeV Ar ions at 450 °C. *J. Nucl. Mater.* **2000**, *283–287*, 259–262. [CrossRef]

MDPI

St. Alban-Anlage 66

4052 Basel

Switzerland

Tel. +41 61 683 77 34

Fax +41 61 302 89 18

www.mdpi.com

Metals Editorial Office

E-mail: metals@mdpi.com

www.mdpi.com/journal/metals

Printed in the USA
CPSIA information can be obtained
at www.ICGtesting.com
LVHW071117291023
762473LV00023B/1166

9 783038 975052